CD-ROM INCLUDED

AVENUE WRAPS

Avenue Wrap-Arounds for ArcObjects

A guide for converting Avenue scripts into VB/VBA code

First Edition

August 2002

The CEDRA Press
Rochester, New York

To those that such authorization is given, no warranty, expressed or implied is made regarding the accuracy, reliability, or functioning of the software and related material, and no responsibility is assumed by the authors or by The CEDRA Corporation for any errors, mistakes, or misrepresentation in the software or related material, or which may occur in its use. Each user of the software and related material agrees that no attempt will be made to hold the authors or The CEDRA Corporation liable or responsible for any errors, mistakes, or misrepresentations as a result of the use of this publication, associated software and related material.

The information in this document is subject to change without notice and does not represent a commitment on the part of the authors or The CEDRA Corporation. The software described in this document is provided under a license agreement. The software may be used only in accordance with the terms of the agreement.

CEDRA-ICEGMA, CEDRA-ITOPO, CEDRA-ISEWERS, CEDRA-IWATER, CEDRA-CARDS, CEDRA-TOPOMAP, CEDRA-POND, CEDRA-SLOPES, CEDRA-RAPID, CEDRA-FOUNDS, all Utility Programs documented in this publication, The CEDRA Corporation, The CEDRA System, CEDRA-SEA, CEDRA-LAND, CEDRA-SAND, CEDRA-WATER, CEDRA-GEOTEK, CEDRA-TRAFF, CEDRA-ENVRON, CEDRA-HYDROL, CEDRA-AVseries, CEDRA-AVcad, CEDRA-AVcogo, CEDRA-AVparcel, CEDRA-AVland, CEDRA-AVsand, CEDRA-AVwater, Avenue Wraps, "Total CADD for Engineers", "CEDRA is Civil Engineering", "Bridging Engineering and GIS", The CEDRA Logo, The CEDRA Corporation Logo, and the logos of each of the program modules of The CEDRA Corporation are trademarks of The CEDRA Corporation, Rochester, New York, United States of America.

ESRI, ArcView, ArcIMS, SDE, and the ESRI globe logo are trademarks of ESRI, registered in the United States and certain other countries; registration is pending in the European Community. ArcObjects, ArcGIS, ArcMap, ArcCatalog, ArcScene, ArcInfo, ArcEdit, ArcEditor, ArcToolbox, 3D Analyst, ArcPress, ArcSDE, GIS by ESRI, and the ArcGIS logo are trademarks and Geographic Network, www.esri.com, and @esri.com are service marks of Environmental Systems Research Institute, Inc., Redlands, California.

Visual Basic is a registered trademark of Microsoft, Inc., Redmond, Washington.

International Standard Book Number: 0-9722631-0-1

Printed in the United States of America

This book is dedicated
to all those
that spent hours after hours writing Avenue code
for ArcView® 3.x
and who now have to convert it
for ArcView® 8.x

About The Authors

Constantine N. Tonias, P. E. is president of The CEDRA Corporation, and overseer of the research and development operations of the corporation. Prior to establishing The CEDRA Corporation, he was director of the computer center for a consulting engineering firm, and responsible for the development and maintenance of structural, hydraulic and other civil engineering computer software including modeling, graphics and business applications. Prior to that he was a research assistant at the interactive graphic's research center of Rensselear Polytechnic Institute, from which he holds a bachelor's and a master's degree in Civil engineering.

Mr. Tonias has been responsible for the development of The CEDRA System, a fully interactive design and drafting system particularly tailored for the civil engineering practice. The system provides the civil engineer with the necessary production tools to carry the design of a project from the basic mapping operations through the geometric, roadway, drainage, and sewer design to the generation of the final contract drawings completely via computer oriented means.

Mr. Tonias has also been responsible for the development of the CEDRA-AVseries software suite which bridges engineering and related operations with the ArcGIS™ geographic information system of Environmental Systems Research Institute of Redlands, California, thus enabling municipalities, and engineers to inventory, model, engineer, construct and manage their infrastructure facilities under one unique software umbrella.

Mr. Tonias' computer experience, in addition to the development of major complex programming systems, includes the conversion of numerous computing programs between various computing hardware and operating system environments. He also is an authorized ArcView® GIS and Avenue instructor and has written in several technical publications and made presentations on the application of computer aided design methods, geographic information system applications, and management of computer aided design operations at various engineering and other seminars and conferences, domestically and abroad.

Elias C. Tonias, P. E. is a technical consultant to The CEDRA Corporation in the development of engineering computing software. He has over forty years of engineering experience in a variety of large engineering projects including highways, bridges, storm water conveyance and detention, wastewater conveyance, tunnels and land development. He was a pioneer in the application of computers in Civil Engineering, and has written and managed the development of numerous software. Having being involved with software development since the very early days of the computer age, he was forced to go through a series of program conversions from computer environment to computer environment.

Mr. Tonias has written extensively on the subject of computer utilization and software development and made presentations domestically and abroad. In addition to his engineering practice, he was for ten years an adjunct professor at the Rochester Institute of Technology. His educational background includes a bachelor's from Rensselear Polytechnic Institute and a master's degree from The Ohio State University, both in Civil engineering.

Preface

Thousands of years ago, humans scratched and etched on rock surfaces, palace and pyramid walls, papyri and other surfaces, animals, birds, plants and various human actions, for communication purposes and to record history. Over the years, the icons of animals, birds, plants and human actions were replaced by small alpha characters we now call letters. Whereas it once took a single icon, of a wild animal being chased by a human with a javelin, it now took a few sheets of papyri to describe the same chase using a collection of these letters. By grouping these letters in a specific manner, the human was able to form different words. Using words, rather than icons, offered the human a more powerful and precise means of communication. As the years went by, humans found it necessary to deal with numbers which were greater than the number of their arms and legs. As such, it became appropriate to augment their alphabet by introducing special characters to represent numbers. Thus, *alphanumerics* were created.

But humans, although they may have had one source of origin, they quickly multiplied and scattered across the surface of planet earth. Travel being not that easy at those times, and telecommunication being nonexistent, groups of humans (tribes, or rather sets) kind of lost communication with each other. In the process, whereas, the iconic depiction of an animal's hunt may have varied from tribe to tribe only by the artistic expertise of its scribe, the alphanumeric symbology and language of communication was distinctly different from tribe to tribe. So that, as the various tribes began to interact, the need for translation from one language, and one alphabet, to another became necessary. This *conversion* process has been going on for thousands of years. Since a language is a product of a tribe's experience and culture over a period of centuries, and possibly millennia, complete and exact translations from one language to another is not that easy, if not impossible.

The above process has evolved over a period of thousands of years, and under rather adverse natural and human-made conditions. In an effort to assist the human, approximately 50 years ago, a new beast, the *computer*, was created. The function of this beast was to perform computations, by manipulating number and letter characters; thousands of times faster than the human could ever do. In order to better utilize the beast, the human had to devise a means of communication with it. Thus, the first languages of computer communication were developed. Note that we said *languages*, and not *language*. One would have thought that the human would have learned from the past experience of thousands of years, the difficulties that arise from having many different languages. But that was not to be, because no matter how smart a human may be, and how accomplished a human may become in searching for new technologies, the human is reluctant to learn from past mistakes.

We first began to communicate with the new beast using a *machine language*, because that was the only language the beast could understand. But machine languages, as you may surmise, are difficult and cumbersome to use. Thus, the human intellect created a new means of communication, the symbolic *languages* (the reader should again note the plural form of the word). However, symbolic languages did not prove to be much less difficult and cumbersome, so new languages were developed. The prominent early languages were Fortran and COBOL, the former primarily for the scientific and engineering communities, while the latter for most others. As expected, there were many flavors of such languages depending upon the computer manufacturer and the entity that developed the language.

Fortran for example *evolved* from Fortran to Fortran IV, Fortran 75, Fortran 97 to Fortran 2000. The reader should note the emphasis on the word evolved, since one could advance from one Fortran level to another with relative ease, as long as one avoided the use of the specific flavors introduced by the manufacturers and language development entities.

In the 1980's, some drastic new computer language developments began to emerge in the open computer market (research and development had started much earlier). One such development was Microsoft's Windows®, that revolutionized the computer utilization, and kind of put all other computer languages in the background. Like any other revolution, this revolution had to evolve until maturity. So we had Window 95®, Windows 98®, Windows 2000®, Windows NT®, Window XP®, and who knows what is to come next.

Well, enough of the reminiscing. Let us come to the specific subject of this book, Avenue Wraps™. In the late 1980's, ArcInfo™, of Environmental Systems Research Institute (ESRI), had established itself as the major geographic information systems application product. In the early 1990's, ArcView®, a younger smaller sister of ArcInfo™, and its programming language Avenue were born. As it would turn out, ArcView® would take the world by storm. During its evolution from ArcView 1.0 to ArcView 2.0, to its current version, 3.2, there are more than an estimated 600,000 ArcView® users, worldwide, as of this writing.

In the year 2000, for the sake of progress as well as other reasons, a new ESRI family of products, ArcGIS™ was publicly announced, becoming available to the open market a year later. ArcGIS™ provided the market with some very exciting new development and functionality, but had one major drawback for the thousands of ArcView 3.x users and Avenue programmers. Microsoft's COM technology and ESRI's ArcObjects™, was the development environment of ArcGIS™. ArcView 8, the ArcView 3.x equivalent within the ArcGIS™ family product provided no direct Avenue compatibility. Thus all Avenue code, for all intents and purposes, became obsolete. This was a revolution, not an evolution. Like most revolutions, the cost is high; the cost of reprogramming that is. Thousands of lines of Avenue code now has to be rewritten, in order to be utilized within the ArcGIS™ software. Furthermore, the conversion does not merely consist of changing the names of Avenue requests and their lists of given and returned arguments; in many instances it is the logic of program portions that has to be modified.

Like most new explorations, it is the pioneers that catch the first arrows. Many of us Avenue program developers had to go through a process of learning from experience. This experience was gained when we first developed applications for the ArcView 3.x environment. Having gone through a process of converting these same ArcView 3.x applications to ArcView 8.x, we thought that a means to facilitate such a conversion should be made available to others. Why should others catch the same arrows we caught? It was for this primary purpose that this book was written. In it, one will find a set of guidelines for converting ArcView® 3.x Avenue code into Microsoft's Visual Basic® and ArcObjects™ code for incorporation into the ArcGIS™ environment. These guidelines take one of three forms:

1. Certain of the conversion guidelines refer to Avenue requests that have a direct Visual Basic® and/or ArcObjects™ command (subroutine or function) counterpart. A direct counterpart implies that the operation is generally the same in both programming environments, even though the name may be somewhat different, and the location of the arguments may follow, rather than precede the request.

2. There are, however, numerous Avenue requests that do not meet the above criteria of a direct counterpart. In order to assist the Avenue programmer to expedite the conversion of Avenue code into Visual Basic® and ArcObjects™ code, various subroutines and functions have been written by The CEDRA Corporation of Rochester, New York, USA which simulate the operation of what are considered to be the most common Avenue requests. These subroutines and functions are jointly referred to as **Avenue Wraps™**.

   The name **Avenue Wraps™** is based upon the word wraparound. When converting code, the easiest conversion work occurs when there is a direct one to one correspondence of the items that need to be converted. In the case here, we have to concern ourselves with the syntax of Avenue code, as well as, the Avenue requests themselves.

   The syntax of Avenue code is different, but similar to that of Visual Basic®. So that the syntax issue can be addressed rather straight forwardly. Avenue requests, on the other hand, are very different from ArcObjects™ code. What has been done with the **Avenue Wraps™** is to provide the Avenue programmer that one to one correspondence, which was mentioned above. In so doing, the conversion work is minimized.

Difficulty of conversion from Avenue to ArcObjects

Had there been direct compatibility	1
With Avenue Wraps	3
Without Avenue Wraps	10

3. In addition to the above **Avenue Wraps™**, there are many others that have been included in this book that do not directly correspond to an Avenue request, but contribute in the overall conversion process.

An attempt has been made to make the **Avenue Wraps™** as generic as possible. Thus, in addition to being of assistance in converting Avenue code into ArcObjects™, these wraps may be used to write totally new programs for ArcObjects™ using Avenue "like" commands. So that, the Avenue programmer can develop applications for ArcGIS™ using an Avenue programming style. In addition to the description of each Avenue Wrap™, numerous code examples are provided in this book on the use of these wraps, as well as, the listing of each **Avenue Wrap™**.

The use of this book assumes a knowledge of Avenue programming, and a basic knowledge, although not extensive expertise in Visual Basic® and ArcObjects™ programming. In using this book, it is recommended that the reader first peruse the entire book, order of chapter reading is not essential, so as to familiarize him or her self as to the contents and the various **Avenue Wraps™**.

Before we conclude this preface to **Avenue Wraps™**, we consider it appropriate to address one aspect of the new computer programming technology, COM. The first programming languages were rather sequential (not that we have forgotten about the GO TO statements) in describing an operation that the computer was to perform. First you do this, then that, then that, and so on; with this and that being as minute as possible. Then came procedures, subroutines and functions, that combined several lines of code and could be called from anywhere in a program by one name with certain variable arguments. Lately, we have the concept of objects, which are capable of having methods and properties, that are entities of programming code and/or graphics which can be referenced at will, not from just within a single program, but from program to program. Perhaps one may consider this as going back in time, where the first computer programs relate to the hieroglyphics of the papyri, and the new object oriented programs to the icons etched on the rocks of caves by our first ancestors.

Constantine N. Tonias, P. E.
Elias C. Tonias, P. E.

Contents

CHAPTER 1

INTRODUCTION TO AVENUE WRAPS

This book, **Avenue Wraps,** presents a set of guidelines for converting ArcView® 3.x Avenue scripts into Microsoft's Visual Basic® and ArcObjects™ code, jointly referred to in this book as "VB code", for incorporation into ArcGIS™ software. It is noted that the words "ArcObjects™ code" as used in this book represents a generic term for the use of ArcObjects™ properties and methods in conjunction with VB code. The reason for using Visual Basic® and not Visual C++®, or any other COM compliant software, is that the syntax of Visual Basic® is much closer to that of Avenue, than Visual C++.

Certain of the conversion guidelines presented in this book refer to Avenue requests that have a direct VB (subroutine or function) or ArcObjects™ counterpart. A direct counterpart implies that the operation is generally the same in both programming environments, even though the name may be somewhat different, and the location of the argument lists may follow rather than precede the request. For the most part, these guidelines are presented in Chapter 2 of this book. Speaking of terminology, the word "program" as used herein is a generic term referring to a main line program, subroutine or function, and is considered synonymous with the Avenue word "script", and "VB macro". Also included in Chapter 2 are the changes that need to be made to the Avenue code that do not utilize Avenue requests, that is, the syntax of the code. For example dealing with "for each" loops, "if" statements, intrinsic functions and so forth.

Unfortunately, there are numerous Avenue requests that do not meet the above criteria of a direct counterpart. In order to assist an Avenue programmer to expedite the conversion of Avenue code to VB code, various subroutines and functions have been written in Visual Basic® that simulate the operation of what are considered to be the most common Avenue requests. The scope and use of these subroutines and functions, jointly referred to as Avenue Wraps™, are presented in the subsequent chapters of this book, and a listing of each Avenue Wrap™ is presented in Appendix D.

The use of this book assumes a knowledge of Avenue programming, and a basic knowledge, although not extensive expertise in Visual Basic® and ArcObjects™ programming. Although primarily intended for those planning on converting Avenue code to ArcObjects™, certain of

DECLARATION OF VARIABLES

the Avenue Wraps™ contained herein will prove helpful to those that do not convert from Avenue, but instead wish to program directly into VB code. In using this book, it is recommended that the reader first peruse the entire book, order of chapter reading is not essential, so as to familiarize him or her self as to the contents and the various Avenue Wraps™.

1.1 Comments on the Declaration of Variables

An attempt has been made to make the Avenue Wraps™ as generic as possible. Each Avenue Wrap™ description is preceded by its corresponding Avenue request, if one exists. The use of an Avenue Wrap™ is basically a direct substitution of an Avenue Wrap™ for an Avenue request. The name of an Avenue Wrap™ is essentially the same as that of the corresponding Avenue request. There are, however, two differences between an Avenue Wrap™ and an Avenue request. The first is that the argument list, for some Avenue Wraps™, will differ slightly with their Avenue counterpart. The second is that, whereas Avenue requests could be concatenated, Avenue Wraps™ cannot.

The description of each Avenue Wrap™ includes the Dim declaration statements of its associated variables, as well as a description of each of the variables. Note that this is a significant difference between Avenue and "VB code". Variables did not have to be declared in Avenue, however with "VB code", variables do. Regarding the Dim statements, the reader is alerted to the following:

It is noted that an Avenue "list" corresponds to a VB "collection", which is different than a VB array. In this book, the words "list" and "collection" are used as meaning the same. Where distinction is necessary with regards to arrays, it is so done.

- Dim statements should appear in the program that calls the subject subroutine or function, or in a preceding program that may be calling the program that calls a subroutine or function. For example consider a subroutine called Force being called by subroutine BBB, which in turn is called by AAA. In this case, the Dim statements for the variables y1, x1, y2, x2, dist, az should appear in the subroutine AAA. So that the declarations would look like this:

```
Public Sub AAA
   Dim x3, y3, y1, x1, y2, x2, dist, az As Double
   ... do something
   Call BBB(x3, y3, y1, x1, y2, x2, dist, az)
End Sub
'
Public Sub BBB(x3, y3, y1, x1, y2, x2, dist, az)
   Dim only whatever variables are local to BBB
```

DECLARATION OF VARIABLES

```
'   ---y1, x1, y2, x2 are the given variables
'   ---dist and az are the returned variables
    Call Force(y1, x1, y2, x2, dist, az)
'   ...do whatever with dist and az
End Sub
'
Public Sub Force(y1, x1, y2, x2, dist, az)
Dim only whatever variables are local to Force
'   ... solve for dist and az
End Sub
```

- Regarding the Dim statement of a function itself, consider the following sample statement, which may appear in a public sub called CCC, that uses the function avAddDoc (an Avenue Wrap™):

```
theLayer = avAddDoc(aDoc)
```

In the text of Appendix D, in which said Avenue Wrap™ is presented, one may get the impression when looking at the function's listing, that in addition to any other variables in the function's argument list, the function name, avAddDoc in this case, should be declared in the calling program as:

```
Dim avAddDoc As Integer
```

This is not the case. When a variable is set equal to a function, it is the variable and not the function name that should appear in a Dim statement within the calling program. In this case, in public sub CCC it is the variable *theLayer* that is declared as:

```
Dim theLayer As Integer     ' correct declaration
theLayer = avAddDoc(aDoc)
```

and not the function name:

```
Dim avAddDoc As Integer     ' incorrect declaration
theLayer = avAddDoc(aDoc)
```

- Certain variable types such as integer, long, double and the like may be declared with more than one declaration on the same line of a Dim statement, such as:

```
Dim opmode, numItems, totItems, xyzRecs As Integer
Dim x1, x2, y1, y2, z1, z2 As Double
Dim aTitle1, aTitle2 As String
Dim anObjt1, anObjt2, anObjt3, anObjt4 As Variant
```

DECLARATION OF VARIABLES

These types of variables are referred to as non-object types. A variable of object type is stored as a 32-bit (4-byte) address, which refers to an object. In addition, objects support methods and properties. ArcObjects™ offers the programmer a lot of different types of objects which enable the programmer to interact with the ArcGIS™ software. The approach taken in this book, when declaring objects, is to do so one per line, such as:

```
Dim pmxDoc As IMxDocument
Dim pPolygon As IPolygon
Dim graphList As New Collection
Dim theFields As New Collection
```

Note that, for the sake of conservation of space, in Appendix D, more than one object may be declared on a single line.

- Certain Avenue Wraps™ utilize global variables. Unlike Avenue, where a global variable was prefixed with the _ character, global variables in "VB code" begin with the letters **ug**, and are initialized in the Avenue Wrap™ avInit, presented in Chapter 3 of this book. Global variables are used to pass information between the Avenue Wraps™ much like a common block would be used in a Fortran program.

The variables of each Avenue Wrap™ calling argument list are described as to their nature, and they are classified as to being either "given" or "returned" arguments within a subroutine or function. There are some Avenue Wraps™ that may not have given and/or returned variables in their argument list. In such cases, the word "nothing" will appear.

1.2 Overview of the Chapter Contents

In the subsequent chapters, Avenue Wrap™ subroutine and function names that appear in bold type, such as **avGetDisplayFlush**, indicate that the complete listing of the subroutine or function is contained in Appendix D. Subroutines or functions that are not shown in bold type are either standard VB or ArcObjects™ code.

The available Avenue Wraps™ are presented in the chapters identified below, and they have been grouped according to function.

- **Chapter 2 - General Conversion Guidelines**
  This chapter presents (a) a set of general syntax guidelines for converting Avenue to VB and ArcObjects™ code, (b) some Avenue Wraps™, and (c) addresses the syntax concerning the following:

- Numbers and arithmetic operations,
- String manipulation,
- Transcendental and other functions,
- Querying and testing variables,
- Lists, arrays and collections, and
- Program flow control such as Do and For loops.
- Data type declaration including a summary table of how various types of variables, views, theme and table names, message box contents, and the like should be declared (dimmed).

OVERVIEW OF CHAPTER CONTENTS

- **Chapter 3 - Project Organization Avenue Wraps™**
  This chapter contains Avenue Wraps™ that are associated with the establishment of an ArcView® project and its views, or are of a general application nature.

- **Chapter 4 - File I/O Avenue Wraps™**
  This chapter addresses Avenue Wraps™ that are concerned with the opening and closing of files, reading of files, writing to files, and other file handling operations including the extraction of data from delimited files.

- **Chapter 5 - Theme and Table Avenue Wraps™**
  This chapter contains Avenue Wraps™ pertaining to the handling of themes and their tables including:
  - Theme and table creation and retrieval,
  - Querying, editing and summarizing tables, and
  - Performing calculations on table cells.

- **Chapter 6 - Feature Selection Avenue Wraps™**
  This chapter is comprised of various Avenue Wraps™ that help create and manipulate selection sets from various groups of features or rows.

- **Chapter 7 - Message and Menu Box Avenue Wraps™**
  This chapter is comprised of the Avenue Wraps™ that help create boxes with messages to the user, or present the user questions for action, provision for input of data, and/or selection from predefined lists.

- **Chapter 8 - Geometric Routines Avenue Wraps™**
  This chapter contains Avenue Wraps™ that enable the programmer to perform various geometric operations such as creating points, line, circles, and polygons, and extracting information thereof.

OVERVIEW OF CHAPTER CONTENTS

- **Chapter 9- User Document Interaction**
  This chapter contains various subroutines and functions that facilitate the graphic interaction ("making picks") between the user and ArcMap™.

- **Chapter 10 - Graphics and Symbols Avenue Wraps™**
  This chapter contains Avenue Wraps™ that enable the programmer to manipulate graphics, assign attributes and symbology, and create and work with graphic text.

- **Chapter 11 - Classification and Legends**
  This chapter contains various subroutines and functions with which the programmer may classify layers, create legends, and work with symbol palettes.

- **Chapter 12 - Utility Macros**
  This chapter provides two subroutines that assist the programmer to (a) mass export VBA code to a specified directory, and (b) import or load VBA code from a specified directory into the current ArcMap™ project. This chapter also contains various subroutines and functions which are used by certain Avenue Wraps™ transparently to the programmer. In addition, these subroutines and functions may be used in the development of individual code to carry out certain geometric and other operations.

- **Appendices**

  A Palette Index Values
  In this appendix, the programmer will find the various symbol index values for each of the 17 standard ArcMap™ palettes.

  B Color Calibration Diagrams
  In this appendix, the calibration charts for the gray scale, CMYK color model, HSV color model and RGB color model will be found.

  C Table of Avenue Request to Avenue Wrap™ Macro Mapping
  The table of this appendix summarizes in alphabetical order the most common Avenue requests and VB or ArcObjects statements, most of which are individually discussed in this book, and identifies the corresponding Avenue Wrap™ and location of its discussion by chapter and section.

  D Avenue Wrap™ Macro Listing
  In this appendix, the complete listing of the various Avenue Wrap™ scripts is presented in alphabetical script name order.

GETTING STARTED

1.3 Getting Started

1.3.1 General Commentary

So we are now ready to start converting Avenue code. Before performing any conversion work, it is recommended that the **Visual Basic Environment (VB and VBA)** and **Visual Basic for Applications Development Environment** sections in the **ArcObjects Developer Help** be read. These sections can be found under the Contents tab, clicking on Getting Started, followed by clicking on Getting Started Start Page, see Figure 1-1. These are not very long sections but they provide valuable information that helps to explain what is presented in the subsequent chapters.

The approach recommended in this book for converting Avenue code into "VB code", is to initially perform the conversion work in the **VBA**, Visual Basic for Applications, environment and then build an extension, if appropriate, in the **VB**, Visual Basic, environment. The reason for doing so is that the **VBA** environment is similar to the Avenue environment in that the programmer can easily test and debug the application directly within ArcMap™. Once the application has been tested and is ready for distribution, the programmer can either (a) package the application as a protected

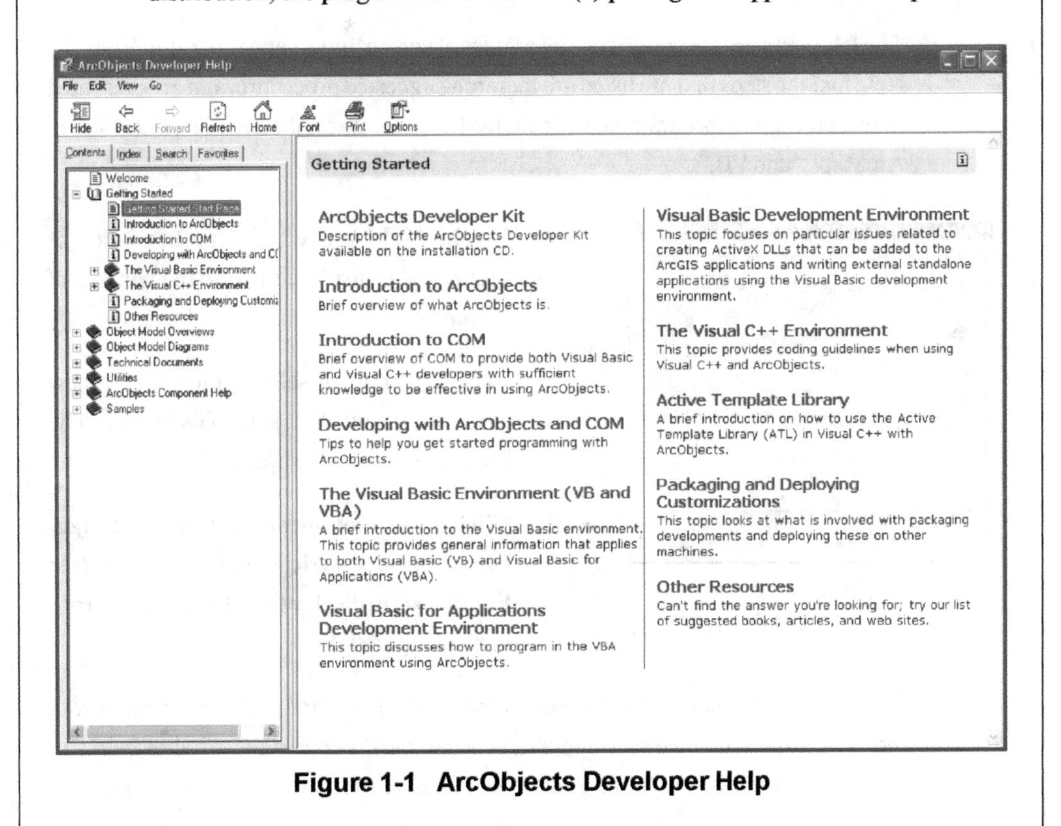

Figure 1-1 ArcObjects Developer Help

GETTING STARTED

project file, similar to an encrypted ArcView-Avenue project file, or (b) build an extension in the **VB**, Visual Basic, environment. Although there will be some duplicitous work (i.e. creating tools, menu items, etc.), any other approach would lead to a longer development/conversion cycle.

When an extension is to be built the programmer will be working in two different, yet somewhat similar, environments, first **VBA** and secondly (once the application has been tested) **VB**. It is suggested that the listings in Appendix D be reviewed, in terms of the variable declaration statements, to see how variables should be declared so that any re-coding can be eliminated. The point is that a programming style should be adopted such that the code can exist in both a **VBA** and **VB** environment.

When a protected project file is to be distributed, the end user working with the protected project file will still be able to customize the project file, but will be unable to modify or view the customizations provided by the developer. The developer in protecting the project file assigns a password that must be entered in order to modify or view his/her customizations. If the password is not properly specified, the end user is unable to modify or view the developer's customizations.

The following pages contain three examples of converting Avenue code into VB code. Note that the steps that are listed are merely a suggested procedure, and are presented here for the novice programmer who for the first time enters the development realm of ArcMap™ and VB.

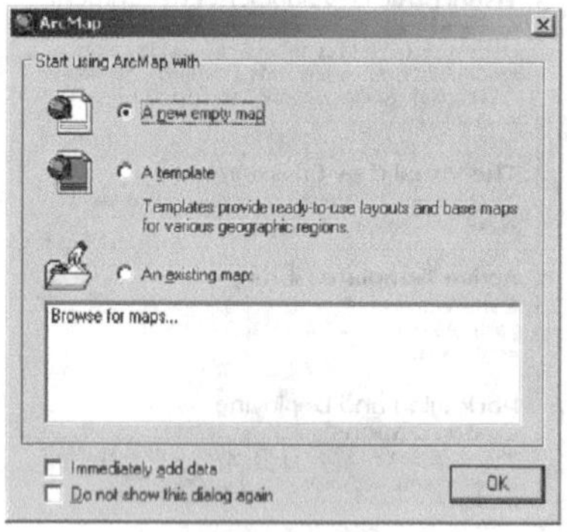

Figure 1-2
Existing Project Browser

➢ 1 **Invoke** ArcMap, at which time the window of Figure 1-2 is displayed within the window of Figure 1-3.

➢ 2 Accept the default selection to create a new empty map, and **click** at the **OK** button. The browser window disappears.

➢ 3 **Click** at the **Tools** menu and then at the **Macros** and **Visual Basic Editor** sub-menus (see Figure 1-4) to display the VBE work environment of Figure 1-5.

The VBE environment is divided into five areas, the menu and tool bars across the top, the Project sub-window in the upper left corner, the Properties sub-window below it, and the main work area (the large dark gray area) to the right.

1.3.2 Converting an Avenue System Script

Converting an Avenue System Script

For our first conversion example we will convert the Avenue system script View.ClearSelect.

➢ 4 **Invoke** ArcView 3.x, and load either an existing project, or create a blank project.

➢ 5 **Click** at the Scripts icon in the Project window, and then at the **New** button of said sub-window (see Figure 1-6A).

➢ 6 **Click** at the **Load System Script** button to display a list of the system scripts, scroll down to **select** the **View.ClearSelect** script, and **click** at the **OK** button. See Figure 1-6(B). The system script is now displayed in the new script window. See Figure 1-6(C). This script is supposed to clear (deselect) any selected features in the active theme within the active document.

We will now translate the above Avenue system script into VB code. Since this script is so small, there is no need to copy it and paste it into a VBA module.

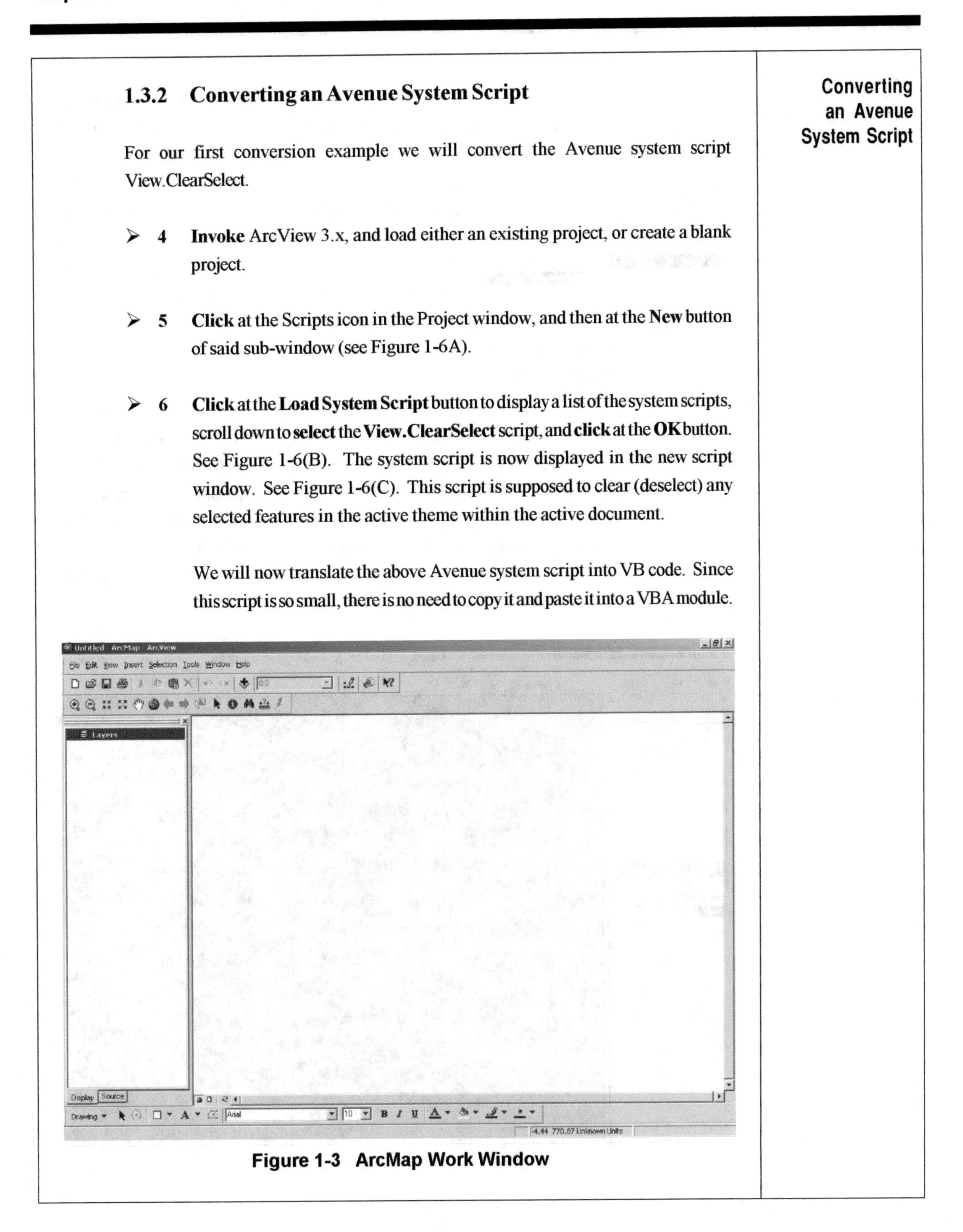

Figure 1-3 ArcMap Work Window

Converting an Avenue System Script

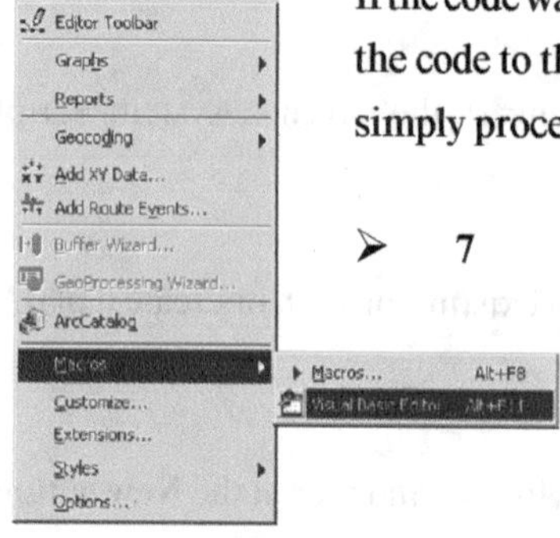

Figure 1-4 Tools Menu and Macros/Visual Basic Editor Sub-Menus

If the code was larger we could use normal Windows functionality to copy the code to the clipboard and paste it in a VBA module. Hence, we will simply proceed to code it into VBA from scratch. So that,

➢ 7 Go back to the VBE work environment of Figure 1-5, and **click** at the **Insert** menu and then at the **Module** sub-menu. *Module1* is now displayed in the Properties sub-window of Figure 1-5, and *Module1 (Code)* is displayed in the work area of the same figure. This process is essentially the same as creating a new script window in *ArcView* 3.x.

➢ 8 In the Properties sub-window of Figure 1-5, **double-click** on top of the name of the module, *Module1*, and replace the name by key **entering** the new name to be assigned to the module, **avViewClearSelectMOD**, followed by **depressing** the **Enter key**. All references to Module1 should have been changed to reflect the new module name. The extension MOD, which appears in the module name, denotes that the file is a module. It is not possible to have a module name that is the same as the name of a subroutine or function.

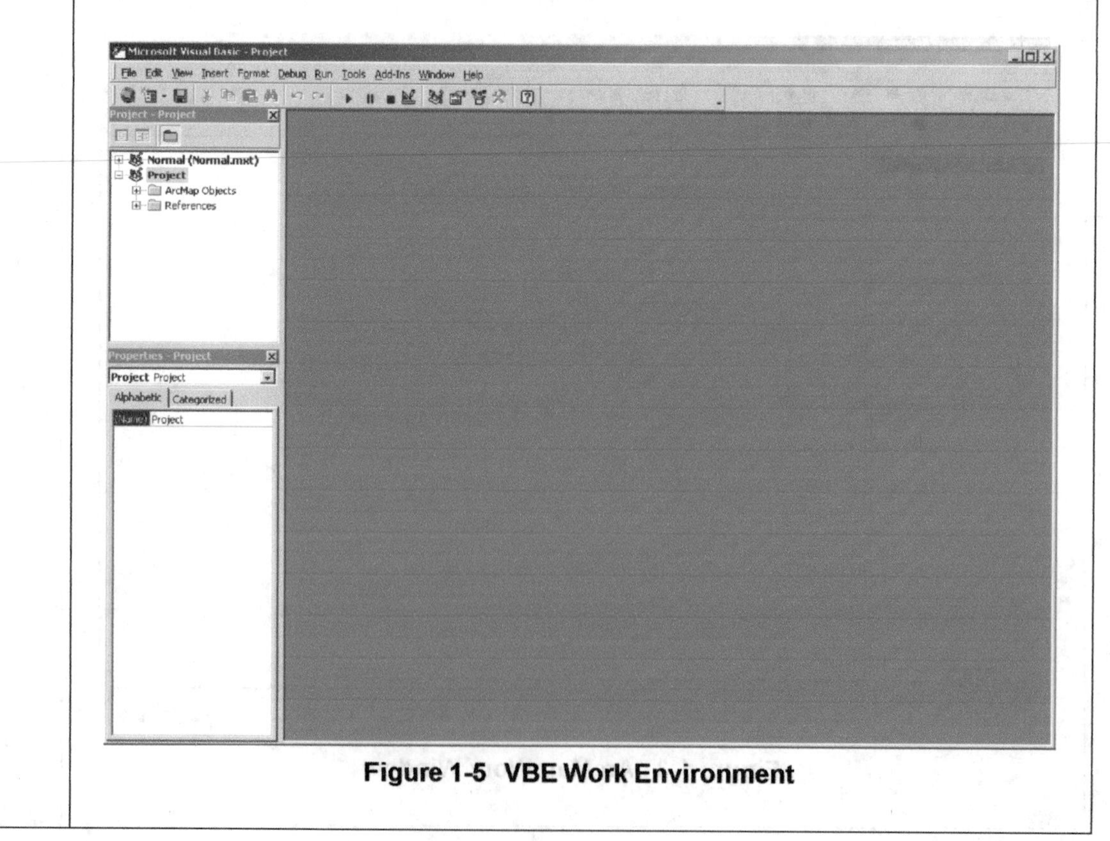

Figure 1-5 VBE Work Environment

Converting an Avenue System Script

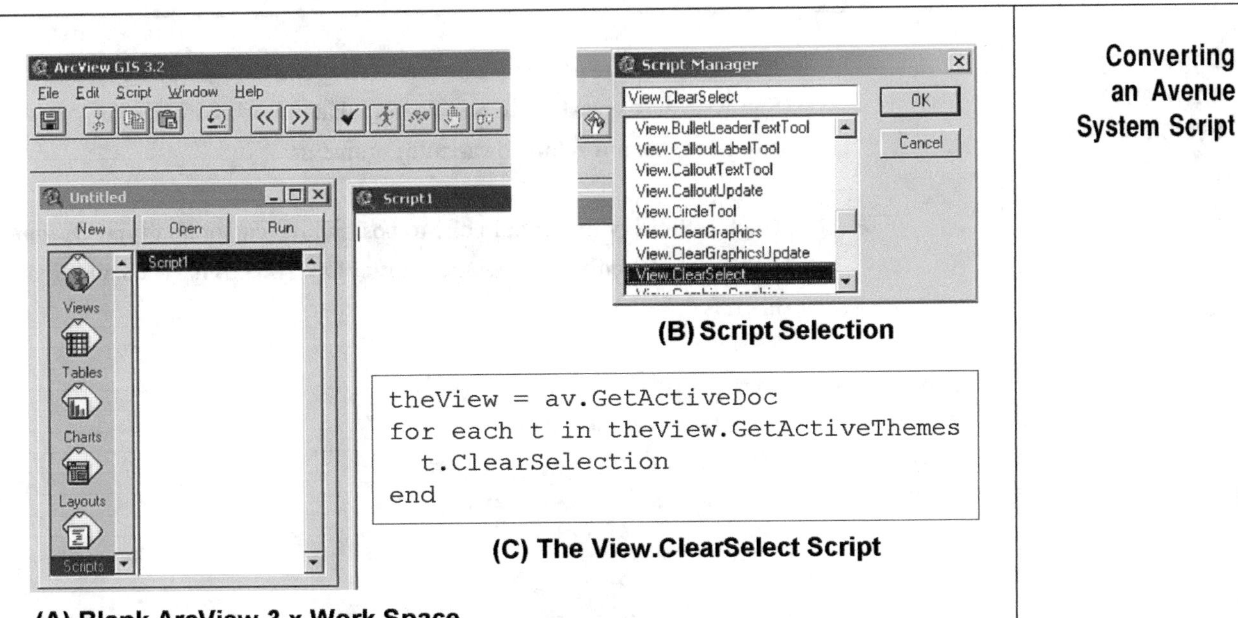

(B) Script Selection

(C) The View.ClearSelect Script

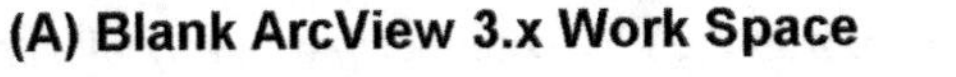

(A) Blank ArcView 3.x Work Space

Figure 1-6 Opening an ArcView 3.x System Script

The script to be converted is to be a public subroutine, so that it may be called by various other procedures (subroutines and functions) to be written later on. Thus, in the VBE work environment of Figure 1-5, we will now create a public subroutine.

Summarizing, scripts in Avenue are referred to as procedures, and are stored in modules, in VBA or VB. Modules, unlike scripts, can contain more than one procedure. The most straight forward approach is to create a module, with a single procedure within it, for every script to be converted.

➤ 9 **Click** in the title bar of the avViewClearSelectMOD window to make the window active. **Click** at the **Insert** menu and then at the **Procedure...** sub-menu. In the data field to the right of the Name: label, **type avViewClearSelect** and **click** at the **OK** button. Note that under the Type and Scope frames the default values denote a public subroutine is to be established.

The following lines of code will appear.

```
Public Sub avViewClearSelect()

End Sub
```

Converting an Avenue System Script

Note that there are no arguments between the two parentheses in the Public Sub avClearSelect() line, thus indicating that this subroutine is not to have any given, nor any returned variable arguments.

➢ 10 **Click** in front of the word Public to position the cursor and **type Option Explicit**, followed by **depressing** the **Enter key, twice**. The code should look like this:

```
Option Explicit

Public Sub avViewClearSelect()

End Sub
```

The statement Option Explicit informs the VBE compiler that all variables are to be explicitly declared. Although an optional statement, it helps tremendously in the development/conversion process.

Now, **click** at the blank line between the second and third lines (between Public Sub avViewClearSelect and End Sub), and **key enter** the statements below. The comment statements explain the code.

```
'
'   ---The Avenue Wrap below gets the active document.
'   ---Refer to section 3.1.3 of this book.
    Call avGetActiveDoc(pMxApp, pmxDoc, pActiveView, pMap)
'
'   ---The Avenue Wrap below creates a collection of the
'   ---various layers (themes) in the active document.
'   ---Refer to section 3.1.4 of this book.
    Call avGetActiveThemes(pmxDoc, ThemesList)
'
'   ---We will now determine the number of layers (themes) in
'   ---the active document
    NumThemes = ThemesList.Count
'
'   ---The Avenue Wrap below within the For Each loop will
'   ---deselect the selected features in each of the themes
'   ---Refer to section 6.2.2 of this book.
```

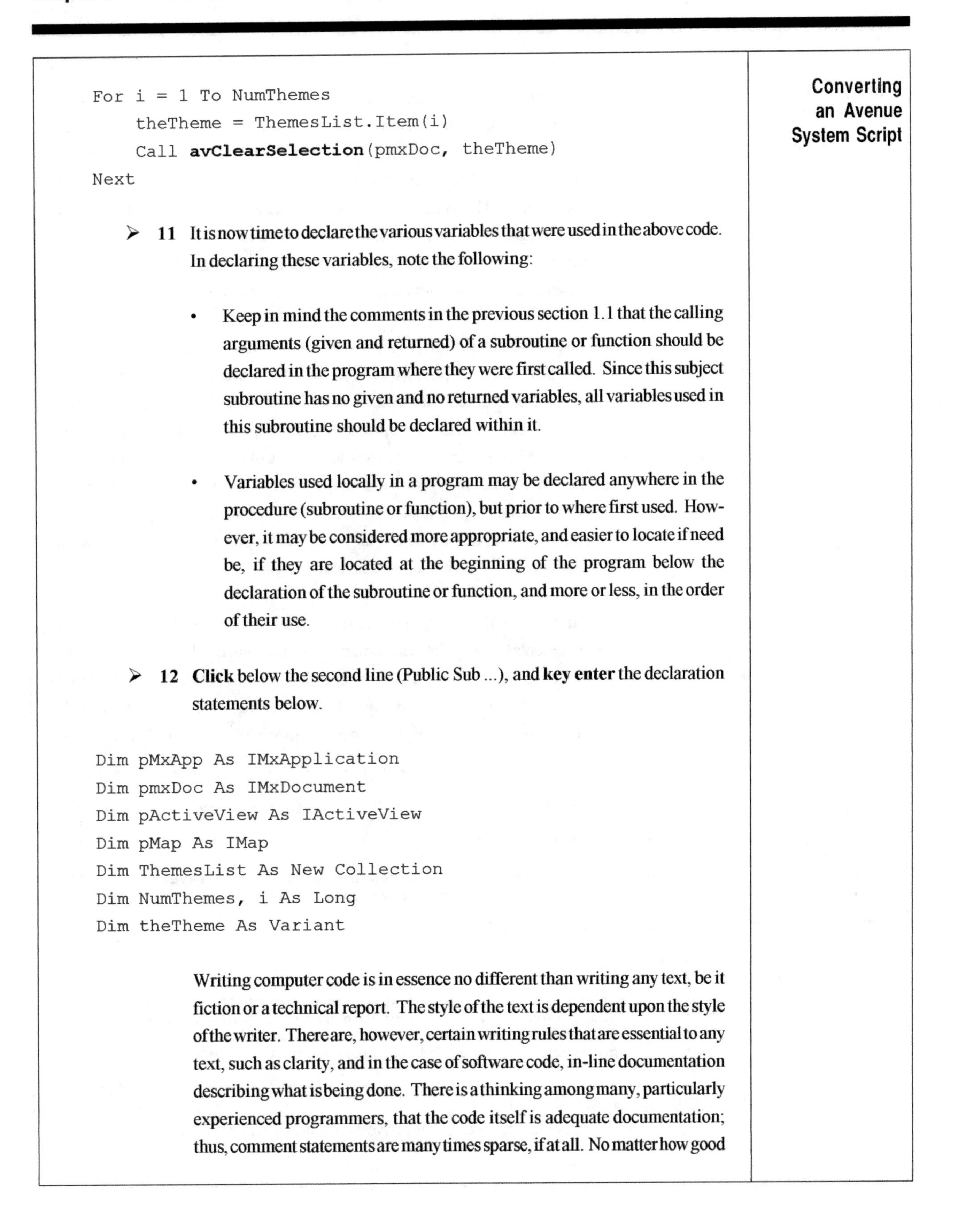

Converting an Avenue System Script

```
For i = 1 To NumThemes
    theTheme = ThemesList.Item(i)
    Call avClearSelection(pmxDoc, theTheme)
Next
```

➤ **11** It is now time to declare the various variables that were used in the above code. In declaring these variables, note the following:

- Keep in mind the comments in the previous section 1.1 that the calling arguments (given and returned) of a subroutine or function should be declared in the program where they were first called. Since this subject subroutine has no given and no returned variables, all variables used in this subroutine should be declared within it.

- Variables used locally in a program may be declared anywhere in the procedure (subroutine or function), but prior to where first used. However, it may be considered more appropriate, and easier to locate if need be, if they are located at the beginning of the program below the declaration of the subroutine or function, and more or less, in the order of their use.

➤ **12** **Click** below the second line (Public Sub ...), and **key enter** the declaration statements below.

```
Dim pMxApp As IMxApplication
Dim pmxDoc As IMxDocument
Dim pActiveView As IActiveView
Dim pMap As IMap
Dim ThemesList As New Collection
Dim NumThemes, i As Long
Dim theTheme As Variant
```

Writing computer code is in essence no different than writing any text, be it fiction or a technical report. The style of the text is dependent upon the style of the writer. There are, however, certain writing rules that are essential to any text, such as clarity, and in the case of software code, in-line documentation describing what is being done. There is a thinking among many, particularly experienced programmers, that the code itself is adequate documentation; thus, comment statements are many times sparse, if at all. No matter how good

Converting an Avenue System Script

a programmer is, the need will come some months, if not years later, when (a) the programmer's memory may not be as good as first thought, and (b) when program logic may not be as easily discernible. It is at this point of time, when properly located comment statements will be more than just welcomed and appreciated. This is not to justify the number of comment statements in the preceding sample code. One may justifiably claim that they are verbose, and some of them redundant. The purpose of their introduction was purely to lucidly describe to a novice as to what is being done.

On the issue of code clarity, it is quite important to describe at the very start of a procedure the purpose or objective of the procedure, and identify its given and returned variables, as well as any special conditions that may pertain to, or required, by the program for its proper execution. If the program is a main line program, a skeleton description of its flow may be apropos. In the "good old days" of computer programming, a detailed flow chart was a requirement, if not a necessity. Nowadays, a concise description in good English, or what may be the language of the programmer, is considered appropriate. Thus it is now time to do just that.

➢ **13** **Click** below the first line (Option Explicit), and **key enter** the comment statements below. Note that comment statements can be created by using the ' character, which may appear anywhere on a data line. Again note that these comments are just a suggestion which the authors utilize in their programming. Others may customize them to their specific needs and desires.

Note that the procedure name and the disk file name may be the same, or different. A 8.3 file name convention is used here.

```
'
' * * * * * * * * * * * * * * * * * * * * * * * * * * * * * * * * *
' *                                                                 *
' *  Name: avViewClearSelect           File Name: avclears.bas     *
' *                                                                 *
' * * * * * * * * * * * * * * * * * * * * * * * * * * * * * * * * *
' *                                                                 *
' *  PURPOSE:  Deselect all selected features in all layers         *
' *            (themes) of an active application (document)         *
' *                                                                 *
' *  GIVEN:    nothing                                              *
' *                                                                 *
' *  RETURN:   nothing                                              *
' *                                                                 *
' *  Dim (in this area the declaration statements of the given      *
' *       and returned variables of the program would be shown      *
' *       as comments.  In the subject example, since there are     *
' *       no given and returned variables, omit these comments)     *
' *                                                                 *
' * * * * * * * * * * * * * * * * * * * * * * * * * * * * * * * * *
'
```

The complete code of the above example is contained in Table 1-1.

Converting an Avenue System Script

Table 1-1 SAMPLE VB CODE

```
Option Explicit
'
' * * * * * * * * * * * * * * * * * * * * * * * * * * * * * * * * *
' *                                                                 *
' *  Name: avViewClearSelect               File Name: avclears.bas  *
' *                                                                 *
' * * * * * * * * * * * * * * * * * * * * * * * * * * * * * * * * *
' *                                                                 *
' *  PURPOSE:  Deselect all selected features in all layers         *
' *            (themes) of an active application (document)         *
' *                                                                 *
' *  GIVEN:    nothing                                              *
' *                                                                 *
' *  RETURN:   nothing                                              *
' *                                                                 *
' *  Dim (in this area the declaration statements of the given      *
' *      and returned variables of the program would be shown       *
' *      as comments.  In the subject example, since there are      *
' *      no given and returned variables, omit these comments)      *
' *                                                                 *
' * * * * * * * * * * * * * * * * * * * * * * * * * * * * * * * * *
'

Public Sub avViewClearSelect()
'
   Dim pMxApp As IMxApplication
   Dim pmxDoc As IMxDocument
   Dim pActiveView As IActiveView
   Dim pMap As IMap
   Dim ThemesList As New Collection
   Dim NumThemes, i As Long
   Dim theTheme As Variant
'
'  ---The Avenue Wrap below gets the active document.
'  ---Refer to section 3.1.3 of this book.
   Call avGetActiveDoc(pMxApp, pmxDoc, pActiveView, pMap)
'
'  ---The Avenue Wrap below creates a collection of the
'  ---various layers (themes) in the active document.
'  ---Refer to section 3.1.4 of this book.
   Call avGetActiveThemes(pmxDoc, ThemesList)
'
'  ---We will now determine the number of layers (themes) in
'  ---the active document
   NumThemes = ThemesList.Count
'
'  ---The Avenue Wrap below within the For Each loop will
'  ---deselect the selected features in each of the themes
'  ---Refer to section 6.2.2 of this book.
   For i = 1 To NumThemes
       theTheme = ThemesList.Item(i)
       Call avClearSelection(pmxDoc, theTheme)
   Next
'
End Sub
```

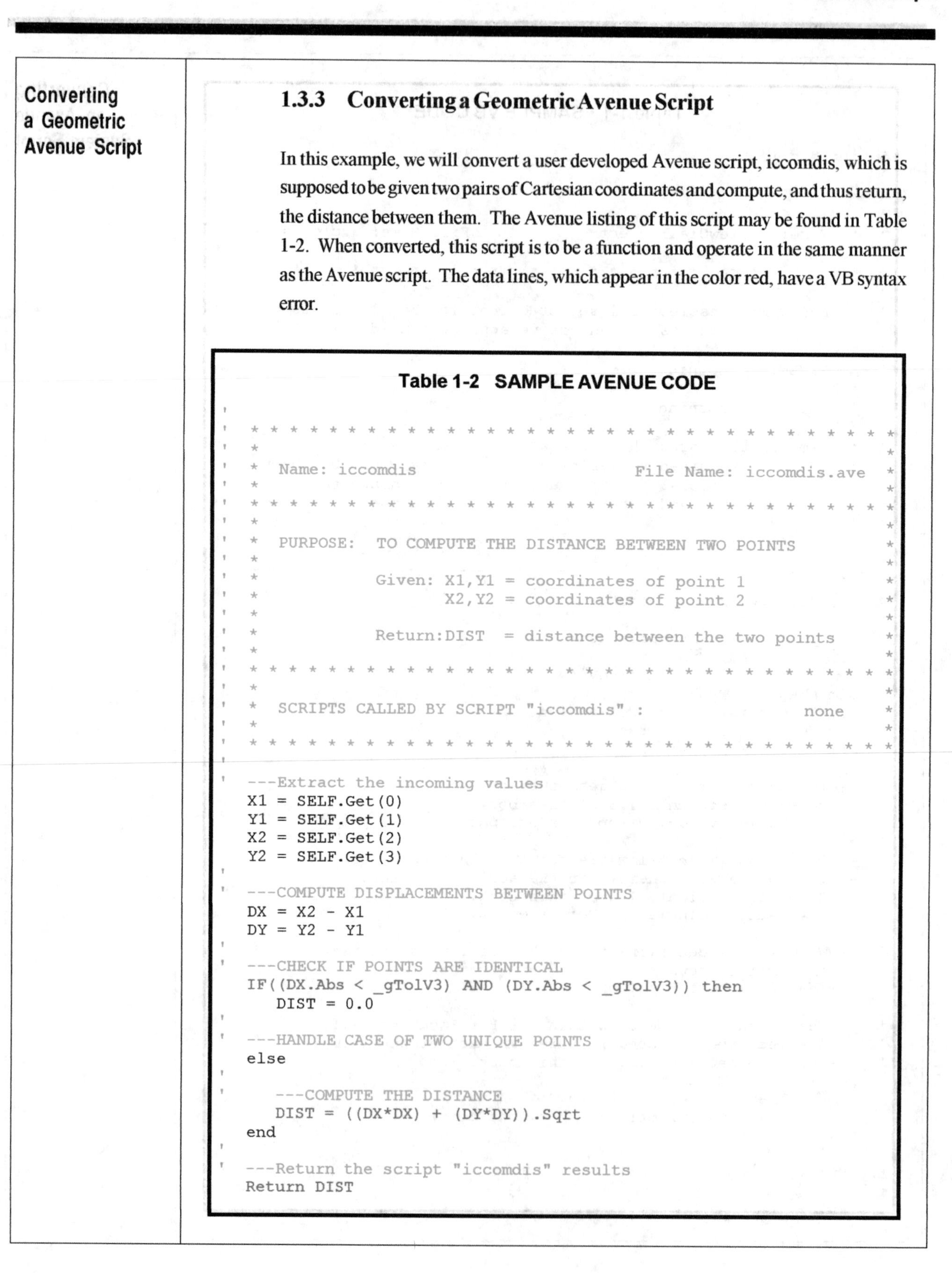

Converting a Geometric Avenue Script

1.3.3 Converting a Geometric Avenue Script

In this example, we will convert a user developed Avenue script, iccomdis, which is supposed to be given two pairs of Cartesian coordinates and compute, and thus return, the distance between them. The Avenue listing of this script may be found in Table 1-2. When converted, this script is to be a function and operate in the same manner as the Avenue script. The data lines, which appear in the color red, have a VB syntax error.

Table 1-2 SAMPLE AVENUE CODE

```
'
' * * * * * * * * * * * * * * * * * * * * * * * * * * * * * * * * *
' *                                                                 *
' *  Name: iccomdis                     File Name: iccomdis.ave    *
' *                                                                 *
' * * * * * * * * * * * * * * * * * * * * * * * * * * * * * * * * *
' *                                                                 *
' *  PURPOSE:  TO COMPUTE THE DISTANCE BETWEEN TWO POINTS           *
' *                                                                 *
' *           Given: X1,Y1 = coordinates of point 1                 *
' *                  X2,Y2 = coordinates of point 2                 *
' *                                                                 *
' *           Return:DIST  = distance between the two points        *
' *                                                                 *
' * * * * * * * * * * * * * * * * * * * * * * * * * * * * * * * * *
' *                                                                 *
' *  SCRIPTS CALLED BY SCRIPT "iccomdis" :              none        *
' *                                                                 *
' * * * * * * * * * * * * * * * * * * * * * * * * * * * * * * * * *
'
' ---Extract the incoming values
X1 = SELF.Get(0)
Y1 = SELF.Get(1)
X2 = SELF.Get(2)
Y2 = SELF.Get(3)
'
' ---COMPUTE DISPLACEMENTS BETWEEN POINTS
DX = X2 - X1
DY = Y2 - Y1
'
' ---CHECK IF POINTS ARE IDENTICAL
IF((DX.Abs < _gTolV3) AND (DY.Abs < _gTolV3)) then
   DIST = 0.0
'
' ---HANDLE CASE OF TWO UNIQUE POINTS
else
'
'    ---COMPUTE THE DISTANCE
   DIST = ((DX*DX) + (DY*DY)).Sqrt
end
'
' ---Return the script "iccomdis" results
Return DIST
```

Converting a Geometric Avenue Script

The first three steps of this conversion process for this script are the same as those of the preceding conversion process, so that once we have entered the VBE work environment, we will commence with Step 4.

➢ 4 **Click** at the **File** menu and then at the **Import** sub-menu of Figure 1-4. The conventional Window file browser window is now displayed.

➢ 5 **Navigate** to the appropriate directory and **select** the desired file, in this example the file is called iccomdis.ave. The contents of Table 1-2 are displayed.

Table 1-3 SAMPLE VB CODE

```
Option Explicit
'
'  * * * * * * * * * * * * * * * * * * * * * * * * * * * * * * * * *
'  *                                                                 *
'  *  Name: iccomdis                        File Name: iccomdis.bas  *
'  *                                                                 *
'  * * * * * * * * * * * * * * * * * * * * * * * * * * * * * * * * *
'  *                                                                 *
'  *  PURPOSE:  TO COMPUTE THE DISTANCE BETWEEN TWO POINTS           *
'  *                                                                 *
'  *  GIVEN:    X1,Y1    = coordinates of the first point            *
'  *            X2,Y2    = coordinates of the second point           *
'  *                                                                 *
'  *  RETURN:   iccomdis = distance between the two points           *
'  *                                                                 *
'  *  Dim X1, Y1, X2, Y2, iccomdis As Double                         *
'  *                                                                 *
'  * * * * * * * * * * * * * * * * * * * * * * * * * * * * * * * * *
'

Public Function iccomdis(X1, Y1, X2, Y2) As Double
'
   Dim DX, DY As Double
'
'  ---COMPUTE DISPLACEMENTS BETWEEN POINTS
   DX = X2 - X1
   DY = Y2 - Y1
'
'  ---CHECK IF POINTS ARE IDENTICAL
   If ((Abs(DX) < ugTolV3) And (Abs(DY) < ugTolV3)) Then
      iccomdis = 0#
'
'  ---HANDLE CASE OF TWO UNIQUE POINTS
   Else
'
'     ---COMPUTE THE DISTANCE
      iccomdis = Sqr((DX * DX) + (DY * DY))
   End If
'
End Function
```

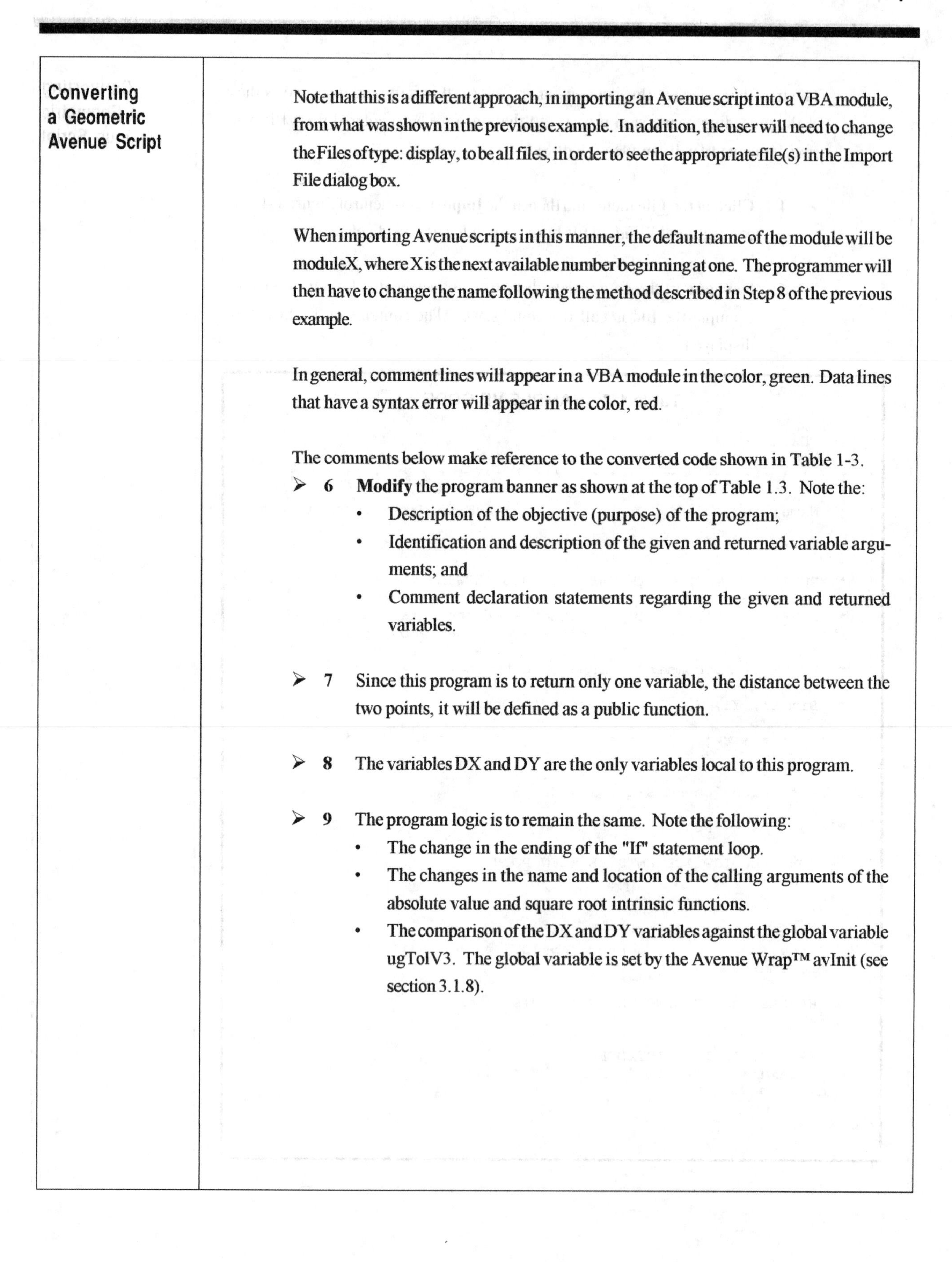

Converting a Geometric Avenue Script

Note that this is a different approach, in importing an Avenue script into a VBA module, from what was shown in the previous example. In addition, the user will need to change the Files of type: display, to be all files, in order to see the appropriate file(s) in the Import File dialog box.

When importing Avenue scripts in this manner, the default name of the module will be moduleX, where X is the next available number beginning at one. The programmer will then have to change the name following the method described in Step 8 of the previous example.

In general, comment lines will appear in a VBA module in the color, green. Data lines that have a syntax error will appear in the color, red.

The comments below make reference to the converted code shown in Table 1-3.

➢ 6 **Modify** the program banner as shown at the top of Table 1.3. Note the:

- Description of the objective (purpose) of the program;
- Identification and description of the given and returned variable arguments; and
- Comment declaration statements regarding the given and returned variables.

➢ 7 Since this program is to return only one variable, the distance between the two points, it will be defined as a public function.

➢ 8 The variables DX and DY are the only variables local to this program.

➢ 9 The program logic is to remain the same. Note the following:

- The change in the ending of the "If" statement loop.
- The changes in the name and location of the calling arguments of the absolute value and square root intrinsic functions.
- The comparison of the DX and DY variables against the global variable ugTolV3. The global variable is set by the Avenue Wrap™ avInit (see section 3.1.8).

1.3.4 Converting Another Geometric Avenue Script

Converting a Geometric Avenue Script

This example is to convert a user developed Avenue script, iccomppt, which is supposed to compare the Cartesian coordinates of a point with those of another given point, and return an indicator flag concerning the comparison, as well as a new pair of coordinates for the first point depending upon whether a match is made or not. The Avenue listing of this script may be found in Table 1-4. When converted, this script is to be a subroutine and operate in the same manner as the Avenue script.

The first three steps of this conversion process for this script are the same as those of the preceding conversion process, thus we will commence with Step 4.

➤ 4 **Click** at the **File** menu and then at the **Import** sub-menu of Figure 1-4. The conventional Window file browser window is now displayed.

➤ 5 **Navigate** to the appropriate directory and **select** the file, iccomppt.ave. The contents of Table 1-4 are displayed.

The comments below make reference to the converted code shown in Table 1-5.

➤ 6 **Modify** the program banner as shown at the top of Table 1.5. Note the:
- Description of the objective (purpose) of the program;
- Identification and description of the given and returned variable arguments; and
- Comment declaration statements regarding the given and returned variables.

➤ 7 Since this program is to return more than one variable, it will be defined as a public subroutine.

➤ 8 Following the definition of the subroutine are the declarations of the various variables local to this subroutine. When a program is small, say a page or two or thereabouts, it is relatively easy to identify all variables and declare them. However, let us be realistic. If a program is large, some if not many of the variables will not be recognized for declaration until the testing period, when the compiler will bring those variables to the programmer's attention. This will occur because the Option Explicit statement appears in the module.

Converting a Geometric Avenue Script

Table 1-4 SAMPLE AVENUE CODE

```
'
' * * * * * * * * * * * * * * * * * * * * * * * * * * * * * * * * *
' *                                                                 *
' * Name: iccomppt                      File Name: iccomppt.ave   *
' *                                                                 *
' * * * * * * * * * * * * * * * * * * * * * * * * * * * * * * * * *
' *                                                                 *
' *  PURPOSE:  CHECK IF A GIVEN POINT MATCHES ANOTHER POINT USING *
' *            A TOLERANCE BASED UPON THE DISPLAY OF THE VIEW      *
' *            Given:  XCORD,YCORD = coord's point to be matched  *
' *                    X2,Y2       = coord's to be matched against*
' *            Return: XCORD,YCORD = input values if NOFND = 0    *
' *                                = X2,Y2 values if NOFND = 1    *
' *                    NOFND = 0 : no match was found              *
' *                          = 1 : match found within tolerance.  *
' *                                                                 *
' * * * * * * * * * * * * * * * * * * * * * * * * * * * * * * * * *
'
' ---Get the active view                                  <<<------
  theView = av.GetActiveDoc
'
' ---Extract the incoming values
  XCORD = SELF.Get(0)
  YCORD = SELF.Get(1)
  X2    = SELF.Get(2)
  Y2    = SELF.Get(3)
'
' ---Obtain the snap tolerance value
  returnList = av.Run("SetViewSnapTol",{theView,X2,Y2})
'
' ---Set the tolerance using the projected coordinate system since
' ---the coordinates being used are in a projected coordinate system
  difxxx = returnList.Get(4)
'
' ---DEVELOP ENCLOSING BOX ABOUT POINT
  XTOLUP = X2 + difxxx
  XTOLDN = X2 - difxxx
  YTOLUP = Y2 + difxxx
  YTOLDN = Y2 - difxxx
'
' ---CHECK IF PICK WITHIN ENCLOSING BOX
  IF((XCORD > XTOLUP) or (XCORD < XTOLDN)) then
     NOFND = 0
'
  elseIF((YCORD > YTOLUP) or (YCORD < YTOLDN)) then
     NOFND = 0
'
' ---POINT MATCHES SPECIFIED POINT
  else
     NOFND = 1
'
'    ---ADJUST GIVEN COORDINATES
     XCORD = X2
     YCORD = Y2
  end
'
' ---Return the script "iccomppt" results
  Return {XCORD,YCORD,NOFND}
```

Converting a Geometric Avenue Script

Table 1-5 SAMPLE VB CODE

```
Option Explicit
'
' * * * * * * * * * * * * * * * * * * * * * * * * * * * * * * * * *
' *                                                                 *
' *  Name: iccomppt                     File Name: iccomppt.bas     *
' *                                                                 *
' * * * * * * * * * * * * * * * * * * * * * * * * * * * * * * * * *
' *                                                                 *
' *  PURPOSE:  CHECK IF POINT IS WITHIN A TOLERANCE, THAT VARIES    *
' *            BASED UPON THE VIEW DISPLAY, OF ANOTHER POINT        *
' *                                                                 *
' *  GIVEN:    XCORD,YCORD = coordinates of point to be checked     *
' *            X2,Y2       = coordinates of the base point          *
' *                                                                 *
' *  RETURN:   XCRD2,YCRD2 = input values if NOFND = 0              *
' *                        = X2,Y2 values if NOFND = 1              *
' *            NOFND       = 0 : no match was found                 *
' *                        = 1 : match was found within the point   *
' *                              snapping tolerance.                *
' *                                                                 *
' *  Dim XCORD, YCORD, X2, Y2, XCRD2, YCRD2 As Double               *
' *  Dim NOFND As Integer                                           *
' *                                                                 *
' * * * * * * * * * * * * * * * * * * * * * * * * * * * * * * * * *
'

Public Sub iccomppt(XCORD, YCORD, X2, Y2, XCRD2, YCRD2, noFnd)
'
   Dim pMxApp As IMxApplication
   Dim pmxDoc As IMxDocument
   Dim pActiveView As IActiveView
   Dim XTOLUP, XTOLDN, YTOLUP, YTOLDN As Double
   Dim viewRect, difxxx, difzzz, difPrj As Double
   Dim thePoint As IPoint
'
'  ---Get the active view
   Set pMxApp = Application
   Set pmxDoc = Application.Document
   Set pActiveView = pmxDoc.ActiveView
'
'  ---Initialize the coordinates to be passed back
   XCRD2 = XCORD
   YCRD2 = YCORD
'
'  ---Obtain the snap tolerance value based upon the current
'  ---view extent
   Call SetViewSnapTol(pmxDoc, X2, Y2, _
                       viewRect, thePoint, difxxx, difzzz, difPrj)
'
'  ---DEVELOP ENCLOSING BOX ABOUT POINT
   XTOLUP = X2 + difPrj
   XTOLDN = X2 - difPrj
   YTOLUP = Y2 + difPrj
   YTOLDN = Y2 - difPrj
```

Converting a Geometric Avenue Script

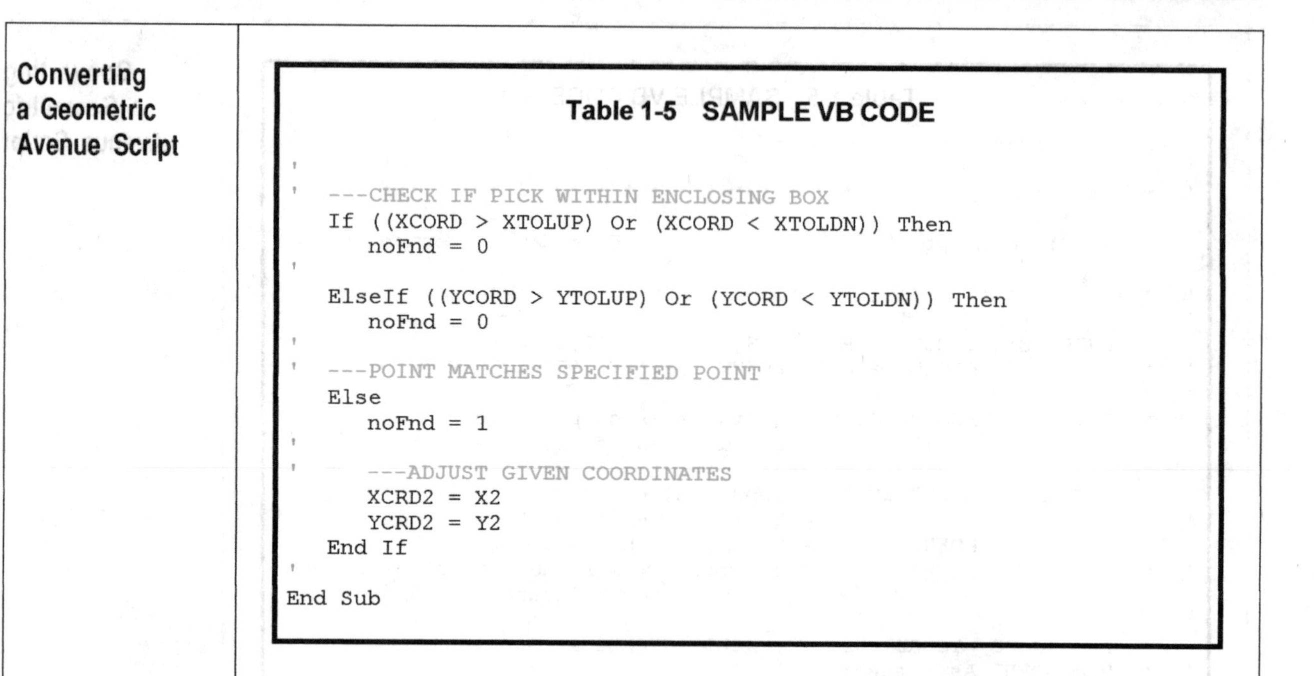

Table 1-5 SAMPLE VB CODE

```
'
'   ---CHECK IF PICK WITHIN ENCLOSING BOX
    If ((XCORD > XTOLUP) Or (XCORD < XTOLDN)) Then
       noFnd = 0
'
    ElseIf ((YCORD > YTOLUP) Or (YCORD < YTOLDN)) Then
       noFnd = 0
'
'   ---POINT MATCHES SPECIFIED POINT
    Else
       noFnd = 1
'
'      ---ADJUST GIVEN COORDINATES
       XCRD2 = X2
       YCRD2 = Y2
    End If
'
End Sub
```

➢ **9** The program logic is to remain the same. Note the following

- The changes in how the subroutine SetViewSnapTol is called. In Avenue, the script, SetViewSnapTol, returned a list, returnList, which contained five items. In the VB code, the subroutine, SetViewSnapTol, passes back the 5 items, not in a list, but as individual items. It is the programmer's discretion as to how to handle the returned arguments in a subroutine call.
- The change in the ending of the "If" statement loop.

CHAPTER 2

General Conversion Guides and Avenue Wraps

This chapter presents a set of general syntax guidelines for converting ArcView® 3.x Avenue code to "VB code", and addresses (a) numeric variables and arithmetic operations, (b) string variables and manipulation thereof, (c) transcendental and other intrinsic functions, (d) query of variables and "If" statements, (e) lists, arrays and collections, (f) data type declaration, definition and conversion, (g) iterative operations such as "Do", "For" and "While" loops, (h) miscellaneous general types of operations, such as getting the current date and time, system alert sound, and summary of declaration of variables, and (i) the use of certain Avenue Wraps of general nature, a list of which is presented below and overleaf.

In the section describing the use of the Avenue Wraps, the user will find the Avenue Wrap's corresponding Avenue request, the description of the input and output (returned) variables, and variable declaration. As a reminder, keep in mind that the variables within the argument list should be declared in the module that first initializes or defines these variables.

It was stated in Chapter 1, and it is worth repeating here, that Avenue requests could be concatenated, by separating each request with a period (.). Avenue Wraps, however, cannot be concatenated. As an example, in Avenue the two requests, GetFTab and FindField could be concatenated in a single statement:

```
aField = aTheme.GetFTab.FindField(aFieldName)
```

However, with Avenue Wraps, each request must appear as a separate statement:

```
Call avGetFTab(pmxDoc,aTheme,aFTab,pFeatClass,pLayer)
aField = aFTab.FindField(aFieldName)
```

The Avenue Wraps of this chapter are listed below in alphabetical order with a short description and the chapter - page number where a full description may be found.

GENERAL CONVERSION GUIDES & AVENUE WRAPS

The source listing of each of the above Avenue Wraps may be found in Appendix D of this book.

2.1 Numbers and Arithmetic Operations

NUMBERS AND ARITHMETIC OPERATIONS

2.1.1 Variable Types and Declarations

In Avenue there is no distinction between types of numeric variables. In VB programming, however, there is a distinction between whole numbers (numbers with no fractional part - no decimal point) and floating point numbers (numbers with fractional part - decimal point), and each one of them is further classified based upon the precision of the associated value. Thus, in VB we have variables that may contain:

- **Integer** numbers that are stored as 16-bit (2-byte) numbers ranging in value from -32,768 to 32,767. Such variables should be declared as:
  ```
  Dim theNumber As Integer
  ```
  Note that in some other programming languages, these variables are referred to as Short integers.
- **Long** integer numbers that are stored as signed 32-bit (4-byte) numbers ranging in value from -2,147,483,648 to 2,147,483,647. Such variables should be declared as:
  ```
  Dim theNumber As Long
  ```
- **Single** precision floating point numbers that are stored as IEEE 32-bit (4-byte), and ranging in value from -3.402823E38 to -1.401298E-45 for negative values and from 1.401298E-45 to 3.402823E38 for positive values. Such variables should be declared as:
  ```
  Dim theNumber As Single
  ```
- **Double** precision floating point numbers that are stored as IEEE 64-bit (8-byte), and ranging in value from -1.79769313486231E308 to -4.94065645841247E-324 for negative values and from 4.94065645841247E-324 to 1.79769313486232E308 for positive values.. Such variables should be declared as:
  ```
  Dim theNumber As Double
  ```

When declaring variables of the same type, more than one variable may appear on the same line, such as:

```
Dim theNumb1, theNumb2, theNumb3 As Double
Dim theNumb4, theNumb5, theNumb6 As Long
```

In view of the above, it is important to distinguish between the four number types when programming in VB. The difference between the short and long integers, and single and double precision is the precision of the numbers and memory requirements. As a general rule, the Avenue Wraps use **long** integers for all counters and loop indices, and short **integers** for all others. As for

NUMBERS AND ARITHMETIC OPERATIONS

floating numbers, Avenue wraps utilized **double** precision variables for all variables associated with geometric related operations. All others are dependent upon their specific application and need.

2.1.2 Arithmetic Operations

There is no difference between Avenue and VB with regards to the most common arithmetic notation (symbology of operations) or sequence of operations, and use of parentheses. Hence, any such code may be ported directly from Avenue to VB. However, there are some subtle differences of which the novice VB programmer should be cognizant. These differences pertain to operators, and to transcendental and other intrinsic functions. The available operators in VB compared to those of Avenue are presented below:

Operation	In Avenue	In VB	Comments
Exponentiation	^	^	
Multiplication	*	*	
Division	/	/	Returns a floating number
Division	Not available	\	Returns an integer number
Addition	+	+	
Subtraction	-	-	

2.1.3 Intrinsic Functions

Whereas in Avenue a function was a request invoked by its name being preceded by its argument list and a period (.), in VB the function is invoked by its name followed by its argument list enclosed in parentheses. Furthermore, there is a difference in some of the function names, and whereas a function may be available in Avenue, it is not so in VB, and vise versa. For examples refer to Table 2-1.

TABLE 2-1 INTRINSIC FUNCTIONS

Function	In Avenue	In VB	Function	In Avenue	In VB
Absolute value	A = B.**Abs**	A = **Abs**(B)	Arccosine	ANG = A.**ACos**	Not available
Arcsine	ANG = A.**ASin**	Not available	Arctangent	ANG = A.**ATan**	ANG = **Atn**(ANG)
Cosine	A = ANG.**Cos**	A = **Cos**(ANG)	e^B	Not available	A = **Exp**(B)
Natural Log	A = **Ln**(b)	A = **Log**(B)	Log of a base	A = **Log**(10)	Not available
Modulo	A = B.**Mod**(C)	Not available	Random number	Not available	A = **Rnd**(B)
Signum	Not available	A = **Sgn**(B)	Sine	A = ANG.**Sin**	A = **Sin**(ANG)
Square root	C = (A^2+B^2).**Sqrt**	C = **Sqr**(A^2+B^2)	Tangent	A = ANG.**Tan**	A = **Tan**(ANG)

NOTES:
1. The angles in all trigonometric functions are in radians
2. For the Arcsine and Arctangent functions, see also the icasinan and icatan functions in Chapter 12

NUMBERS AND ARITHMETIC OPERATIONS

TABLE 2-2
ROUNDING AND TRUNCATION OF NUMBERS

Given B =	if A is to be	In Avenue Use	In VB Use	
10.2	10	A = B.**Floor**	B = **Int**(A)	or
		A = B.**Round**	B = **Fix**(A)	or
		A = B.**Truncate**		
10.5	10	A = B.**Floor**	B = **Int**(A)	or
		A = B.**Round**	B = **Fix**(A)	or
		A = B.**Truncate**		
10.8	10	A = B.**Floor**	B = **Int**(A)	or
		A = B.**Truncate**	B = **Fix**(A)	
10.2	11	A = B.**Ceiling**	B = **Int**(A) + 1	or
			B = **Fix**(A) + 1	
10.5	11	A = B.**Ceiling**	B = **Int**(A) + 1	or
			B = **Fix**(A) + 1	
10.8	11	A = B.**Ceiling**	B = **Int**(A) + 1	or
		A = B.**Round**	B = **Fix**(A) + 1	
-10.2	-10	A = **Ceiling**(B)	B = **Fix**(A)	or
		A = B.**Round**		or
		A = B.**Truncate**		
-10.5	-10	A = **Ceiling**(B)	B = **Fix**(A)	or
		A = B.**Round**		or
		A = B.**Truncate**		
-10.8	-10	A = **Ceiling**(B)	B = **Fix**(A)	or
		A = B.**Truncate**		
-10.2	-11	A = B.**Floor**	B = **Int**(A)	
-10.5	-11	A = B.**Floor**	B = **Int**(A)	
-10.8	-11	A = B.**Floor**	B = **Int**(A)	or
		A = B.**Round**		

NUMBERS AND ARITHMETIC OPERATIONS

2.1.4 Rounding and Truncation of Numbers

Table 2-2 identifies the use, and hence the comparison between the various functions that are available in Avenue and VB.

2.1.5 String Messages

There are several Avenue Wraps contained in Chapter 6 that enable the programmer to display various types of message boxes, or to display messages in the status bar, all of which require the input of a message box title or heading, and/or an instruction. These may be specified as direct text in the Avenue Wrap subroutine or function, or in the form of a variable. In either case, there are four conversion issues that should be kept in mind. Note that these issues represent generic string manipulation rules and they are not specific to the message boxes.

When copying Avenue source code and pasting it on a VB procedure, if there are two plus signs (++) in a statement line, and there are no other conversion errors in that statement, or they have been corrected, one of the two plus signs will disappear. Hence, first take care of the two pluses.

- **Concatenation**: Two strings may be concatenated to form one by use of the plus (+) sign. This is the same in both Avenue and VB. However, if a space is required between the two strings, in Avenue the programmer could introduce two consecutive plus signs (++) to denote an extra space. This is not possible in VB. If an additional space is desired it must be so introduced between double quotes if it is to separate two numeric variables, or be incorporated at the end of the preceding string, or at the start of the subsequent string.

- **Number Conversion** In Avenue, a number was converted to a string with the request AsString. In VB, such conversion is typically made with the function CStr.

- **New Message Line** In Avenue, a new line was introduced in a message string by introducing the characters +NL+ between two strings. In VB, this is done by introducing the function Chr(13) within two plus signs.

- **Program Continuation Lines** In Avenue, the program was able to break the code and continue it in the next line. This is not so in VB. To continue a statement onto the next line, a space and an underscore (_) must appear at the end of the line to be continued.

As an example of the above consider the Avenue code below and its conversion to VB. Note: (a) the conversion of ++ to + and the introduction of the space character(s) in the hard-coded strings, (b) the substitution of

NUMBERS AND ARITHMETIC OPERATIONS

CStr for .AsString, (c) the substitution of Chr(13) for NL, and (d) the introduction of " _"to continue the statement on another line.

With Avenue

```
MsgBox.Warning("The lengths"++D1.AsString++
"and"++D2.AsString+NL+"are invalid", aTitle)
```

With Avenue Wraps

```
Call avMsgBoxWarning("The lengths " + CStr(D1) + _
" and " + CStr(D2) + Chr(13) + "are invalid", aTitle)
```

MANIPULATION OF STRING VARIABLES

TABLE 2-3
STRING MANIPULATION FUNCTIONS

	In Avenue	In VB
Concatenate two strings	aString1+ aString2	aString1+ aString2
Concatenate two strings separated by a space	String1++ aString2	aString1+ " " + aString2
Capitalize the first letter of a word in a string when words are separated with input character	String1.**BasicProper** (chr)	Not available
Remove from the start and end of a string the specified characters	String1.**BasicTrim** (L, R)	**avBasicTrim**(String1,L,R)
Extract the word at the specified position (0 to N-1), where N is the number of words (space delimited)	String1.**Extract** (Position)	Not available
Change all string characters to lower case	String1.**LCase**	**LCase**(String1)
Return the specified left most characters	String1.**Left** (nChr)	**Left**(String1, nChr)
Extract a string, starting at a specified offset, the specified number of characters	String1.**Middle** (Off, nChr)	**Mid**(String1, Off, nChr)
Capitalize the first letter of a word in a string when words are separated with blank space	String1.**Proper**	Not available
Place a string within a pair of quotes	String1.**Quote**	Not available
Return the specified left most characters	String1.**Right** (nChr)	**Right** (String1,nChr)
Introduce at the positions shown in the cntrList, the str characters	String1.**Split**(cntrList,str)	Not available
Replace in a string all occurrences of a1 with a2	String1.**Substitute** (a1,a2)	**Replace**(String1, a1, a2)
Replace in a string all occurrences of a1 with a2	String1.**Translate** (a1,a2)	Not available
Remove leading & trailing spaces	String1.**Trim**	**Trim**(String1)
Change all string characters to upper case	String1.**UCase**	**UCase**(String1)
Remove the pair of quotes from a string within a pair of quotes	String1.**Unquote**	Not available

MANIPULATION OF STRING VARIABLES

2.2 Manipulation of String Variables

2.2.1 String Manipulation Requests and Functions

There are several text string manipulation requests in Avenue, most all of which have to be converted to VB code. The sole exception is the concatenation of two strings with a plus sign (+) to create a single new string. Three of the string manipulation requests have been addressed in the preceding section. Shown in Table 2-3 are the various Avenue string manipulation requests and their counterparts, if any in VB. In addition to the requests of Table 2-3, the following are considered as requests of rather common use:

- To determine the number of characters in a string:

The **Avenue** request is: `nChars = theString.Count`

The **VB** function is: `nChars = Len(theString)`

with the variables declared as:

```
Dim nChars As Long
Dim theString As String
```

TABLE 2-4
BOOLEAN QUERYING OF VARIABLES AND IF STATEMENTS

The concatenation of more than one if condition in an "if" statement is the same in both Avenue and VB

In Avenue	In VB

- Querying whether a string variable is a number

```
If(theString.IsNumber)Then          If(IsNumeric(theString))Then
    ...do something                     ...do something
End                                 End If
```

- Querying whether a string variable is not a number

```
If(theString.IsNumber.Not)Then      If(Not IsNumeric(theString))Then
    ...do something                     ...do something
End                                 End If
```

- Querying whether a string variable has not been defined

```
If(theString.IsNull)Then            If(IsNull(theString))Then
    ...do something                     ...do something
End                                 End If
```

- Querying whether a string variable has been defined

```
If(theString.IsNull.Not)Then        If(Not IsNull(theString))Then
    ...do something                     ...do something
End                                 End if
```

- Querying whether a string variable has not been defined

```
If(theString = Nil)Then             If(IsNull(theString) Then
```

MANIPULATION OF STRING VARIABLES

2.2.2 Querying Variables and If Statements

At times it is necessary to query a variable in order to determine the type of the variable, and/or to change a variable from one type to another. For example, one may wish to change a number into a string so that it may be incorporated in a message box, or convert a number, which is in the form of a string into a number, so as to perform arithmetic operations. The latter often occurs when all data (text and numbers) of an application are read into the program as strings. Regarding the conversion of numbers, stored as strings, to variables of number type, attention has to be paid as to the type of number, integer, long, single or double. In Avenue there is no distinction between these types, they are all converted in the same manner. This is not so in VB.

Since in Avenue the "If", "For" and "While" statements all terminate with an "End" statement, it may be a good idea if the first thing to be done when converting to VB is to convert all "End" statements of an "If" statement to "End If".

Generally, querying of variables is done with the "If...Then...Elseif...Else...End" statement in Avenue or with the "If...Then...Elseif...Else...End If" statement in VB, with the Elseif and Else parts being optional in either Avenue or VB. Note that the operation of the If statement is the same in both Avenue and VB. The only difference between them being the ending statement.

In Avenue, an "If" query terminates with the word "End", while in VB it terminates with the words "End If". A compilation error will be displayed if an "If" statement does not terminate with "End If".

When querying a variable, although most of the times a positive (true) response is expected, at times a negative response is desired (false). Also, the variable theString must have been declared as a string or variant.

The most common commands regarding queries of variables and "If" statements are contained in Table 2-4, including positive and negative tests.

2.2.3 Converting and Initializing Variables

At times it is necessary to convert a variable from one type to another for most of the same reasons stated in the preceding section on querying variables.

- Setting a string variable to be undefined. This is usually performed at the beginning of a program to initialize a variable.

In Avenue	**In VB**
`theString = Nil`	`theString = Null`

MANIPULATION OF STRING VARIABLES

Alternatively to the example above, in both Avenue and VB, a variable may be initialized to some number of string, which is bound to indicate an initialization such as:

A=999999.0	which in VB will display as 999999#
B=0.0	which in VB will display as 0#
I=-1	which in VB will also display as -1
theString = " "	which in VB will also display as " "

- Converting to a number a string which has been proven to be a number. Note the distinction in VB between the four aforementioned types of numbers.

In Avenue

```
theNumber = theString.AsNumber
```

In VB

`theNumber = CInt(theString)` to change the string into an integer in the range of -32,768 to 32,767; fractions are rounded to the nearest integer.

`theNumber = CLng(theString)` to change the string into a long integer in the range of -2,147,483,648 to 2,147,483,647; fractions are rounded to the nearest integer.

`theNumber = CSng(theString)` to change the string into a single precision floating number in the range of -3.402823E38 to -1.401298E-45 for negative values; 1.401298E-45 to 3.402823E38 for positive values.

`theNumber = CDBL(theString)` to change the string into a double precision floating number in the range of -1.79769313486231E308 to -4.94065645841247E-324 for negative values; 4.94065645841247E-324 to 1.79769313486232E308 for positive values,

`theVariant = CVar(aVariable)` to change a string, integer, long, single or double number into a variant

LISTS, ARRAYS AND COLLECTIONS

TABLE 2-5
LISTS AND COLLECTIONS

In Avenue	In VB	
◗ To **create** or initialize a collection		
`aList = List.Make`	`Call CreateList(aList)`	
◗ To **append** an item to a collection (see below for inserting an item)		
`aList.Add(aVal)`	`aList.Add(aVal)`	(#)
	`aList.Add aObj`	(##)
◗ To **get** (extract) an item from a collection		
`aVal = aList.Get(j)`	`aVal = aList.Item(j)`	
◗ To **remove** an item from a collection		
`aList.Remove(j)`	`aList.Remove(j)`	
◗ To **insert** an item at the beginning of a collection		
`aList.Insert(aVal)`	`aList.Add (aVal), before:=1`	(#)
	`aList.Add aObj, before:=1`	(##)
◗ To **insert** an item within a collection - j in this case denotes the position (subscript) after which the item aVal is to be inserted		
`aList.Insert(aVal)`	`aList.Add(aVal), after:=j`	(#)
`aList.Shuffle(aVal,j)`	`aList.Add aVal, after:=j`	(##)
◗ To **replace** an item within a collection		
`aList.Set(j,aVal)`	`aList.Add(aVal), after:=j`	(#)
	`aList.Remove(j)`	
	`aList.Add aVal, after:=j`	(##)
	`aList.Remove(j)`	
◗ To **count** the number of items in a collection		
`nItems = aList.Count`	`nItems = aList.Count`	
◗ To **clear** a collection		
`aList.Empty`	`Call CreateList(aList)`	
◗ To **clone** a collection		
`aList1 = aList2.Clone`	`Set aList1 = aList2`	

Notes:

- For the Avenue Wrap CreateList refer to the end of this chapter and Appendix D
- (#) — Use this for non-objects (strings, numbers, etc.)
- (##) — Use this for objects (Collections and ArcObjects)
- j is measured from base 0 — j is measured from base 1, so that, all Avenue index values used in collections will need to be incremented by one

2.3 Lists, Arrays and Collections

LISTS, ARRAYS AND COLLECTIONS

2.3.1 Definitions

In Avenue, a grouping of items such as variables, themes, tables, views and others could constitute a list. In VB, lists are referred to as collections. In addition to collections, the user is able to utilize arrays, much the same way as one would in Fortran or C. In VB, arrays are declared based upon the type of data they are to contain, while collections are declared as themselves. Thus, the corresponding Dim statements for the following samples would be:

`Dim iAry(5) As Integer`	iAry is a one dimensional array.
`Dim jAry(2,30) As Long`	jAry is a two dimensional array.
`Dim kAry(2,4,6) As Single`	kAry is a three dimensional array.
`Dim mAry() As Double`	mAry is a dynamic array.
`Dim aCol As Collection`	aCol is declared to be a non-initialized collection (Nothing).
`Dim aCol As New Collection`	aCol is declared to be, and initiated as a zero-length or empty collection.

Note that collections are one dimensional only. Reference is made to the Avenue Wrap CreateList of this chapter which may be used to initialize and empty a collection. This Avenue Wrap is not applicable to arrays.

2.3.2 Working with Arrays

Working with arrays in VB is quite similar to working with arrays in Fortran or C. One may assign values to array cells, or extract values from such cells by referring to the array and the desired cell index. The programmer should note that the default base index of an array in VB is zero (0) and not one (1). However, it can be changed to one (1), if so desired, by introducing in the declaration section of the module the statement:

```
Option base 1
```

Another way to control the issue of array subscripts is to specify the low and upper bounds of the subscripts. For example, the declaration:

```
Dim iArray(15) As Integer
```

denotes a one dimensional array with 15 cells between 0 and 14, or between 1 and 15 if Option base 1 had been specified, while the declaration:

```
Dim iArray(3 To 7) As Integer
```

denotes a one dimensional array with 5 cells between 3 and 7. In the latter case there are no 0, 1 and 2 cells.

LISTS, ARRAYS AND COLLECTIONS

Note that you cannot use the "Add" and "Count" commands with an array. Some of the operations associated with arrays include:

- To extract a value from a cell use an equation:

```
aVal = theArray(j)
```

  where j must have been previously declared to be an integer or a long variable within the range limits of the Dim statement that declared theArray.

- To assign a value to a cell use an equation:

```
theArray(j) = aVal
```

  where j must have been previously declared to be an integer or a long variable within the range limits of the Dim statement that declared theArray.

2.3.3 Working with Collections

Working with collections is not quite the same as working with arrays. The differences are:

- A collection may contain variables of different type.

- The base index of a collection is one (1) and not zero (0), the opposite of arrays.

The most common commands regarding Avenue lists and VB collections, and the differences between them are contained in Table 2-5.

2.3.4 Sorting of Collections

In Avenue the request "Sort" is used to sort a list. In VB, the Avenue Wrap "SortTwoLists" may be used to sort one or two collections, but not arrays. When sorting two collections, SortTwoLists treats the two collections as a two dimensional array to be sorted under one sort key, that being the first collection. The use of this Avenue Wrap is presented later on in this chapter, while the source code of "SortTwoLists" may be found in Appendix D.

2.3.5 Copying of Collections

There are two Avenue Wraps that do not have Avenue counterparts, and which allow the programmer to copy one collection into another. The CopyList enables the programmer to copy a non-object collection into another non-object collection, while CopyList2 enables the programmer to copy an object collection into another object collection. The use of these two

LISTS, ARRAYS AND COLLECTIONS

Avenue Wraps is presented later on in this chapter, while their source listing may be found in Appendix D.

ITERATIVE OPERATIONS

2.4 Iterative Operations

2.4.1 The Iterative Statements

In Avenue there are only two iterative operation statements, the "For Each ... End" statement and the "While ... End" statement. In VB there are three, the "Do", the "For" and the "While ... Wend", of which:

- the "Do" has four variations, the "Do While ... Loop", the "Do Until ... Loop", the "Do ... Loop While", and the "Do ... Loop Until", and

- the "For" has two variations, the "For ... Next", and the "For Each ... Next".

2.4.2 Converting the Avenue "For Each ... End" Statement

In Avenue, this statement is comprised of the following lines:

```
For Each Rec in theList
    ... do something with Rec
End
```

where Rec is an object in theList and theList is a list.

To convert this statement, the programmer must be cognizant of what theList is comprised. If theList contains:

- Objects, then the programmer should use the following

  ```
  For Each Rec in theList
      ... do something with Rec
  Next Rec
  ```

  where Rec is an object in theList and declared accordingly, and theList is a collection of objects. In this example, theList contains objects of the same type.

- Variables, then the programmer should use the following

  ```
  For iRec = 1 To theList.Count
      Rec = theList.Item(iRec)
      ... do something with Rec
  Next iRec
  ```

  where iRec should be declared as an integer or long, theList may be a collection or array, and Rec as a variant. Note that in the above example, iRec is both a counter and an index to theList.

ITERATIVE OPERATIONS

Alternatively, the user may elect to compute the index to theList for which something is to be done.

```
K = 5
For I = iLow To iHigh
    K = K + 1
   ... do something with theList(K)
Next I
```

In using the above variables I, K, iLow and iHigh, the programmer should keep in mind that in a collection the base reference to an Avenue list is zero (0), while the base reference to a VB collection is one (1).

2.4.3 Converting the Avenue "While ... Wend" Statement

In Avenue, this statement is comprised of the following lines:

```
While Expression
   ... do something as long the Expression is true
End
```

In VB, the programmer may use any one of the four "Do" iterative statement variations depending on how the programmer wishes to set the conditional expression to be evaluated. For example, consider the following:

```
DoOver = True
Do While DoOver
   ... do something
   If (something) Then
      ... do some other things
   Else
      DoOver = False
   End If
Loop
```

Regarding the four variations of the "Do" statement, the programmer should note that:

- The "While" condition performs the operations between "Do" and "Loop" for as long as the conditional expression (DoOver in the above example) is true, while the "Until" condition performs said operations until said condition is met.

ITERATIVE OPERATIONS

- By placing the conditional test at the top with the "Do" loop, the subsequent statements are executed up to the "Loop" statement only if the condition is true. Thus the possibility exists that said subsequent statements may never be executed. By placing the conditional test at the bottom with the "Loop" said subsequent statements will be executed at least once.

In addition to one of the above four variations of the "Do" statement, the programmer may elect to use the "While ... Wend" statement, which represents a more direct one to one conversion between Avenue and VB, and has only one difference, the substitution of "Wend" for "End" in the ending statement of the iterative operation. While this may at first seem to be preferential, it does not provide as good of a structured approach as the "Do" statement, particularly if an early exit of the iterative process is desired.

2.4.4 Early Exit of an Iterative Statement

At times it becomes desirable to exit an iterative process earlier than provided by the conditions of the iterative processes. In Avenue, the programmer could exit an iterative process earlier than dictated by the conditions of a "For" or "While" statement by introducing the "Break" statement. In VB, the user has the following options:

- In any of the iterative processes, the user may terminate a subroutine or function without completing the entire iteration process by introducing the "Exit Sub", or "Exit Function" statement respectively.

- In the two variations of the "For" statement, the programmer may terminate the iterative process, and proceed to continue with the next statement after the "Next" statement by introducing the "Exit For" statement line.

- In the four variations of the "Do" statement, the programmer may terminate the iterative process, and proceed to continue with the next statement after the "Loop" statement by introducing the "Exit Do" statement line.

- The only way to prematurely exit a "While ... Wend" iterative process is with a "GoTo" statement (see the following section about *Advancing to the Next Iteration*).

ITERATIVE OPERATIONS

2.4.5 Advancing to the Next Iteration

At times it its desirable to skip to the next iteration from somewhere within the code of the iteration process. In Avenue, this can be accomplished with the "Continue" statement. Such a statement and function is not available in VB. One way to get around this problem is to restructure the code of the iteration routine perhaps with properly constructed "If" statements. Another way is with the use of the "GoTo" statement. As an example consider the following:

```
    DoOver = True
    K = 1
    While DoOver
       Do something that involves modification of K
       If (K > 0) Then
          ... do something else with K
       Elseif (K = 0) Then
          Exit Sub      ' If K=0 exit the subroutine
       Elseif (K < 0) Then
          GoTo Line 1   ' If K<0 skip remaining steps,
       End If           ' but do not exit subroutine
          ... continue doing something
Line 1                  ' Come here when K<0
    Wend
```

Note that a "GoTo" statement can be used in other instances, and more than once, in which case, different line numbers or text should be used. It is recommended that the use of this statement be a last resort case, because it does not create a well structured code, and can become confusing during the debugging stage.

MISCELLANEOUS OPERATIONS

2.5 Miscellaneous Operations

2.5.1 Current Time and Date

At times it is desirable to retrieve from the computing system the time and date that a program is being executed. This may be done as follows:

In Avenue

```
D  = Date.Now
d1 = D.SetFormat("d MMMM yyyy hhh m s").AsString
d2 = D.SetFormat("d MMMM yyyy").AsString
d3 = D.SetFormat("hhh m s").AsString
```

The second line (d1) above will get the date and time, the third line (d2) will get the date only, and fourth line (d3) will get the time only. The string appearing in the SetFormat statement may vary from what is shown above to meet a specific user format for the date and/or time.

In VB

```
aDate1 = Date
aDate2 = Now
aDate3 = FormatDateTime(aDate1,K)
MsgBox aDate1   prints   5/14/2002
MsgBox aDate2   prints   5/14/20029:28:11AM
MsgBox aDate3   prints   5/14/20029:28:11AM        ifK=0
                         Tuesday, May 14, 2002     ifK=1
                         5/14/2002                 ifK=2
                         9:28:11AM                 ifK=3
                         9:28                      ifK=4
```

2.5.2 System Beep

Usually when an error occurs during the execution of a program, or if an erroneous data is key entered in a form, it is a good idea for the program to issue a warning sound or beep. This is done as follows:

In Avenue

```
System.Beep
```

In VB

```
Beep
```

2.5.3 Variable Declarations

Although some of the following may have been addressed elsewhere in this book, it is felt that it is worth repeating. Before proceeding any further, it is necessary to distinguish between the

MISCELLANEOUS OPERATIONS

words "declare" and "define", and derivatives thereof. Each variable and object used in a program must first be **declared** as to its type (variant, integer, string, etc.). This is done with the Dim statement. Table 2-6 contains a summary of how various type of variables and objects should be declared. The list of declarations in this table is not by any means the complete list of declarations. Only the ones that are considered as the most common are presented therein.

TABLE 2-6
VB DECLARATION OF COMMON OBJECTS AND VARIABLES

Object/variable	Declaration Statement
Document	Dim pDoc As IMxDocument
Map	Dim pMap As IMap
Layer (theme)	Dim pLayer As ILayer
Table (FTab)	Dim aFTab As IFields
Collection	Dim aList As New Collection
Selection	Dim aSel As ISelectionSet
Point	Dim pPoint As IPoint
Lines	Dim pLine As IPolyline
Polygon	Dim pPolygon As IPolygon
Variant	Dim aNumber As Variant
Integer	Dim aNumber As Integer
Single precision	Dim aNumber As Single
Double precision	Dim aNumber As Double
String	Dim aString As String

In Avenue, all variables used in a script have to be **defined** or **initialized** prior to their use. That is, one could not say

```
A = B + 5.9
```

Unless B had been previously been assigned a value. Likewise, the statement below would be invalid

```
theFTab = theTheme.GetFTab
```

unless theTheme had previously been defined as a theme. However, there are variables and objects that for some reason need to be defined as null objects or empty variables. This implies that it will be desirable to also know

TABLE 2-7
LIST OF NULL DEFINITION IN VB

Object or variable	To define an object or variable as null or empty	To query whether an object or variable is null or empty
All objects	Set anObject = Nothing	If (anObject Is Nothing) Then
Variants	aVariant = Null	If (IsNull(aVariant)) Then
Strings	aString = Null	If (IsNull(aString)) Then
All numbers	not applicable	If (aNumber = 0) Then when a number has not been initialized

MISCELLANEOUS OPERATIONS

whether an object or a variable has been defined or not. In Avenue, the key word for such querying is "Nil". In VB, the corresponding word is "Null" for non-objects, and "Nothing" for objects, see Table 2-7 for a summary.

In VB we have the word "Empty" and the function "IsEmpty", which are associated with variables only, and not with objects. Thus, if we wish, for some reason or another to not define or initialize a variable, we can write:

```
B = Empty
... do something and then later on ask
If (Not IsEmpty(B)) Then
   A = B + 5.9
End If
```

This has no counterpart in Avenue.

2.5.4 Script Execution

In Avenue, the programmer was able to execute another script by using the av.Run statement. With VB code, the programmer executes another script by calling a subroutine or a function, depending upon how the other script has been implemented. Functions return one and only one value, while subroutines can return many, or none, values. For example:

In Avenue

```
myList = List.Make
myList.Add( TRUE )
returnValue = av.run( "script2", myList )
```

In VB with script2 implemented as a Subroutine

```
Dim myList As New Collection
Dim returnValue As Variant
Call CreateList(myList)
myList.Add (TRUE)
Call script2 (myList, returnValue)
```

In VB with script2 implemented as a Function

```
Dim myList As New Collection
Dim returnValue As Variant
Call CreateList(myList)
myList.Add (TRUE)
returnValue = script2 (myList)
```

Note, all references to the SELF statement must be replaced by putting the variables created with the SELF statement in the argument list of the subroutine or function.

GENERAL AVENUE WRAPS

2.6 General Avenue Wraps

2.6.1 Function avBasicTrim

This function enables the programmer to remove from a given string the specified leading and/or trailing characters.

The corresponding Avenue request is:

```
newString = theString.BasicTrim(LeadChar, TrailChar)
```

The call to this Avenue Wrap is:

newString = **avBasicTrim**(theString, LeadChar, TrailChar)

avBasicTrim

GIVEN:	theString	= the given string to be trimmed
	LeadChar	= the characters to be removed at the start of the given string
	TrailChar	= the characters to be removed at the end of the given string
RETURN:	newString	= the resultant string

The given and returned variables should be declared where first called as:

```
Dim theString, LeadChar, TrailChar As String
Dim newString As String
```

2.6.2 Function avClone

This function enables the programmer to make a new object by copying an existing object.

The corresponding Avenue request is:

```
theNewObject = theObject.Clone
```

The call to this Avenue Wrap is:

Set theNewObject = **avClone**(theObject)

avClone

GIVEN:	theObject	= object which is to be copied
RETURN:	theNewObject	= copy of the object

The given and returned variables should be declared where first called as:

```
Dim theObject As IUnknown
Dim theNewObject As IClone
```

GENERAL AVENUE WRAPS

2.6.3 Subroutine avExecute

This subroutine enables the programmer to execute a system level command. In using this subroutine, note that once the command has been issued, the statements that follow the call to avExecute will be immediately executed, there is no waiting for the system command to finish its processing. In order to pause ArcMap until said command is completed, one possibility is to perform a loop checking for the existence of a file, which could be created when said command has finished processing (see example below).

The corresponding Avenue request is:

System.Execute (aCommand)

avExecute

The call to this Avenue Wrap is:

Call **avExecute**(aCommand)

GIVEN: aCommand = the command to be executed

RETURN: nothing

The given and returned variables should be declared where first called as:
Dim aCommand As String

The code below is an example of how to invoke a program from within a VBA module. The program "Adjust" reads a file called "inFile" and will create a "dummy" file called "outFile" when its processing is complete.

```
Public Sub Test
Dim aCmnd, inFile, outFile As String
   .....
   Perform some operations to create "inFile"
   .....
   aCmnd = "c:\Dir1\SubDir3\Adjust.exe " + inFile
   Call avExecute(aCmnd)
   Do While (True)
      If (avFileExists(outFile)) Then
         Exit Do
      End If
   Loop
   .....
End Sub
```

GENERAL AVENUE WRAPS

2.6.4 Function avGetEnvVar

This function enables the programmer to get the full path for an environment variable. Below are examples of what is returned for what is given:

Given	**Return**
ARCHOME	C:\ARCGIS\ARCEXE81
TMP	C:\WINDOWS\TEMP
ABC	yields an empty string (""), assuming the ABC does not exist

The corresponding Avenue request is:

```
theEnvVar = System.GetEnvVar (aPath)
```

The call to this Avenue Wrap is:

theEnvVar = **avGetEnvVar**(aPath)

avGetEnvVar

GIVEN: aPath = name of the environment variable to be processed

RETURN: theEnvVar = full path name associated with the variable

The given and returned variables should be declared where first called as:

```
Dim aPath, theEnvVar As String
```

2.6.5 Subroutine avRemoveDupStrings

This function enables the programmer to remove duplicate strings or numbers from a list (collection). In addition, the programmer can specify whether the strings in the list are to be treated as case sensitive or case insensitive. That is, are upper and lower case characters to be treated the same. If they are *not* to be treated the same, this is referred to as being *case sensitive*.

The corresponding Avenue request is:

```
aList.RemoveDuplicates
```

The call to this Avenue Wrap is:

Call **avRemoveDupStrings**(aList, caseFlag)

avRemove DupStrings

GIVEN: aList = list of strings or numbers to be modified
caseFlag = flag denoting the case sensitivity of the list
True = case sensitive, False = insensitive

RETURN: nothing

GENERAL AVENUE WRAPS

The given and returned variables should be declared where first called as:
Dim aList As New Collection
Dim caseFlag As Boolean

2.6.6 Subroutine CopyList

This subroutine enables the programmer to copy a collection into another collection, and then initialize (clear) the original collection (the collection that was copied). Note that this subroutine operates only on non-object collections, collections containing variants, numbers and strings. To copy an object collection into another object collection the programmer must use the CopyList2 Avenue Wrap, which is presented later on.

The corresponding Avenue request is:
There is no corresponding Avenue request.

CopyList

The call to this Avenue Wrap is:
Call **CopyList**(origList, newList)

GIVEN: origList = list to be copied and then cleared

RETURN: newList = copy of the original list

The given and returned variables should be declared where first called as:
Dim origList As New Collection
Dim newList As New Collection

2.6.7 Subroutine CopyList2

This subroutine enables the programmer to copy a collection into another collection, and then initialize or clear the original collection (the one that was copied). Note that these collections contain objects, not variables such as strings, numbers and so forth. To copy a non-object collection into another non-object collection use the CopyList Avenue Wrap described above.

The corresponding Avenue request is:
There is no corresponding Avenue request.

CopyList2

The call to this Avenue Wrap is:
Call **CopyList2**(origList, newList)

GENERAL AVENUE WRAPS

GIVEN: origList = Object list to be copied and then cleared

RETURN: newList = Object copy of the original list

The given and returned variables should be declared where first called as:

```
Dim origList As New Collection
Dim newList As New Collection
```

2.6.8 Subroutine CreateList

CreateList

This subroutine enables the programmer to create a collection which contains either objects or variables, and initialize it to be an empty collection.

The corresponding Avenue request is:

```
newList = List.Make
```

The call to this Avenue Wrap is:

Call **CreateList**(newList)

GIVEN: nothing

RETURN: newList = the new empty collection

The given and returned variables should be declared where first called as:

```
Dim newList As New Collection
```

2.6.9 Function Dformat

This function enables the programmer to format for output a number according to a Fortran Fa.b format. In using this function, note the following regarding the given arguments of the function list:

- The given number (theNumber) to be formatted may be an integer or long number, or a single or double precision number.
- If the number of digits to the right of the decimal point (DigitsRight) is zero (0), the output string will contain only the whole number part of a floating number (no decimal point will be included).
- If the number of digits to the right of the decimal point (DigitsRight) is one (1) or higher, the output string will contain a decimal point and that many zeros to the right of the decimal point.
- If the number will not fit within the specified data field length as specified by TotalDigits, then the data field will be expanded to accommodate the number, automatically.

GENERAL AVENUE WRAPS

Dformat

The corresponding Avenue request is:

There is no corresponding Avenue request.

The call to this Avenue Wrap is:

theString = **Dformat**(theNumber, TotalDigits, DigitsRight)

GIVEN: theNumber = the number to be formatted
TotalDigits = the total number of characters, including leading spaces, decimal point and decimal digits, in the string to be passed back
DigitsRight = digits to the right of the decimal point

RETURN: theString = string representing the number in the specified format

The given and returned variables should be declared where first called as:

```
Dim theNumber As Integer     or as
Dim theNumber As Long        or as
Dim theNumber As Single      or as
Dim theNumber As Double
Dim TotalDigits, DigitsRight As Integer
Dim theString As String
```

2.6.10 Subroutine SortTwoLists

This subroutine enables the programmer to sort one or two different lists (collections, not arrays). When sorting two lists, the sorting of the second list corresponds to the sort of the first list. That is, SortTwoLists treats the two lists (List1 and List2) as a two dimensional array (List1 constituting the first column and List2 the second) to be sorted under one sort key, that of List1.

In using this subroutine, note the following:

- The order of the lists passed in are changed by this script to reflect the effects of the sort (List1 and List2 are modified by this subroutine).
- If only one list is to be sorted, the collection List2 can be an empty list, or passed in as Nothing.
- If Null is specified for aMssg, a progress bar will not be displayed during the sorting process.

GENERAL AVENUE WRAPS

There is no corresponding Avenue request for sorting two lists. The corresponding Avenue request to sort one list is:

```
aList.Sort(anOrder)
```

SortTwoLists

The call to this Avenue Wrap is:

Call **SortTwoLists**(List1, List2, aMssg, anOrder)

GIVEN:

List1	=	first list of items to be sorted
List2	=	second list of items to be sorted, if only one list to be sorted specify as NOTHING
aMssg	=	progress bar message, if no message is desired specify as NULL
anOrder	=	the sort order as a Boolean: True = ascending, and False = Descending

RETURN: nothing

The given and returned variables should be declared where first called as:

```
Dim List1 As New Collection
Dim List2 As New Collection
Dim aMssg As Variant
Dim anOrder As Boolean
```

GENERAL AVENUE WRAPS

CHAPTER 3

APPLICATION - DOCUMENT AVENUE WRAPS

This chapter contains various Avenue Wraps to assist a programmer in converting Avenue scripts that are associated with the establishment of an ArcView® 3.x application, specifically, its views, retrieval of active themes, and manipulation of the view display area including panning, zooming and refreshing, into "VB code".

The Avenue Wraps within this chapter are listed below in alphabetical order with a short description along with the chapter - page number where a full description can be found.

PROJECT ORGANIZATION

AVENUE WRAPS

The source listing of each of the above Avenue Wraps may be found in Appendix D of this publication.

3.1 Application - Document Related Avenue Wraps

APPLICATION - DOCUMENT AVENUE WRAPS

3.1.1 Function avAddDoc

This function enables the programmer to add a layer or table to the currently active map.

The corresponding Avenue request is:

```
aProject.AddDoc (aDoc)
```

The call to this Avenue Wrap is:

theLayer = **avAddDoc**(aDoc)

avAddDoc

GIVEN: aDoc = the document to be added

RETURN: theLayer = error flag
(0 denotes no error encountered)
(1 denotes an error has been encountered)

The given and returned variables should be declared where first called as:

```
Dim aDoc As IUnknown
Dim theLayer As Integer
```

3.1.2 Function avFindDoc

This function enables the programmer to obtain the index into the Table of Contents for a layer or table.

The corresponding Avenue request is:

```
aDoc = av.FindDoc (aDocName)                    or
aDoc = av.GetProject.FindDoc(aDocName)
```

The call to this Avenue Wrap is:

theIndex = **avFindDoc**(aDocName)

avFindDoc

GIVEN: aDocName = name of layer or table to be found

RETURN: theIndex = index into Table of Contents of the layer or table; if
= -1 the layer or table was not found

The given and returned variables should be declared where first called as:

```
Dim aDocName As Variant
Dim theIndex As Long
```

APPLICATION - DOCUMENT AVENUE WRAPS

3.1.3 Subroutine avGetActiveDoc

This subroutine enables the programmer to get the currently active document, view and focus map.

The corresponding Avenue request is:

```
aDoc = av.GetActiveDoc
```

avGetActiveDoc

The call to this Avenue Wrap is:

Call **avGetActiveDoc**(pMxApp, pmxDoc, pActiveView, pMap)

GIVEN: nothing

RETURN: pMxApp = the application
pmxDoc = the document
pActiveView = the active view
pMap = the focus map

The given and returned variables should be declared where first called as:

```
Dim pMxApp As IMxApplication
Dim pmxDoc As IMxDocument
Dim pActiveView As IActiveView
Dim pMap As IMap
```

3.1.4 Subroutine avGetActiveThemes

This subroutine enables the programmer to get a list of the currently active or selected layers (themes).

The corresponding Avenue request is:

```
ThemesList = aView.GetActiveThemes
```

avGetActiveThemes

The call to this Avenue Wrap is:

Call **avGetActiveThemes**(pmxDoc, ThemesList)

GIVEN: pmxDoc = the active view

RETURN: ThemesList = list of themes

The given and returned variables should be declared where first called as:

```
Dim pmxDoc As IMxDocument
Dim ThemesList As New Collection
```

3.1.5 Subroutine avGetName

This subroutine enables the programmer to get the application (project) name that appears in the upper left corner of the overall application window. Refer to the Avenue Wrap avSetName for setting the name of the application.

The corresponding Avenue request is:

```
aTitle = av.GetName
```

The call to this Avenue Wrap is:

Call **avGetName**(aTitle)

avGetName

GIVEN: nothing

RETURN: aTitle = the name of the application appearing in the upper left corner of the application window

The given and returned variables should be declared where first called as:

```
Dim aTitle As String
```

3.1.6 Subroutine avGetProjectName

This subroutine enables the programmer to get the name of the current document.

The corresponding Avenue request is:

```
aFileName = aProject.GetFileName
```

The call to this Avenue Wrap is:

Call **avGetProjectName**(aTitle)

avGetProjectName

GIVEN: nothing

RETURN: aTitle = the name of the current document, which will include the .mxd extension (i.e. sample.mxd)

The given and returned variables should be declared where first called as:

```
Dim aTitle As String
```

APPLICATION - DOCUMENT AVENUE WRAPS

3.1.7 Subroutine avGetWinFonts

This subroutine enables the programmer to get a list (collection) of the fonts that are installed on the computer.

The corresponding Avenue request is:

```
fontList = aFontManager.ReturnFamilies
```

avGetWinFonts

The call to this Avenue Wrap is:

Call **avGetWinFonts**(fontList)

GIVEN: Nothing

RETURN: fontList = list of fonts installed on the computer

The given and returned variables should be declared where first called as:

```
Dim fontList As New Collection
```

3.1.8 Subroutine avGetWorkDir

This subroutine enables the programmer to get the current working directory.

The corresponding Avenue request is:

```
theWorkDir = aProject.GetWorkDir
```

avGetWorkDir

The call to this Avenue Wrap is:

Call **avGetWorkDir**(theWorkDir)

GIVEN: Nothing

RETURN: theWorkDir = current working directory

The given and returned variables should be declared where first called as:

```
Dim theWorkDir As String
```

3.1.9 Subroutine avInit

This subroutine enables the programmer to initialize the various global variables that are used by the Avenue Wraps. The programmer may add additional custom global variables into the subroutine, if desired. In an ArcMap project file, avInit can be called in the OpenDocument event for the MxDocument function within the ThisDocument module.

The corresponding Avenue request is:

There is no corresponding Avenue request.

The call to this Avenue Wrap is:

Call **avInit**()

avInit

GIVEN: Nothing

RETURN: Nothing

The given and returned variables should be declared where first called as:

```
Public ugAVvers, ugbtTreePath, ugAVcedinstall As String
...........
...........
```

3.1.10 Subroutine avMoveTo

This subroutine enables the programmer to reposition a window object within the overall display (view).

The corresponding Avenue request is:

av.MoveTo(aLeft, aTop)

The call to this Avenue Wrap is:

Call **avMoveTo**(aDocName, aLeft, aTop)

avMoveTo

GIVEN: aDocName = the window object
aLeft = distance from the left side of the screen
aTop = distance from the top of the screen

RETURN: Nothing

The given and returned variables should be declared where first called as:

```
Dim aDocName As IUknown
Dim aLeft, aTop As Long
```

APPLICATION - DOCUMENT AVENUE WRAPS

3.1.11 Subroutine avObjGetName

This subroutine enables the programmer to get the name of an object, such as a layer (theme) or table. Refer to the Avenue Wrap avObjSetName for setting the name of an object.

The corresponding Avenue request is:

```
aName = aObj.GetName
```

avObjGetName

The call to this Avenue Wrap is:

Call **avObjGetName**(aObj, aName)

GIVEN: aObj = the object to be processed

RETURN: aName = the name assigned to the object

The given and returned variables should be declared where first called as:

```
Dim aObj As IUnknown
Dim aName As String
```

3.1.12 Subroutine avObjSetName

This subroutine enables the programmer to set the name of an object, such as a layer (theme). Refer to the Avenue Wrap avObjGetName for getting the name of an object.

The corresponding Avenue request is:

```
aObj.SetName(aName)
```

avObjSetName

The call to this Avenue Wrap is:

Call **avObjSetName**(aObj,aName)

GIVEN: aObj = the object to be named
aName = the name to be assigned to the object

RETURN: nothing

The given and returned variables should be declared where first called as:

```
Dim aObj As IUnknown
Dim aName As String
```

APPLICATION - DOCUMENT AVENUE WRAPS

3.1.13 Subroutine avRemoveDoc

This subroutine enables the programmer to remove the specified layer (theme) or table from the table of contents (TOC), without deleting any files from the disk.

The corresponding Avenue request is:

aProject.RemoveDoc(aDoc)

The call to this Avenue Wrap is:

avRemoveDoc

Call **avRemoveDoc**(aDocName)

GIVEN:	aDocName	= name of the layer (theme) or table to be removed from the TOC
RETURN:	Nothing	if the layer (theme) or table cannot be found in the TOC, the TOC remains unchanged and no error message is displayed

The given and returned variables should be declared where first called as:

Dim aDocName As String

3.1.14 Subroutine avResize

This subroutine enables the programmer to re-size a window object by specifying its name, width and height.

The corresponding Avenue request is:

av.Resize(aWidth, aHeight)

The call to this Avenue Wrap is:

avResize

Call **avResize**(aDoc, aWidth, aHeight)

GIVEN:	aDocName	= the window object
	aWidth	= the new width of the window
	aHeight	= the new height of the window
RETURN:	Nothing	

The given and returned variables should be declared where first called as:

Dim aDocName As IUnknown

Dim aWidth, aHeight As Long

APPLICATION - DOCUMENT AVENUE WRAPS

3.1.15 Subroutine avSetName

This subroutine enables the programmer to set the application (project) name that appears in the upper left corner of the overall application window. Refer to the Avenue Wrap avGetName for getting the name of the application.

The corresponding Avenue request is:

av.SetName(aTitle)

avSetName

The call to this Avenue Wrap is:

Call **avSetName**(aTitle)

GIVEN: aTitle = the name of the application appearing in the upper left corner of the application window

RETURN: nothing

The given and returned variables should be declared where first called as:

Dim aTitle As String

3.1.16 Subroutine avSetWorkDir

This subroutine enables the programmer to set the current working directory.

The corresponding Avenue request is:

aProject.SetWorkDir (theWorkDir)

avSetWorkDir

The call to this Avenue Wrap is:

Call **avSetWorkDir**(theWorkDir)

GIVEN: theWorkDir = the new working directory

RETURN: Nothing

The given and returned variables should be declared where first called as:

Dim theWorkDir As String

3.2 Theme - Layer Related Avenue Wraps

THEME - LAYER AVENUE WRAPS

3.2.1 Subroutine avGetLayerIndx

This subroutine enables the programmer to get the index of a layer or table

The corresponding Avenue request is:

There is no corresponding Avenue request.

The call to this Avenue Wrap is:

theIndex = **avGetLayerIndx**(theTheme)

avGetLayerIndx

GIVEN: theTheme = the layer or table to be processed

RETURN: theIndex = index of the layer or table

The given and returned variables should be declared where first called as:

Dim theTheme As Variant
Dim theIndex As Long

3.2.2 Subroutine avGetLayerType

This subroutine enables the programmer to get the type of a layer or table

The corresponding Avenue request is:

There is no corresponding Avenue request.

The call to this Avenue Wrap is:

Call **avGetLayerType**(pSelected, aName, aType)

avGetLayerType

GIVEN: pSelected = the data layer object to be processed

RETURN: aName = name of the data layer object (uppercase)
aType = type of data layer object

0 = unknown 1 = stand-alone table
2 = raster layer 3 = tin layer
4 = annotation layer 5 = feature layer
6 = CAD annotation layer 7 = CAD layer

The given and returned variables should be declared where first called as:

Dim pSelected As IUnknown
Dim aName As Variant
Dim aType As Integer

THEME - LAYER AVENUE WRAPS

3.2.3 Subroutine avGetSelectedExtent

This subroutine enables the programmer to get the rectangle that encloses (a) the selected set of a layer (theme), or (b) the entire layer (theme) if the selected set is empty (there are no selected features)

The corresponding Avenue request is:

```
theRect = anFTheme.GetSelectedExtent
```

avGetSelected Extent

The call to this Avenue Wrap is:

Call **avGetSelectedExtent**(pmxDoc, theTheme, theRect)

GIVEN: pmxDoc = the active view
theTheme = the layer (theme) to be processed

RETURN: theRect = the enclosing rectangle encompassing the selected features in a theme, or the entire theme

The given and returned variables should be declared where first called as:

```
Dim pmxDoc As IMxDocument
Dim theTheme As Variant
Dim theRect As IEnvelope
```

3.2.4 Subroutine avGetThemes

This subroutine enables the programmer to get a list of all layers (themes) or tables in a document.

The corresponding Avenue request is:

```
ThemesList = aView.GetThemes
```

avGetThemes

The call to this Avenue Wrap is:

Call **avGetThemes**(pmxDoc, opmode, ThemesList)

GIVEN: pmxDoc = the active view
opmode = mode of operation: 0 = find all layers
1 = find only feature layers 2 = find all tables

RETURN: ThemesList = list of themes

The given and returned variables should be declared where first called as:

```
Dim pmxDoc As IMxDocument
Dim opmode As Integer
Dim ThemesList As New Collection
```

THEME - LAYER AVENUE WRAPS

3.2.5 Subroutine avGetVisibleCADLayers

This subroutine enables the programmer to get a list of the visible CAD annotation layers and visible CAD layers in a document.

The corresponding Avenue request is:

There is no corresponding Avenue request.

The call to this Avenue Wrap is:

avGetVisibleCAD Layers

Call **avGetVisibleCADLayers**(pmxDoc, vThemesList)

GIVEN: pmxDoc = the active view

RETURN: vThemesList = list of visible CAD layers

The given and returned variables should be declared where first called as:

```
Dim pmxDoc As IMxDocument
Dim vThemesList As New Collection
```

3.2.6 Subroutine avGetVisibleThemes

This subroutine enables the programmer to get a list of the visible layers (themes) in a document.

The corresponding Avenue request is:

vThemesList = aView.GetVisibleThemes

The call to this Avenue Wrap is:

avGetVisibleThemes

Call **avGetVisibleThemes**(pmxDoc, vThemesList)

GIVEN: pmxDoc = the active view

RETURN: vThemesList = list of visible themes

The given and returned variables should be declared where first called as:

```
Dim pmxDoc As IMxDocument
Dim vThemesList As New Collection
```

DISPLAY AVENUE WRAPS

3.3 Display Related Avenue Wraps

3.3.1 Subroutine avDisplayInvalidate

This subroutine enables the programmer to refresh or redraw the active display.

The corresponding Avenue request is:

aDisplay.Invalidate(aFlag)

avDisplay Invalidate

The call to this Avenue Wrap is:

Call **avDisplayInvalidate**(aFlag)

GIVEN: aFlag = when to redraw (true: refresh at the next refresh; false: refresh immediately)

RETURN: nothing the display is refreshed as per aFlag

The given and returned variables should be declared where first called as:

Dim aFlag As Boolean

3.3.2 Subroutine avGetDisplay

This subroutine enables the programmer to get the current focus map display (get the display of the active view).

The corresponding Avenue request is:

pScreenDisplay = aView.GetDisplay

avGetDisplay

The call to this Avenue Wrap is:

Call **avGetDisplay**(pActiveView,pScreenDisplay,pDT)

GIVEN: pActiveView = the focus map active view

RETURN: pScreenDisplay = the screen display

pDT = the screen display transformation

The given and returned variables should be declared where first called as:

Dim pActiveView As IActiveView

Dim pScreenDisplay As IScreenDisplay

Dim pDT As IDisplayTransformation

DISPLAY AVENUE WRAPS

3.3.3 Subroutine avGetDisplayFlush

This subroutine enables the programmer to make certain that the display is up-to-date (flush the display).

The corresponding Avenue request is:

aView.GetDisplay.Flush

The call to this Avenue Wrap is:

Call **avGetDisplayFlush**

avGetDisplay Flush

GIVEN: Nothing

RETURN: Nothing

3.3.4 Subroutine avPanTo

This subroutine enables the programmer to center the display about a point

The corresponding Avenue request is:

aDisplay.PanTo (thePoint)

The call to this Avenue Wrap is:

Call **avPanTo**(pmxDoc, thePoint)

avPanTo

GIVEN: pmxDoc = the active view document
thePoint = point about which the display is to be centered

RETURN: Nothing The display is centered about the given point

The given and returned variables should be declared where first called as:

Dim pmxDoc As IMxDocument
Dim thePoint As IPoint

3.3.5 Function avReturnVisExtent

This function enables the programmer to get the extent of the current active display (a rectangle in map units that encloses the current active display).

The corresponding Avenue request is:

aRect = aDisplay.ReturnVisExtent

The call to this Avenue Wrap is:

Set aRect = **avReturnVisExtent**(pDT)

avReturn VisExtent

DISPLAY AVENUE WRAPS

GIVEN: pDT = the screen display transformation

RETURN: aRect = the current extent of the view

The given and returned variables should be declared where first called as:
Dim pDT As IDisplayTransformation
Dim aRect As IEnvelope

3.3.6 Subroutine avSetExtent

This subroutine enables the programmer to set the extent of the current active display (a rectangle in map units that encloses the current active display).

The corresponding Avenue request is:
aDisplay.SetExtent (newRect)

avSetExtent

The call to this Avenue Wrap is:
Call **avSetExtent**(pActiveView, pDT, newRect)

GIVEN: pActiveView = the active view
pDT = the screen display transformation
newRect = view extent rectangle

RETURN: Nothing Defines or establishes the extent of the display

The given and returned variables should be declared where first called as:
Dim pActiveView As IActiveView
Dim pDT As IDisplayTransformation
Dim newRect As IEnvelope

3.3.7 Subroutine avZoomToSelected

This subroutine enables the programmer to zoom (a) to the extent of the selected set of features in a theme, or (b) to the extent of all selected features in the map.

The corresponding Avenue request is:
There is no corresponding Avenue request.

avZoomToSelected

The call to this Avenue Wrap is:
Call **avZoomToSelected**(pmxDoc, theTheme)

DISPLAY AVENUE WRAPS

GIVEN: pmxDoc = the active view
theTheme = theme to be processed, if NULL is specified all selected features in the map will be zoomed to

RETURN: nothing The display zooms

The given and returned variables should be declared where first called as:
Dim pmxDoc As IMxDocument
Dim theTheme As Variant

3.3.8 Subroutine avZoomToTheme

This subroutine enables the programmer to zoom to the extent of all features in a theme, even though some may be selected.

The corresponding Avenue request is:
There is no corresponding Avenue request.

The call to this Avenue Wrap is:
Call **avZoomToTheme**(pmxDoc, theTheme)

avZoomToTheme

GIVEN: pmxDoc = the active view
theTheme = the theme to be processed

RETURN: nothing The display zooms

The given and returned variables should be declared where first called as:
Dim pmxDoc As IMxDocument
Dim theTheme As Variant

3.3.9 Subroutine avZoomToThemes

This subroutine enables the programmer to zoom to the extent of a group of themes.

The corresponding Avenue request is:
There is no corresponding Avenue request.

The call to this Avenue Wrap is:
Call **avZoomToThemes**(pmxDoc, thmList)

avZoomToThemes

GIVEN: pmxDoc = the active view
thmList = list of themes to be processed

DISPLAY AVENUE WRAPS

RETURN: nothing The display zooms

The given and returned variables should be declared where first called as:

Dim pmxDoc As IMxDocument
Dim thmList As New Collection

3.3.10 Subroutine ChangeView

This subroutine enables the programmer to alter the display of the view by (a) changing the scale factor, (b) panning in the X and/or Y coordinate direction, (c) redefining the extent, or (d) centering the display about a point.

The corresponding Avenue request is:

There is no corresponding Avenue request.

ChangeView

The call to this Avenue Wrap is:

Call **ChangeView**(pmxDoc, opmode, scl, panX, panY, usrRect, _
iok, newRect)

GIVEN: pmxDoc = the active view
opmode = mode of operation
1 : zoom scale factor to be applied
2 : panning values to be applied
3 : a new extent to be defined
4 : center display about a point
scl = scale factor to be applied to view
panX = distance in world units to pan along x axis or display center X coordinate if opmode = 4
panY = distance in world units to pan along y axis or display center Y coordinate if opmode = 4
usrRect = user-defined view extent rectangle

RETURN: iok = error flag (0 = no error, 1 = error)
newRect = view extent rectangle

The given and returned variables should be declared where first called as:

Dim pmxDoc As IMxDocument
Dim opmode As Integer
Dim scl, panX, panY As Double
Dim usrRect As IUnknown
Dim iok As Integer
Dim newRect As IEnvelope

3.4 Progress Run and Status Bar Display Avenue Wraps

PROGRESS RUN & STATUS BAR AVENUE WRAPS

3.4.1 Subroutine avClearStatus

This subroutine enables the programmer to clear the status bar area.

The corresponding Avenue request is:

av.ClearStatus

The call to this Avenue Wrap is:

Call **avClearStatus**

avClearStatus

GIVEN: Nothing

RETURN: Nothing

3.4.2 Subroutine avShowStopButton

This subroutine enables the programmer to display the Cancel (Stop in Avenue) button in the progress bar. Use of this command will result in the progress bar appearing in the middle of the display, and not in the status bar area (see Figure 3-2).

The corresponding Avenue request is:

av.ShowStopButton

The call to this Avenue Wrap is:

Call **avShowStopButton**

avShow StopButton

GIVEN: Nothing

RETURN: Nothing

3.4.3 Class Calc_Callback

Calc_Callback

This class module consists of a number of procedures enabling the programmer to initialize, report on, and terminate the reporting of the progress bar for the Calculate method (see avCalculate Avenue Wrap in Chapter 5). This class module is necessary because a standard ArcMap progress bar cannot be displayed concurrently with the avCalculate calculation process.

This module can be modified or used as a guide to display a progress bar for other custom operations that may have a similar callback function.

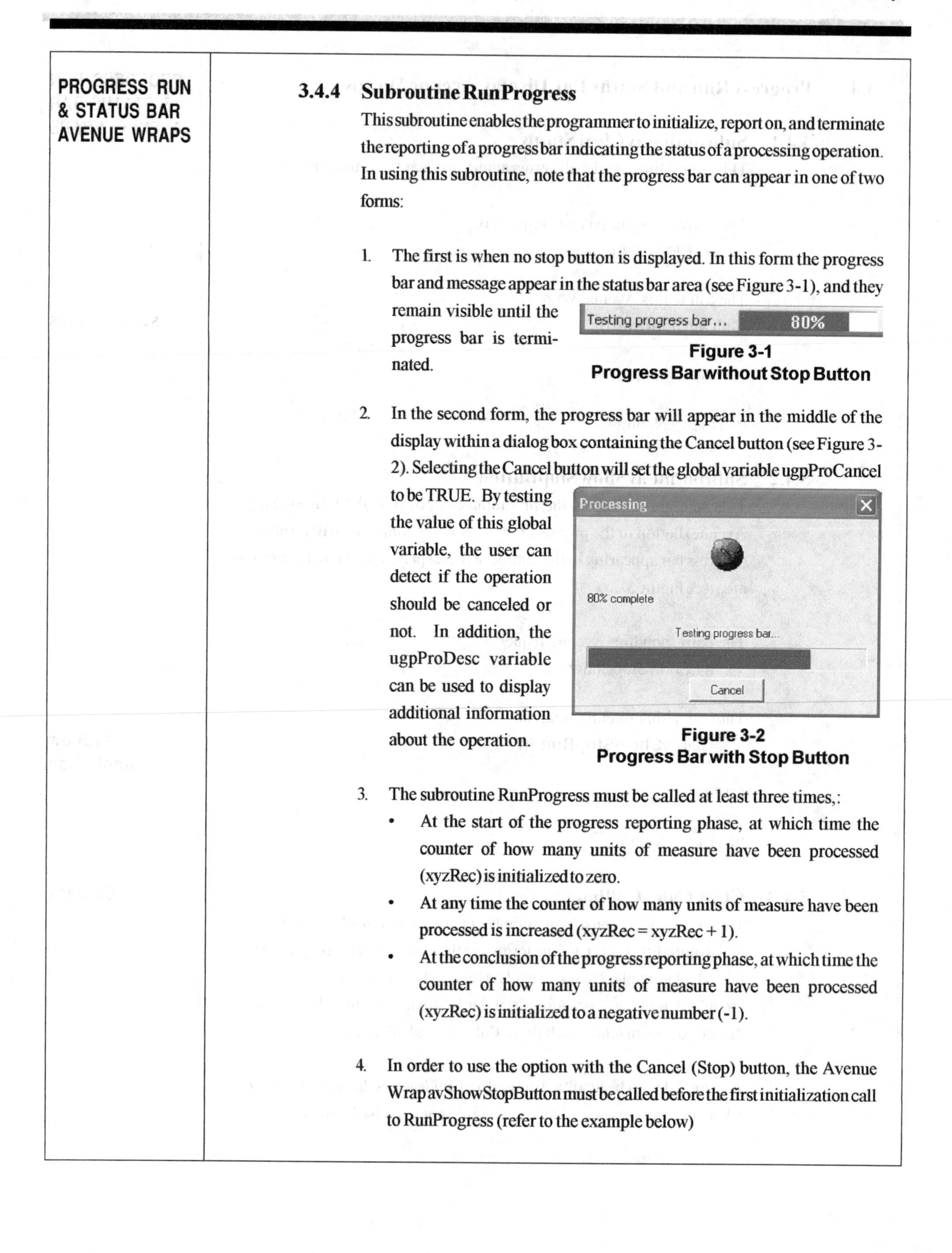

PROGRESS RUN
& STATUS BAR
AVENUE WRAPS

3.4.4 Subroutine RunProgress

This subroutine enables the programmer to initialize, report on, and terminate the reporting of a progress bar indicating the status of a processing operation. In using this subroutine, note that the progress bar can appear in one of two forms:

1. The first is when no stop button is displayed. In this form the progress bar and message appear in the status bar area (see Figure 3-1), and they remain visible until the progress bar is terminated.

Figure 3-1
Progress Bar without Stop Button

2. In the second form, the progress bar will appear in the middle of the display within a dialog box containing the Cancel button (see Figure 3-2). Selecting the Cancel button will set the global variable ugpProCancel to be TRUE. By testing the value of this global variable, the user can detect if the operation should be canceled or not. In addition, the ugpProDesc variable can be used to display additional information about the operation.

Figure 3-2
Progress Bar with Stop Button

3. The subroutine RunProgress must be called at least three times,:
   - At the start of the progress reporting phase, at which time the counter of how many units of measure have been processed (xyzRec) is initialized to zero.
   - At any time the counter of how many units of measure have been processed is increased (xyzRec = xyzRec + 1).
   - At the conclusion of the progress reporting phase, at which time the counter of how many units of measure have been processed (xyzRec) is initialized to a negative number (-1).

4. In order to use the option with the Cancel (Stop) button, the Avenue Wrap avShowStopButton must be called before the first initialization call to RunProgress (refer to the example below)

PROGRESS RUN & STATUS BAR AVENUE WRAPS

The corresponding Avenue request is:

There is no direct corresponding Avenue request for this Avenue Wrap. To accomplish the task of this Avenue Wrap requires several Avenue statements.

The call to this Avenue Wrap is:

RunProgress

Call **RunProgress**(xyzRec, totalRecs, aMessage)

GIVEN: xyzRec = the current unit of measure of progress. If it is:

= 0 : initiate the progress report phase

> 0 : report on the progress; number of units of measure processed up to this call of this statement.

< 0 : terminate the progress report phase

totalRecs = the total unit of measure of progress

aMessage = identification of the progress reporting

RETURN: nothing

The given and returned variables should be declared where first called as:

```
Dim xyzRec, totalRecs As Long
Dim aMessage As Variant
```

SAMPLE CODE FOR PROGRESS RUN

3.5 Sample Code for Progress Run

```
'
' ---
' ---Sample illustrating how to display and update a progress bar
' ---without a Cancel button (see Figure 3-1)
' ---
'
  Dim i, j As Long
  Dim aMsg As Variant
'
' ---Define the upper limit of the loop
  j = 30000
'
' ---Define the progress bar message
  aMsg = "Testing progress bar..."
'
' ---Initialize the progress bar display
  Call RunProgress(0, j, aMsg)
'
' ---Begin the loop and increment the progress bar
  For i = 1 To j
      Call RunProgress(i, j, aMsg)
  Next
'
' ---Terminate the display of the progress bar
  Call RunProgress(-1, j, aMsg)
'
' ---
' ---Sample illustrating how to display and update a progress bar
' ---with a Cancel button (see Figure 3-2)
' ---
'
' ---Display the cancel button
  Call avShowStopButton
'
' ---Initialize the progress bar display
  Call RunProgress(0, j, aMsg)
'
' ---Begin the loop and increment the progress bar
  For i = 1 To j
      Call RunProgress(i, j, aMsg)
'     ---Check if the cancel button was selected
      If (ugpProCancel) Then
         Exit For
      End If
  Next
'
' ---Terminate the display of the progress bar
  Call RunProgress(-1, j, aMsg)
'
' ---Inform user how may loop iterations were performed
  MsgBox "Performed " + CStr(i) + " loops"
```

CHAPTER 4

FILE INPUT AND OUTPUT AVENUE WRAPS

FILE INPUT & OUTPUT AVENUE WRAPS

This chapter contains various Avenue Wraps to assist a programmer in converting Avenue scripts to VB code that pertain to file handling such as opening and closing files, listing and questioning their existence, and ASCII file reading and writing, as well as extracting data from a file being read. In addition to the various Avenue Wraps, the reader will also find in this chapter certain VB statements that constitute a counter part to Avenue requests in the reading of ASCII files. To introduce a graphic progress status bar display during the reading process of a file, the reader is directed to the RunProgress Avenue Wrap described in Chapter 2. A sample code for writing to a file and reading from a file is presented at the end of this chapter.

The Avenue Wraps of this chapter are listed below in alphabetical order with a short description and the chapter - page number where a full description may be found.

The source listing of each of the above Avenue Wraps may be found in Appendix D of this publication.

Also contained in this chapter are various VB statements and their avenue counterparts that pertain to the following:

4.1 Directory Utility Related Avenue Wraps

DIRECTORY UTILITIES

4.1.1 Function avDirExists

This function enables the programmer to determine if a directory exists or not. If the directory to be checked is not a sub-directory of the current directory, then the complete path name must be given.

The corresponding Avenue request is:

There is no corresponding Avenue request.

The call to this Avenue Wrap is:

DirExists = **avDirExists**(DirName)

avDirExists

GIVEN: DirName = the directory to be checked for existence

RETURN: DirExists = existence flag (true = yes, false = no)

The given and returned variables should be declared where first called as:

Dim DirName As String
Dim DirExists As Boolean

4.1.2 Subroutine avListFiles

This subroutine enables the programmer to get a list of all files in a directory.

The corresponding Avenue request is:

There is no corresponding Avenue request.

The call to this Avenue Wrap is:

Call **avListFiles**(aDir, filType, filList)

avListFiles

GIVEN: aDir = the directory to be scanned

filType = the type of files to be searched for. Enter either the enumerator (i.e. vbNormal, or its numeric equivalent, i.e. 0) below to specify:

vbNormal	0	files with no attributes
vbReadOnly	1	Read-only files in addition to files with no attributes
vbHidden	2	Hidden files in addition to files with no attributes

DIRECTORY UTILITIES

VbSystem	4	System files in addition to files with no attributes
vbVolume	8	Volume label; if any other attribute is given vbVolume is ignored
vbDirectory	16	Directories or folders in addition to files with no attributes

RETURN: filList = list of files that were found as specified by the filType argument

The given and returned variables should be declared where first called as:

```
Dim aDir As String
Dim filType As Integer
Dim filList As New Collection
```

4.2 File Existence and Deletion Related Avenue Wraps

FILE EXISTENCE AND DELETION

4.2.1 Function avDeleteDS

This function enables the programmer to delete from the hard drive a data set such as a shapefile or dBase file. In using this function, note the following:

(a) Since a data set may be comprised of more than one file, this function will delete from the hard drive all files associated with the data set file name.

(b) The data set must not appear in the Table of Contents. If it does, the error flag will be set to 1. Use the subroutine avRemoveDoc to remove the data set from the Table of Contents before calling this function.

(c) If the input (given) file name contains an extension such as .shp, .mdb, .dbf, etc., the extension will be stripped off, without generating an error message. All files having the input (given) file name to the left of the data set extension will be deleted from the hard drive.

(d) If the input (given) file name does not contain a complete path name, the current working directory will be used. An example of aFileName for a:

shapefile is	c:\project\test\l_0ln
access database is	c:\project\test\montgomery
dBase file is	c:\project\test\table

(e) If the input (given) file name does not pertain to a data set, the function will search for a single standard disk file that matches the given file name.

The corresponding Avenue request is:

There is no corresponding Avenue request.

The call to this Avenue Wrap is:

errFlag = **avDeleteDS**(aFileName)

avDeleteDS

GIVEN: aFileName = name of the data set to be deleted. Refer to notes (a) through (e) above.

RETURN: errFlag = error flag. See note (b) above.
0 denotes an error has not been encountered
1 denotes an error has been encountered

The given and returned variables should be declared where first called as:

```
Dim aFileName As String
Dim errFlag As Integer
```

FILE EXISTENCE AND DELETION

4.2.2 Function avFileDelete

This function enables the programmer to delete from the hard drive a file. In using this function, note the following:

(a) The input (given) file name must not appear in the Table of Contents. If it does, the error flag will be set to 1. Use the subroutine avRemoveDoc to remove the data set from the Table of Contents before calling this function.

(b) If the input (given) file name contains an extension such as .shp, .mdb, .dbf, etc., the extension will be stripped off, without generating an error message.

(c) If the input (given) file name does not contain a complete path name, the current working directory will be used.

(d) If the input (given) file name pertains to a data set, only one of the files bearing the name of the input (given) file name will be deleted. Hence, make certain that this input (given) file name is not that of a data set.

The corresponding Avenue request is:

```
File.Delete(aFileName)
```

avFileDelete

The call to this Avenue Wrap is:

errFlag = **avFileDelete**(aFileName)

GIVEN: aFileName = name of the file to be deleted. Refer to notes (a) through (d) above.

RETURN: errFlag = error flag. See note (a) above.
0 denotes an error has not been encountered
1 denotes an error has been encountered

The given and returned variables should be declared where first called as:

```
Dim aFileName As String
Dim errFlag As Integer
```

FILE EXISTENCE AND DELETION

4.2.3 Function avFileExists

This function enables the programmer to determine whether a file exists or not.

The corresponding Avenue request is:

```
existFlag = File.Exists(aFileName)
```

The call to this Avenue Wrap is:

existFlag = **avFileExists**(aFileName)

avFileExists

GIVEN: aFileName = name of file to be checked. If the file is not in the current folder, a complete path name must be specified.

RETURN: existFlag = existence status.
True denotes the file exists
False denotes the file does not exist

The given and returned variables should be declared where first called as:

```
Dim aFileName As String
Dim existFlag As Boolean
```

FILE NAME MANIPULATION

4.3 File Name Manipulation Related Avenue Wraps

4.3.1 Function avGetBaseName

This function enables the programmer to get the base name that appears in a path name. Below are examples of what is returned for what is given:

given	return	given	return
c:\test\vb\aFile.shp	aFile.shp	c:\test\vb\aFile	aFile
c:\test\vb\	vb	c:\test\vb	vb
c:\a	a	aFile.txt	aFile.txt
c:\	empty string ("")		

The corresponding Avenue request is:

```
theBaseName = aFileName.GetBaseName
```

avGetBaseName

The call to this Avenue Wrap is:

theBaseName = **avGetBaseName**(aPathName)

GIVEN: aPathName = the path name to be processed.

RETURN: theBaseName = base name appearing in a path name including the filename extension, if one is present in the base name

The given and returned variables should be declared where first called as:

```
Dim aPathName As String
Dim theBaseName As String
```

4.3.2 Function avGetBaseName2

This function enables the programmer to get the base name that appears in a path name, minus any extension that may appear in said path name. Below are examples of what is returned for what is given:

given	return	given	return
c:\test\vb\aFile.shp	aFile	c:\test\vb\aFile	aFile
c:\test\vb\	vb	c:\test\vb	vb
c:\a	a	aFile.txt	aFile
c:\	empty string ("")		

The corresponding Avenue request is:

There is no corresponding Avenue request.

avGetBaseName2

The call to this Avenue Wrap is:

theBaseName = **avGetBaseName2**(aPathName)

FILE NAME MANIPULATION

GIVEN: aPathName = the path name to be processed.

RETURN: theBaseName = base name appearing in a path name without the file name extension

The given and returned variables should be declared where first called as:

```
Dim aPathName As String
Dim theBaseName As String
```

4.3.3 Function avGetExtension

This function enables the programmer to get the extension that appears in a path name. Below are examples of what is returned for what is given:

given	return	given	return
c:\test\vb\aFile.shp	shp	c:\test\vb\aFile	empty string ("")
c:\test\vb\	empty string ("")	c:\test\vb	empty string ("")
c:\a	empty string ("")	aFile.shp	shp
c:\	empty string ("")		

The corresponding Avenue request is:

```
theExtension = aFileName.GetExtension
```

The call to this Avenue Wrap is:

theExtension = **avGetExtension**(aPathName)

avGetExtension

GIVEN: aPathName = the path name to be processed.

RETURN: theExtension = the extension in the path name

The given and returned variables should be declared where first called as:

```
Dim aPathName As String
Dim theExtension As String
```

4.3.4 Function avGetPathName

This function enables the programmer to get the path name of the file that appears in a string. Below are examples of what is returned for what is given:

given	return	given	return
c:\test\vb\aFile.shp	c:\test\vb	c:\test\vb\aFile	c:\test\vb
c:\test\vb\	c:\test	c:\test\vb	c:\test
c:\a	c:\	aFile.shp	empty string ("")
c:\	empty string ("")		

FILE NAME MANIPULATION

The corresponding Avenue request is:

There is no corresponding Avenue request.

avGetPathName

The call to this Avenue Wrap is:

theGetPathName = **avGetPathName**(aPathString)

GIVEN: aPathName = the path name to be processed.

RETURN: thePathName = the path name appearing in a string minus the last component in the string

The given and returned variables should be declared where first called as:

Dim aPathString As String
Dim thePathName As String

4.3.5 Function avSetExtension

This function enables the programmer to set the file extension in a base name, or change the extension if a full path name is given.

The corresponding Avenue request is:

aFileName.SetExtension (aExt)

avSetExtension

The call to this Avenue Wrap is:

newPathName = **avSetExtension**(aPathName,aExt)

GIVEN: aPathName = a base name, or a full path name to be processed
aExt = extension to be set on the base name. It should not contain a period, just the desired three character extension.

RETURN: NewPathName = the new base name, or full path name with the specified extension applied

The given and returned variables should be declared where first called as:

Dim aPathName, aExt As String
Dim newPathName As String

FILE OPEN AND CLOSE

4.4 File Opening and Closing Related Avenue Wraps

4.4.1 Subroutine avLineFileClose

This subroutine enables the programmer to close an open file. Note that in Avenue a LineFile is a class that refers to a file comprised of lines of characters, and not graphic lines.

The corresponding Avenue request is:

```
aFile.Close
```

The call to this Avenue Wrap is:

Call **avLineFileClose**(aFile)

avLineFileClose

GIVEN: aFile = text stream object to be processed to be the same as that which was specified when this file was opened (made) with the avLineFileMake function (see below)

RETURN: nothing

The given and returned variables should be declared where first called as:

Dim aFile — aFile is declared internally as an object. Do not include in this declaration an "As" type.

4.4.2 Function avLineFileMake

This function enables the programmer to open a file connection for reading and/or writing. Note that in Avenue a LineFile is a class that refers to a file comprised of lines of characters, and not graphic lines. Note the "Set" statement in front of the returned variable from this function.

The corresponding Avenue request is:

```
aLineFile = LineFile.Make (aFileName, aFilePerm)
```

The call to this Avenue Wrap is:

Set aLineFile = **avLineFileMake**(aFileName, aFilePerm)

avLineFileMake

GIVEN: aFileName = the name of the file to be opened

aFilePerm = the type of file connection desired

READ	to open file for reading
WRITE	to open file for writing
APPEND	to open file for appending

RETURN: aLineFile = text stream object that can be used for reading and/or writing operations

FILE OPEN AND CLOSE

The given and returned variables should be declared where first called as:

Dim aFileName As String
Dim aFilePerm As String
Dim aLineFile — theLineFile is declared internally as an object. Do not include in this declaration an "As" type.

4.4.3 ASCII File Read/Write Example

Presented below is a sample program that demonstrates the opening and closing of files, as well as reading from and writing to them (refer to section 4.5 for reading from and writing to ASCII files). This program makes two passes. During the first pass, the file cedFile does not exist, so it is created for writing purposes and two lines are written to it. During the second pass, the cedFile file exists, and the previously written two lines are read and written in a message box. At the end of the second pass the program deletes the file.

```
Dim aFileName, aString As String
Dim cedFile
Dim k, iok As Integer
'
aFileName = "c:\temp\l_zxcv"
For k = 1 To 2
    Set cedFile = avLineFileMake(aFileName, "READ")
    If (cedFile Is Nothing) Then               ' ---Pass 1
       MsgBox aFileName + " does not exist"
       Set cedFile = avLineFileMake(aFileName, "WRITE")
       cedFile.WriteLine ("First data line")
       cedFile.WriteLine ("Second data line")
       MsgBox "2 lines written to the file"
    Else
       MsgBox aFileName + " exists"         ' ---Pass 2
       aString = cedFile.ReadLine
       MsgBox "Line 1 in File: " + aString
       aString = cedFile.ReadLine
       MsgBox "Line 2 in File: " + aString
    End If
    Call avLineFileClose(cedFile)
Next k
If (avFileExists(aFileName)) Then
   iok = avFileDelete(aFileName)             ' ---Delete file
End If
```

WRITING TO AND READING FROM FILES

4.5 Writing to and Reading from Files Statements and Related Avenue Wraps

4.5.1 Writing to an ACII File

The common Avenue request to write a series of lines into an ASCII file (LineFile) is the use of the WriteElt request within an iterative operation such as a "For", "While" or "If" statement. The corresponding VB statement is "WriteLine", which can be used within a "Do", "For", "While" or "If" statement. Thus we have:

The corresponding Avenue request is:

```
aLineFile.WriteElt (aLine)
```

The VB statement is:

```
aLineFile.WriteLine (aLine)
```

GIVEN: aLine = text string of data to be written to the file. It could be a single variable, or a collection of variables concatenated into a single string.

aLineFile = text stream object to be written to

RETURN: nothing aLine is written to aLineFile

The given and returned variables should be declared where first called as:

```
Dim aLine As String
Dim aLineFile
```

aFile is declared internally as an object. Do not include in this declaration an "As" type.

Example 1: To write into a file a series of numbers, one per data line. This example assumes that the file, aFile, was previously opened.

```
Dim X As Double
Dim i As Long
Dim aLine As String
Dim aFile
X = 0                             ' ---Initialize X
For i = 1 To 10
    X = (X + 1) * i               ' ---Modify X
    aLine = CStr(X)               ' ---Change to string
    aFile.WriteLine (aLine)       ' ---Write X to file
Next i
```

WRITING TO AND READING FROM FILES

Example 2: To write into a file nine data lines containing five, comma separated values, per data line.

```
Dim Delim, aLine As String
Dim j, i As Long
Dim X As Double
Dim aLineFile
'
' ---Define comma as the separator
Delim = ","
' ---You may use any other character as separator.
' ---For a tab use Chr(9), i.e. Delim = Chr(9)
j = 1                      ' ---Initialize lines written
Do While j < 10
   aLine = ""                        ' ---Initialize string
   X = 0                                ' ---Initialize X
   For i = 1 To 5
       X = (X + 1) * i                      ' ---Modify X
       aLine = aLine + CStr(X) ' ---Add to string
       If (i <> 5) then
          aLine = aLine + Delim         ' ---Add comma
       End If
   Next i
   aLineFile.WriteLine (aLine)
   j = j + 1                ' ---Increment lines written
Loop
```

4.5.2 Reading from an ASCII File

The common Avenue request to read a series of lines from an ASCII file (LineFile) is the use of the ReadElt request within an iterative operation such as a "For", "While" or "If" statement. The corresponding VB statement is "ReadLine", which can be used within a "Do", "For", "While" or "If" statement. The line to be read from the file may contain one item per line, or several separated from each by a known delimiter character. Thus we have:

The corresponding Avenue request is:

aLine = aLineFile.ReadElt

The VB statement is:

```
aLine = aLineFile.ReadLine
```

GIVEN: aLineFile = text stream object to be read from

RETURN: aLine = the text string corresponding to the data that was read from the file. It could contain only one item, or it could contain several concatenated items, which are separated by a distinct character (delimiter).

The given and returned variables should be declared where first called as:

Dim aLineFile — aFile is declared internally as an object. Do not include in this declaration an "As" type.

Dim aLine As String

4.5.3 Length of an ASCII File

When reading an ASCII file, it is important to know the length of the ASCII file. In Avenue the programmer was able to determine the number of lines that comprised a file. In VB however, the programmer can only determine the length of the file in terms of bytes.

The corresponding Avenue request is:

```
totalRecs = aLineFile.GetSize
```

The VB statement is:

```
totalRecs = FileLen (aFileName)
```

GIVEN: aFileName = the name of the ASCII file

RETURN: totalRecs = number of lines (Avenue) or bytes (VB) in the ASCII file

The given and returned variables should be declared where first called as:

Dim aFileName As String

Dim totalRecs As Long

WRITING TO AND READING FROM FILES

4.5.4 End of an ASCII File

When reading an ASCII file, in addition to its length, it also is important to know when the end of the ASCII file has been reached. In Avenue the programmer could read a line from a file and then determine whether the file's end was reached or not. In VB however, the programmer must first determine whether the file's end has been reached, or not, before proceeding to read a line from the file.

The corresponding Avenue request is:

```
ReadOver = 0
While ReadOver = 0
   aLine = aLineFile.ReadElt
   if (aLine = nil) then
      ReadOver = 1
   Else
      ... do something with data that was read
   End
Wend
```

The VB statement is:

```
ReadOver = 0
While ReadOver = 0
   If (aLineFile.AtEndOfStream) Then
      ReadOver = 1
   Else
      aLine = aLineFile.ReadLine
      ... do something with data that was read
   End If
Wend
```

GIVEN: aLineFile = text stream object to be read from

RETURN: aLine = the text string corresponding to the data that was read from the file. It could contain only one item, or it could contain several concatenated items, which are separated by a distinct character (delimiter).

WRITING TO AND READING FROM FILES

The given and returned variables should be declared where first called as:

Dim ReadOver As Integer
Dim aLine As String
Dim aLineFile — aFile is declared internally as an object. Do not include in this declaration an "As" type.

4.5.5 Rewind an ASCII File

At times it is necessary to rewind a file so that it may be read more than once. Although there are statements to go to the beginning of a file in both Avenue ("GoToBeg") and VB ("Seek"), one may also close and then reopen the file to position the file pointer at the beginning of the file.

4.5.6 Subroutine avAsTokens

This subroutine enables the programmer to extract from a delimited string a list of string items. Although this is a generic string manipulation subroutine, it is included in this section of the book because it is felt that it would be used the most in extracting information from a data line, which was read from a file.

The subroutine avAsTokens reads the given text string, the word delineator and the indicator whether to change all characters to upper or lower case characters, and then:

1. Removes the leading and trailing blank spaces from said given text string;
2. Creates a list of words from the contents of said given text string, based upon the given word delineator, excluding therefrom any blank spaces that may be present between the words; and
3. Computes the number of words in the returned list.

The corresponding Avenue request is:

theList = theString.AsTokens (delString)

The call to this Avenue Wrap is:

avAsTokens

Call **avAsTokens**(theString, delString, UpperLower, _
theList, nWords)

GIVEN:	theString	= the input string, or line read from a file
	delString	= the word delineator in theString
	UpperLower	= U - change all characters to upper case
		= L - change all characters to lower case
		= any other character - no change

WRITING TO AND READING FROM FILES

RETURN: theList = the returned list of words
nWords = number of words (items) extracted

The given and returned variables should be declared where first called as:

Dim theString, delString, UpperLower As String
Dim theList As New Collection
Dim nWords As Integer

Example: As sample code for the use of the AvAsTokens Avenue Wrap consider the example below. In this example, we are opening and reading an ASCII file that must have at least two items per data line and may have:

- Comment lines, which are denoted with a slash and asterisk (/*) in columns 1 and 2, that should be disregarded
- Blank lines
- Have the items on a data line separated by either a comma, blank space or a tab.

```
Dim aFileName As String
Dim cedFile
Dim totalRecs, xyzRec, iLinLeft, nChrs As Long
Dim aMsg As Variant
Dim buf1, buf2, tmpStrng, theDChr As String
Dim xyzTokens As New Collection
Dim datalineItems As Long
'
' ---Set the file name and open the file
aFileName = "c:\temp\l_zxcv"
Set cedFile = avLineFileMake(aFileName, "READ")
'
' ---Find the number of bytes in the file
totalRecs = FileLen(aFileName)
'
' ---Initialize number of lines in the file read
xyzRec = 0
'
' ---Initialize status area and display message
aMsg = "Reading the file " + aFileName + "..."
Call RunProgress(xyzRec, totalRecs, aMsg)
```

WRITING TO AND READING FROM FILES

```
'
' ---Skip over comment lines appearing at top of file
' ---A valid data line will be a non-comment line
' ---Start reading file, checking if EOF reached
For iLinLeft = 1 To totalRecs
'
'    ---Check if the end of file encountered
    If (cedFile.AtEndOfStream) Then
       Exit For
    End If
'
'    ---Grab line from input file and store in a buffer
    buf1 = cedFile.ReadLine
'
'    ---Increment total number of bytes read, and
'    ---update the progress bar accounting for the end
'    ---of line and new line characters
    xyzRec = xyzRec + Len(buf1) + 2
    Call RunProgress(xyzRec, totalRecs, aMsg)
'
'    ---Tim any leading and/or trailing blank spaces
    buf2 = Trim(buf1)
'
'    ---Check if a non-comment data line was read
    nChrs = Len(buf2)
    If (nChrs > 1) Then
'
'       ---Get the first two characters in the data line
       tmpStrng = Mid(buf2, 1, 2)
       If (tmpStrng <> "/*") Then
'
'          ---A valid data line has been found.
'          ---Set flag to check for the comma delimiter
          theDChr = ","
'
'          ---Extract the list of items on the data line
'          ---using the comma delineator character
          Call avAsTokens(buf2, theDChr, "N", _
                          xyzTokens, datalineItems)
```

WRITING TO AND READING FROM FILES

```
'
'           ---Not enough items on the data line, try
'           ---using the space delimiter
            If (datalineItems <= 1) Then
               theDChr = " "
               Call avAsTokens(buf2, theDChr, "N", _
                                xyzTokens, datalineItems)
'
'              ---Not enough items on the data line, try
'              ---using the tab delimiter character
               If (datalineItems <= 1) Then
                  theDChr = Chr(9)
                  Call avAsTokens(buf2, theDChr, "N", _
                                   xyzTokens, datalineItems)
               End If
            End If
'
'           ---Handle case when enough items specified
            If (datalineItems > 1) Then
            End If
         End If
      End If
Next iLinLeft
'
' ---Clear the status bar area
Call RunProgress(-1, totalRecs, aMsg)
```

4.6 File Dialog Windows Related Avenue Wraps

FILE DIALOG WINDOWS

4.6.1 Subroutine avFileDialogPut

This subroutine enables the programmer to get a file by browsing through the various files of the computer. As a result of incorporating this subroutine into an application, the user will:

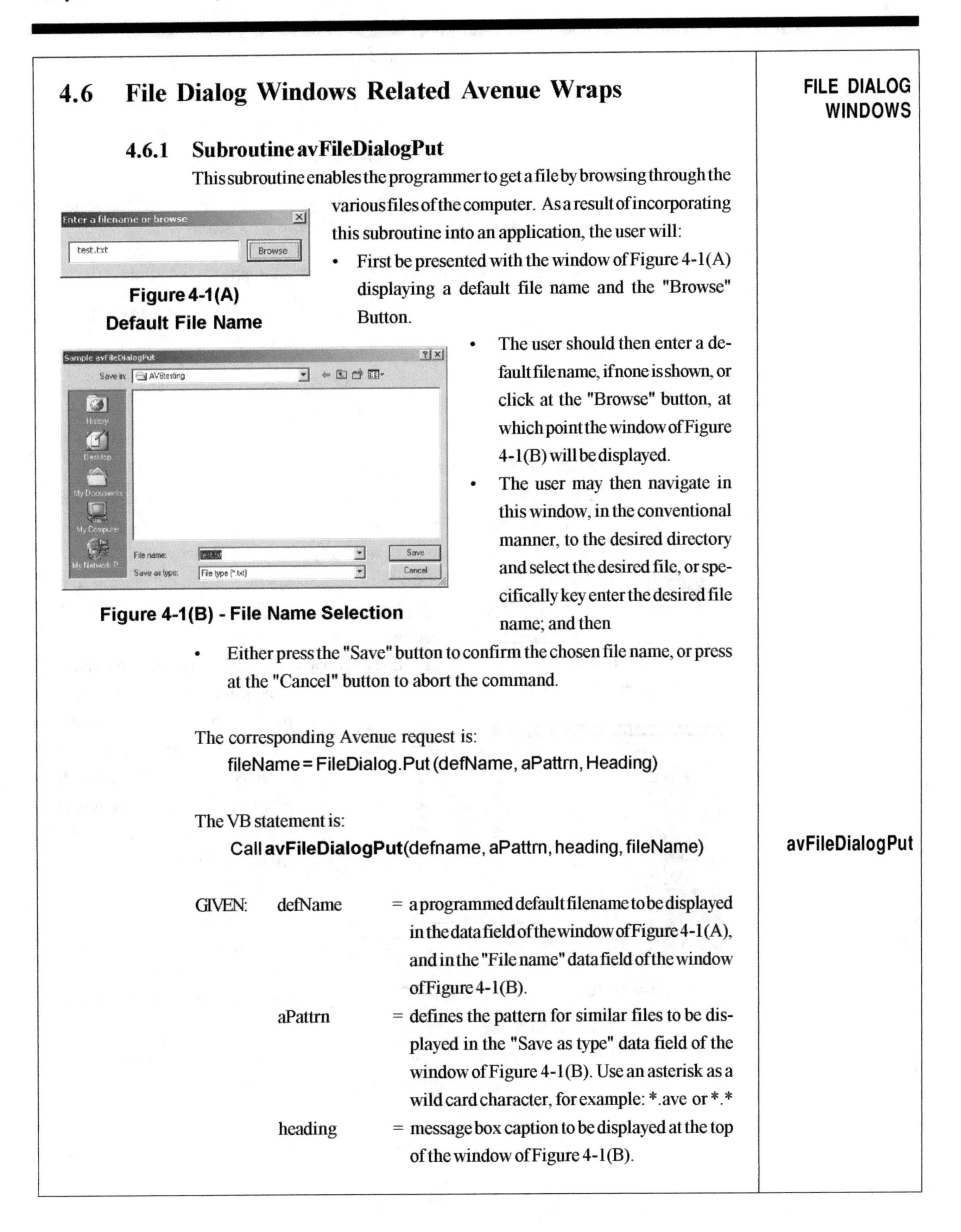

Figure 4-1(A)
Default File Name

Figure 4-1(B) - File Name Selection

- First be presented with the window of Figure 4-1(A) displaying a default file name and the "Browse" Button.
- The user should then enter a default file name, if none is shown, or click at the "Browse" button, at which point the window of Figure 4-1(B) will be displayed.
- The user may then navigate in this window, in the conventional manner, to the desired directory and select the desired file, or specifically key enter the desired file name; and then
- Either press the "Save" button to confirm the chosen file name, or press at the "Cancel" button to abort the command.

The corresponding Avenue request is:

fileName = FileDialog.Put (defName, aPattrn, Heading)

The VB statement is:

Call **avFileDialogPut**(defname, aPattrn, heading, fileName)

avFileDialogPut

GIVEN: defName = a programmed default filename to be displayed in the data field of the window of Figure 4-1(A), and in the "File name" data field of the window of Figure 4-1(B).

aPattrn = defines the pattern for similar files to be displayed in the "Save as type" data field of the window of Figure 4-1(B). Use an asterisk as a wild card character, for example: *.ave or *.*

heading = message box caption to be displayed at the top of the window of Figure 4-1(B).

FILE DIALOG WINDOWS

RETURN: fileName = name of file to be created. If the user Cancels the command, fileName will be set to a blank character (a single space).

The given and returned variables should be declared where first called as:
Dim defName, aPattrn, heading, fileName As String

The sample code for creating the windows of Figures 4-1(A) and 4-1(B) is:

```
'   ---Test the avFileDialogPut macro
Dim fileName As String
Call avFileDialogPut("test.txt", "*.txt", _
                     "Sample avFileDialogPut", _
                     fileName)
```

4.6.2 Subroutine avFileDialogReturnFiles

This subroutine enables the programmer to get a collection of one or more file names by browsing through the various files of the computer. As a result of incorporating this subroutine into an application, the user will:

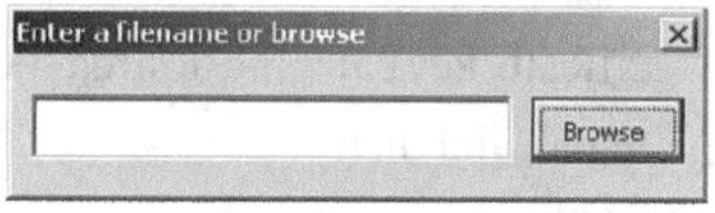

Figure 4-2(A)
Default File Name

- First be presented with the window of Figure 4-2(A) displaying a default file name and the "Browse" button.
- The user then may, or may not enter a default file name, none will be shown, and click at the "Browse" button. If a file name is key entered, it will be accepted and the program will return to execute the next step. If a file name is not entered, the window of Figure 4-2(B) will be dsplayed.
- The user may then navigate in this window, in the conventional manner, to the desired directory and select the desired file(s), or specifically key enter the desired file name(s); and then
- Either press the "Save" button to save the chosen file name(s), or press at the "Cancel" button to abort the command.

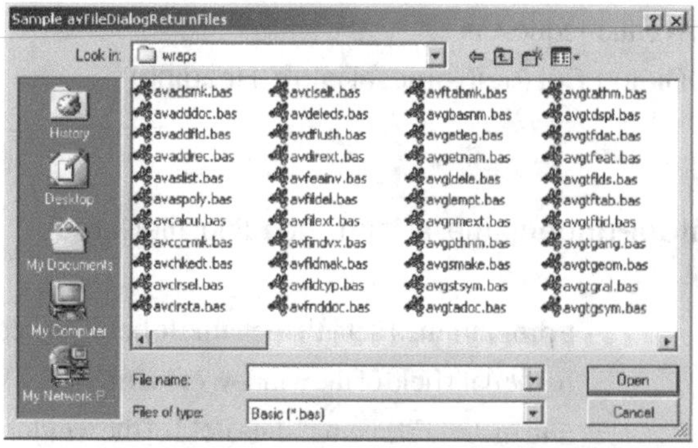

Figure 4-2(B) - File Name Selection

FILE DIALOG WINDOWS

The corresponding Avenue request is:

```
fileList = FileDialog.ReturnFiles (patrns, labels, heading,
                                   defIndex)
```

The VB statement is:

Call **avFileDialogReturnFiles**(patrns, labels, heading, _
defIndex, fileList)

avFileDialog ReturnFiles

GIVEN: patrns = collection of file patterns, in the form of "*.txt" for example. This collection is for internal purposes, and it will not be displayed in either of the two dialog windows of this Avenue Wrap. See the example below.

labels = collection of labels, in the form of "Text (*.txt)" for example, and corresponding to the collection of patterns. See the example below. These labels will be contained in the Files of Type data field of the window of Figure 4-2(B).

heading = message box caption to be displayed at the top of the window of Figure 4-2(B)

defIndex = numeric index, starting at 1, into the Patrns (and Labels) collection denoting the default pattern to be displayed in the Files of Type data field of the window of Figure 4-2(B)

RETURN: fileList = list of file names, of string type, that were selected by the user. If the user Cancels the command this list will be empty.

The given and returned variables should be declared where first called as:

```
Dim patrns As New Collection
Dim labels As New Collection
Dim heading As String
Dim defIndex As Long
Dim fileList As New Collection
```

FILE DIALOG WINDOWS

The sample code for creating the windows of Figures 4-2(A) and 4-2(B) is:

```
' ---Test the avFileDialogReturn macro
Dim Patrns As New Collection
Dim Labels As New Collection
Dim defIndex As Long
Dim aHeading As String
Dim fileList As New Collection
Dim aNumbStr, aName As String
Dim i As Long
Patrns.Add ("*.txt")
Patrns.Add ("*.bas")
Patrns.Add ("*.ave")
Labels.Add ("Text (*.txt)")
Labels.Add ("Basic (*.bas)")
Labels.Add ("Avenue (*.ave)")
defIndex = 1
aHeading = "Sample avFileDialogReturnFiles"
Call avFileDialogReturnFiles(Patrns, _
                             Labels, _
                             aHeading, _
                             defIndex, _
                             fileList)
' ---Check if any files were specified
If (fileList.Count > 0) Then
'   ---Display number of selected files
   aNumbStr = CStr(fileList.Count)
   MsgBox aNumbStr + " files specified"
'   ---List the files that were selected, one at
'   ---a time
   For i = 1 to fileList.Count
       aName = fileList.Item(i)
       MsgBox "File " + CStr(i) + " is " + aName
   Next i
' ---Handle case when command is aborted
Else
   MsgBox "No files specified"
End If
```

CHAPTER 5

THEME AND TABLE AVENUE WRAPS

This chapter contains Avenue Wraps pertaining to the handling of themes and their tables including (a) layer (theme) and table creation, retrieval, visibility control and manipulation, (b) creation and extraction of fields and attributes thereof, (c) extraction of lists of records, editing of records and features, and storing of values and shapes, (d) redrawing of features. (e) querying and summarizing tables, (f) performing calculations on table cells, and (g) creation of shapefiles and personal geodatabases.

An example has been included at the end of this chapter. The example demonstrates how to create a (a) shapefile, and (b) a table, as well as, how to edit, query and summarize the table.

The Avenue Wraps of this chapter are listed below in alphabetical order with a short description and the chapter - page number where a full description may be found.

The source listing of each of the above Avenue Wraps may be found in Appendix D of this publication.

LAYERS (THEMES) and TABLES

5.1 Layer (Theme) and Table Related Avenue Wraps

5.1.1 Subroutine avCheckEdits

This subroutine enables the programmer to perform checks on the editing of data. This routine first determines whether the editor is in an edit state or not. If it is not in an edit state, this routine does nothing. If it is in an edit state, it checks to see if the given data set is currently being edited. If it is not, the routine saves the edits on the data set currently being edited, and starts the editor on the given data set. If the given data set is currently being edited, the routine does nothing.

The corresponding Avenue request is:

There is no corresponding Avenue request.

The call to this Avenue Wrap is:

Call **avCheckEdits**(pEditor, pDataSet)

avCheckEdits

GIVEN: pEditor = the ArcMap Editor extension

pDataSet = the data set to be processed. If the word NOTHING is specified, and if the editor is in an edit state, the editor is stopped, and any edits that may have been made are saved. If the editor is in not in an edit state, the routine does nothing.

RETURN: Nothing

The given and returned variables should be declared where first called as:

```
Dim pEditor As IEditor
Dim pDataSet As IDataset
```

5.1.2 Function avFTabMakeNew

This function enables the programmer to create a new shapefile, the name and type (class) of which are specified by the programmer as given arguments. In using this function, note the following:

1. Regarding the name of the shapefile:

   Some examples of a valid shapefile name include:

   c:\project\test\l_0ln or c:\project\test\l_0ln.shp

   l_0ln or l_0ln.shp

LAYERS (THEMES) and TABLES

2. The shapefile name may or may not contain the extension .shp.
3. If the name does not contain a complete path name, the current working directory will be used.
4. Regarding the shapefile type (class), it may be one of the following:

POINT	MULTIPOINT	POLYLINE	POLYGON
POINTM	MULTIPOINTM	POLYLINEM	POLYGONM
POINTZ	MULTIPOINTZ	POLYLINEZ	POLYGONZ

5. This function creates three fields called FID, SHAPE and ID.
6. Subsequently to this function, the function avAddDoc may be used to add the shapefile to the map, if need be.
7. If the shapefile to be created exists on the disk, the routine will abort, and the existing shapefile will not be overwritten

The corresponding Avenue request is:

theNewFTab = FTab.MakeNew (aFileName, aClass)

avFTabMakeNew

The call to this Avenue Wrap is:

Set theNewFTab = **avFTabMakeNew**(aFileName, aClass)

GIVEN: aFileName = name of the shapefile to be created (refer to notes 1, 2 and 3 above)

aClass = type of shapefile to be created (refer to note 4 above)

RETURN: theNewFTab = feature layer object that is created

The given and returned variables should be declared where first called as:

Dim aFileName, aClass As String
Dim theNewFTab As IFeatureLayer

5.1.3 Function avGetShapeType

This function enables the programmer to get the default shape type for a theme. In using this function, note the following:

1. The shapefile type (class), may be one of the following:

POINT	MULTIPOINT	POLYLINE	POLYGON
POINTM	MULTIPOINTM	POLYLINEM	POLYGONM
POINTZ	MULTIPOINTZ	POLYLINEZ	POLYGONZ

2. The Avenue request corresponding to the subject function operates on the FTab of a theme, while this VB function operates on the layer (theme).

LAYERS (THEMES) and TABLES

The corresponding Avenue request is:

```
anFTab = theTheme.GetFTab
theShapeType = anFTab.GetShapeClass
```

The call to this Avenue Wrap is:

theShapeType = **avGetShapeType**(pmxDoc, theTheme)

avGetShapeType

GIVEN: pmxDoc = the active view
theTheme = the theme to be processed

RETURN: theShapeType = the default shape type of the theme

The given and returned variables should be declared where first called as:

```
Dim pmxDoc As IMxDocument
Dim theTheme As Variant
Dim theShapeType As esriGeometryType
```

5.1.4 Subroutine avGetThemeExtent

This subroutine enables the programmer to get the smallest rectangle enclosing a layer (theme).

The corresponding Avenue request is:

There is no corresponding Avenue request.

The call to this Avenue Wrap is:

Call **avGetThemeExtent**(pmxDoc, theTheme, theRect)

avGetThemeExtent

GIVEN: pmxDoc = the active view
theTheme = the theme to be processed

RETURN: theRect = the smallest rectangle enclosing all of the features in theTheme

The given and returned variables should be declared where first called as:

```
Dim pmxDoc As IMxDocument
Dim theTheme As Variant
Dim theRect As IEnvelope
```

LAYERS (THEMES) and TABLES

5.1.5 Subroutine avInvalidateTOC

This subroutine enables the programmer to refresh the display of the Table of Contents.

The corresponding Avenue request is:

```
aView.InvalidateTOC(theName)
```

avInvalidateTOC

The call to this Avenue Wrap is:

Call **avInvalidateTOC**(theName)

GIVEN: theName = name of the theme or table in the Table of Contents to be refreshed. If NULL is specified, the entire Table of Contents will be refreshed.

RETURN: nothing

The given and returned variables should be declared where first called as:

```
Dim theName As Variant
```

5.1.6 Function avIsEditable

This function enables the programmer to determine if a layer (theme) or table is editable or not. Note that in:

- Avenue, the editability status is asked of an object, FTab or VTab,
- Avenue Wrap, the editability status is asked of a theme, or table name.

The corresponding Avenue request is:

```
theAnsw = theFTab.IsEditable
```

avIsEditable

The call to this Avenue Wrap is:

theAnsw = **avIsEditable**(theName)

GIVEN: theName = name of theme or table for which its editability status is to be checked

RETURN: theAnsw = editability status of the layer or table (true = is editable, false = is not editable)

The given and returned variables should be declared where first called as:

```
Dim theName As Variant
Dim theAnsw As Boolean
```

LAYERS (THEMES) and TABLES

5.1.7 Function avIsFTheme

This function enables the programmer to determine whether a layer (theme) is of the feature layer type, or not

The corresponding Avenue request is:

```
theAnsw = aTheme.Is(FTheme)
```

The call to this Avenue Wrap is:

theAnsw = **avIsFTheme**(theName)

avIsFTheme

GIVEN: theName = name of input object for which its feature layer type is to be determined

RETURN: theAnsw = flag denoting whether the input object is a feature layer (theme) or not (true = it is, false = it is not a feature layer)

The given and returned variables should be declared where first called as:

```
Dim theName As Variant
Dim theAnsw As Boolean
```

5.1.8 Function avIsVisible

This function enables the programmer to determine if an object is visible or not. Note that in:

- Avenue, the visibility status is asked of an object (aTheme),
- Avenue Wrap, the visibility status is asked of a theme name (theName).

The corresponding Avenue request is:

```
theAnsw = aTheme.IsVisible
```

The call to this Avenue Wrap is:

theAnsw = **avIsVisible**(theName)

avIsVisible

GIVEN: theName = name of input object for which its visibility status is to be determined

RETURN: theAnsw = visibility status of the layer (true = is editable, false = is not editable)

The given and returned variables should be declared where first called as:

```
Dim theName As Variant
Dim theAnsw As Boolean
```

LAYERS (THEMES) and TABLES

5.1.9 Subroutine avSetEditable

This subroutine enables the programmer to start or stop the editing of a layer (theme) or table. In using this subroutine, note the following:

- For layers, the editing is not terminated (the editor is not stopped), but instead any buffered writes are simply flushed. This allows the user to undo an edit.
- For tables the editing is terminated.
- To terminate the editing on layers use the subroutine avStopEditing.

The corresponding Avenue request is:

```
aVTab.SetEditable(eStatus)                    '---FTab or VTab
```

avSetEditable

The call to this Avenue Wrap is:

```
Call avSetEditable(pmxDoc, theTheme, eStatus)
```

GIVEN:
- pmxDoc = the active view
- theTheme = name of the theme or table to be processed
- eStatus = editing status. Specify
  True to start editing   False to stop editing

RETURN: nothing

The given and returned variables should be declared where first called as:

```
Dim pmxDoc As IMxDocument
Dim theTheme As Variant
Dim eStatus As Boolean
```

5.1.10 Subroutine avSetEditableTheme

This subroutine enables the programmer to stop the editing of a layer (theme) or set the type of task to be performed. This subroutine operates only on layers (themes), and can be used to display the handles of a feature. In using this subroutine note that the global variable ugSketch is used to keep track of whether a sketch session is active or not. If the value of ugSketch = 0, a sketch session is not active, if ugSketch = 1, a sketch session is active.

The corresponding Avenue request is:

```
theAnsw = aView.SetEditableTheme(aTheme)
```

LAYERS (THEMES) and TABLES

The call to this Avenue Wrap is:

Call **avSetEditableTheme**(pmxDoc, theTheme, theType)

GIVEN: pmxDoc = the active view
theTheme = name of the theme to be processed, if NULL, the editor will be stopped saving any edits that may have been made
theType = the type of task to be performed. Specify 0 to stop sketch session, 1 to modify feature, 2 to create new feature, or NULL to do nothing

RETURN: nothing

avSetEditable Theme

The given and returned variables should be declared where first called as:

Dim pmxDoc As IMxDocument
Dim theTheme As Variant
Dim theType As Variant

5.1.11 Subroutine avSetVisible

This subroutine enables the programmer to set the visibility status of an object.

The corresponding Avenue request is:

aTheme.SetVisible (aStatus)

The call to this Avenue Wrap is:

Call **avSetVisible**(theName, aStatus)

avSetVisible

GIVEN: theName = name of input object for which its visibility status is to be defined
aStatus = the visible state of the input object (true = visible, false = not visible)

RETURN: nothing

The given and returned variables should be declared where first called as:

Dim theName As Variant
Dim aStatus As Boolean

LAYERS (THEMES) and TABLES

5.1.12 Subroutine avStartOperation

This subroutine enables the programmer to start an operation within an edit session.

The corresponding Avenue request is:

There is no corresponding Avenue request.

The call to this Avenue Wrap is:

avStartOperation

Call **avStartOperation**

GIVEN: nothing

RETURN: nothing

5.1.13 Subroutine avStopEditing

This subroutine enables the programmer to terminate the editing of a layer (theme) or table. This subroutine stops the editor committing any edits that may have been made to the layer (theme) or table, thus prohibiting the undo of said edits.

The corresponding Avenue request is:

aTable.StopEditing

The call to this Avenue Wrap is:

avStopEditing

Call **avStopEditing**

GIVEN: nothing

RETURN: nothing

The given and returned variables should be declared where first called as:

5.1.14 Subroutine avStopOperation

This subroutine enables the programmer to stop an operation within an edit session.

The corresponding Avenue request is:

There is no corresponding Avenue request.

The call to this Avenue Wrap is:

Call **avStopOperation**(oprMssg)

avStopOperation

GIVEN: oprMssg = edit operation message that will appear to the right of the Undo menu item under the Edit menu item

RETURN: nothing

The given and returned variables should be declared where first called as:

Dim oprMssg As Variant

NOTE: When the editor is stopped it is not possible to use the Undo command under the Edit menu item, so that, if the Undo command is to be used, the Editor must be active (in use). The subroutine avStopEditing merely signals that an edit operation has been completed and that the operation should be added to the Undo list. It does not terminate the edit session and as such the editor is left in an active state and the user is able to employ the Undo command, if need be.

5.1.15 Subroutine avThemeInvalidate

This subroutine enables the programmer to redraw either the entire display or only that of a theme.

The corresponding Avenue request is:

aTheme.Invalidate(rdStatus)

The call to this Avenue Wrap is:

Call **avThemeInvalidate**(pmxDoc, theTheme, rdStatus)

avThemeInvalidate

GIVEN: pmxDoc = the active view
theTheme = name of theme to be processed
rdStatus = redraw status. Specify:
True to redraw entire view, or
False to redraw the theme only

RETURN: nothing

LAYERS (THEMES) and TABLES

The given and returned variables should be declared where first called as:

Dim pmxDoc As IMxDocument
Dim theTheme As Variant
Dim rdStatus As Boolean

5.1.16 Function avVTabMakeNew

This function enables the programmer to create a new dBase or text file type table. In using this function, note the following:

1. In specifying the name of the new table, if the name does not contain a complete path name, the current working directory will be used. Some examples of a table name include:

   c:\project\test\atable   c:\project\test\atable.dbf
   atable   atable.dbf

   The name may or may not contain the extension .dbf or .txt
2. Two fields called OID and ID will be created by this routine.
3. The function avAddDoc can be used to add the table to the map, if need be.
4. If the table to be created exists on disk, the routine will abort, and the existing table will not be overwritten.

The corresponding Avenue request is:

theNewTable = VTab.MakeNew (aFileName, aClass)

The call to this Avenue Wrap is:

avVTabMakeNew

Set theNewTable = **avVTabMakeNew**(aFileName, aClass)

GIVEN: aFileName = name of the table to be created (see Note 1).
aClass = type of table to be created. Specify: dBase or TEXT

RETURN: theNewTable = table object that is created

The given and returned variables should be declared where first called as:

Dim aFileName, aClass As String
Dim theNewTable As ITable

LAYERS (THEMES) and TABLES

5.1.17 Function FindLayer

This function enables the programmer to find a layer (theme) in a map.

The corresponding Avenue request is:

aTheme = aView.FindTheme (theName)

The call to this Avenue Wrap is:

Set theLayer = **FindLayer**(aMap, theName)

FindLayer

GIVEN: aMap = map to be searched
theName = name of the layer to be found

RETURN: theLayer = the layer in the map

The given and returned variables should be declared where first called as:

Dim aMap As IMap
Dim theName As Variant
Dim theLayer As ILayer

5.1.18 Function FindTheme

This function enables the programmer to find a layer (theme) in a map. This is similar to FindLayer with the exception, it does not return an ILayer object but rather, it returns a variable of Variant type. If the layer to be found can not be found, the returned value will be a single blank character. When a layer is found, the global variable, ugLayerIndx, will be set to the index value of the layer, so that the statement Set theLayer = pMap.Layer(ugLayerIndx) could be used to get an ILayer object. Note that pMap is an IMap object.

The corresponding Avenue request is:

aTheme = aView.FindTheme (theName)

The call to this Avenue Wrap is:

theTheme = **FindTheme**(aMap, theName)

FindTheme

GIVEN: aMap = map to be searched
theName = name of layer to be found

RETURN: theTheme = the layer in the map

The given and returned variables should be declared where first called as:

Dim aMap As IMap
Dim theName As Variant
Dim theTheme As Variant

5.2 Theme or Table Attribute Field Related Avenue Wraps

ATTRIBUTE FIELDS

5.2.1 Function avAddFields

This function enables the programmer to add attribute fields in a layer (theme) or table. In using this function, note the following:

1 In order to add fields into a layer or table, the editor can not be in an edit state. Thus this function will stop the editor, saving any changes that may have been made, prior to adding the fields.

2. In both Avenue and Avenue Wraps, the items in the given collection (list) are objects, not strings. Thus, before calling this function, the given argument, theFields, must be populated with items declared as iField. The Avenue Wrap avFieldMake may be used to create the iField items.

The corresponding Avenue request is:

```
anFTab.AddFields (theFields)
```

The call to this Avenue Wrap is:

errFlag = **avAddFields**(pmxDoc, theTheme, theFields)

avAddFields

GIVEN:	pmxDoc	= the active view
	theTheme	= the theme or table to be processed
	theFields	= list of fields to be added (see Note 2 above)
RETURN:	errFlag	= error flag (0 = no error, 1 = error)

The given and returned variables should be declared where first called as:

```
Dim pmxDoc As IMxDocument
Dim theTheme As Variant
Dim theFields As New Collection
Dim errFlag As Integer
```

5.2.2 Function avFieldGetType

This function enables the programmer to get the type of field that a field object is. In using this function, one of the numbers shown below will be returned to indicate the type of field object that was processed:

0 : Small Integer	1 : Long Integer
2 : Single-precision float	3 : Double-precision float
4 : String	5 : Date
6 : Long Integer denoting the OID	7 : Geometry
8 : Blob	

ATTRIBUTE FIELDS

avFieldGetType

The corresponding Avenue request is:

theFieldType = aField.GetType

The call to this Avenue Wrap is:

theFieldType = **avFieldGetType**(pField)

GIVEN: pField = field object to be processed

RETURN: theFieldType = numeric value denoting type of field (see above)

The given and returned variables should be declared where first called as:

Dim pField As iField
Dim theFieldType As esriFieldType

5.2.3 Function avFieldMake

avFieldMake

This function enables the programmer to create a field that can be added to a layer (theme) or table. In using this function, note the following:

1. Specify the key word below for the argument aFieldType to denote the indicated type of field object:

BYTE	Small Integer	CHAR	String
DATE	Date	DECIMAL	Single
DOUBLE	Double	FLOAT	Single
ISODATE	Date	ISODATETIME	Date
ISOTIME	Date	LOGICAL	String
LONG	Integer	MONEY	Double
SHORT	Small Integer	BLOB	Blob
VCHAR	String		

2. This routine can not be used to create a geometry field.

The corresponding Avenue request is:

theNewField = Field.Make(aName, aFieldType, nchr, ndr)

The call to this Avenue Wrap is:

Set theNewField = **avFieldMake**(aName, aFieldType, nchr, ndr)

GIVEN: aName = name of field to be created
aFieldType = type of field to be created (see Note 1 above)
nchr = total character width of field including decimal point and negative sign, if they are to appear in the field
ndr = number of digits to the right of the decimal point. Specify 0 for non-numeric fields

RETURN: theNewField = field object that was created

ATTRIBUTE FIELDS

The given and returned variables should be declared where first called as:
Dim aName, aFieldType As String
Dim nchr, ndr As Long
Dim theNewField As IFieldEdit

5.2.4 Subroutine avGetFields

This subroutine enables the programmer to get a list of attribute field names for a layer (theme) or table. These are not the alias names for the fields.

The corresponding Avenue request is:

theList = aVTab.GetFields

The call to this Avenue Wrap is:

Call **avGetFields**(theVTab, theList)

avGetFields

GIVEN: theVTab = field list for the theme or table

RETURN: theList = list of field names for an attribute table

The given and returned variables should be declared where first called as:
Dim theVTab As IFields
Dim theList As New Collection

5.2.5 Function avGetPrecision

This subroutine enables the programmer to get the decimal precision for a field. This is the number of digits to the right of the decimal point. This function always returns zero for fields contained in a personal geodatabase.

The corresponding Avenue request is:

aNumb = aField.GetPrecision

The call to this Avenue Wrap is:

aNumb = **avGetPrecision**(theVTab, fieldIndex)

avGetPrecision

GIVEN: theVTab = field list for the theme or table
fieldIndex = index of the field to be processed

RETURN: aNumb = decimal precision for the field

ATTRIBUTE FIELDS

The given and returned variables should be declared where first called as:

```
Dim theVTab As IFields
Dim fieldIndex As Long
Dim aNumb As Long
```

5.2.6 Function avRemoveFields

This function enables the programmer to remove attribute fields from a layer (theme) or table. In using this function, note the following:

1. In order to remove fields from a layer or table, the editor can not be in an edit state. Thus, this routine will stop the editor, saving any changes that may have been made, prior to removing the fields
2. If an invalid index value appears in the list, -1, it will be ignored (an error is not generated)
3. Do not use this routine to delete the SHAPE field
4. The items in the given argument list, theFields, are numeric index values (not objects) for the fields to be deleted.

The corresponding Avenue request is:

aVTab.RemoveFields(theFields)

avRemoveFields

The call to this Avenue Wrap is:

errFlag = **avRemoveFields**(pmxDoc, theTheme, theFields)

GIVEN:	pmxDoc	= the active view
	theTheme	= the theme or table to be processed
	theFields	= list of fields to be removed (see Note 4 above)
RETURN:	errFlag	= error flag (0 = no error, 1 = error detected)

The given and returned variables should be declared where first called as:

```
Dim pmxDoc As IMxDocument
Dim theTheme As Variant
Dim theFields As New Collection
Dim errFlag As Integer
```

5.3 Theme or Table Record Related Avenue Wraps

LAYER (THEME) and TABLE RECORDS

5.3.1 Function avAddRecord

This function enables the programmer to add a record into a layer (theme) or table.

The corresponding Avenue request is:

theRecordID = aVTab.AddRecord

The call to this Avenue Wrap is:

theRecordID = **avAddRecord**(pmxDoc, theTheme)

avAddRecord

GIVEN: pmxDoc = the active view
theTheme = the theme or table to be processed

RETURN: theRecordID = the ID of the record that was added. If a record can not be added it will be -1.

The given and returned variables should be declared where first called as:

Dim pmxDoc As IMxDocument
Dim theTheme As Variant
Dim theRecordID As Long

5.3.2 Subroutine avGetFTab

This subroutine enables the programmer to get the attribute table, feature class and associated layer (theme) for a specified theme. Note that if a table, rather than a theme, is specified, the values for the theFeatureClass and theLayer arguments will be set to Nothing, while theFTab object will reflect the attributes for the table.

The corresponding Avenue request is:

theFTab = aTheme.GetFTab

The call to this Avenue Wrap is:

Call **avGetFTab**(pmxDoc, theTheme, _
theFTab, theFeatureClass, theLayer)

avGetFTab

GIVEN: pmxDoc = the active view
theTheme = the theme or table to be processed

LAYER (THEME) and TABLE RECORDS

RETURN: theFTab = the attribute table for the theme
theFeatureClass = the feature class for the theme
theLayer = the associated layer for the theme

The given and returned variables should be declared where first called as:

```
Dim pmxDoc As IMxDocument
Dim theTheme As Variant
Dim theFTab As IFields
Dim theFeatureClass As IFeatureClass
Dim theLayer As IFeatureLayer
```

5.3.3 Subroutine avGetFTabIDs

This subroutine enables the programmer to get a list of the object identification numbers (OIDs) for a layer (theme).

The corresponding Avenue request is:

There is no corresponding Avenue request.

avGetFTabIDs

The call to this Avenue Wrap is:

Call **avGetFTabIDs**(pmxDoc, theTheme, theRecsList)

GIVEN: pmxDoc = the active view
theTheme = the theme or table to be processed

RETURN: theRecsList = the list of OIDs for the layer (theme)

The given and returned variables should be declared where first called as:

```
Dim pmxDoc As IMxDocument
Dim theTheme As Variant
Dim theRecsList as New Collection
```

5.3.4 Function avGetNumRecords

This function enables the programmer to get the number of records in a layer (theme), or table.

The corresponding Avenue request is:

numRecs = aVTab.GetNumRecords

avGetNumRecords

The call to this Avenue Wrap is:

numRecs = **avGetNumRecords**(pmxDoc, theTheme)

GIVEN: pmxDoc = the active view
theTheme = the theme or table to be processed

RETURN: numRecs = number of records in the theme or table

The given and returned variables should be declared where first called as:

Dim pmxDoc As IMxDocument
Dim theTheme As Variant
Dim numRecs As Long

5.3.5 Subroutine avGetVTab

This subroutine enables the programmer to get the attribute table for a layer (theme) or table.

The corresponding Avenue request is:

theVTab = aTable.GetVTab

The call to this Avenue Wrap is:

Call **avGetVTab**(pmxDoc, theTheme, theVTab)

avGetVTab

GIVEN: pmxDoc = the active view
theTheme = the theme to be processed

RETURN: theVTab = the attribute table for the theme or table

The given and returned variables should be declared where first called as:

Dim pmxDoc As IMxDocument
Dim theTheme As Variant
Dim theVTab As IFields

5.3.6 Subroutine avGetVTabIDs

This subroutine enables the programmer to get a list of the object identification numbers (OIDs) for a table.

The corresponding Avenue request is:

There is no corresponding Avenue request.

The call to this Avenue Wrap is:

Call **avGetVTabIDs**(pmxDoc, theTable, theRecsList)

avGetVTabIDs

LAYER (THEME) and TABLE RECORDS

GIVEN: pmxDoc = the active view
theTable = the table to be processed

RETURN: theRecsList = the list of object identification numbers (OIDs) for the table

The given and returned variables should be declared where first called as:

```
Dim pmxDoc As IMxDocument
Dim theTable As Variant
Dim theRecsList As New Collection
```

5.3.7 Subroutine avSetValue

This subroutine enables the programmer to store a value in a specific field of a specific row of a layer (theme) or table. In using this subroutine, note the following:

1. Do not use this subroutine to store geometry in the SHAPE field. Use this subroutine to store attribute information only.
2. To store geometry in the SHAPE field, use the subroutine avSetValueG.
3. While in Avenue the same request may be used to write attribute information and geometry, there are two distinct Avenue Wrap requests because of there are two distinct interfaces in ArcObjects.
4. While the Avenue request operates on an FTab or VTab, the Avenue Wrap operates on a layer (theme) or table name.

avSetValue

The corresponding Avenue request is:

```
aVTab.SetValue (aField, aRecord, anObj)
```

The call to this Avenue Wrap is:

Call **avSetValue**(pmxDoc, theTheme, aField, aRecord, anObj)

GIVEN: pmxDoc = the active view
theTheme = the theme or table to be processed
aField = field to be written to
aRecord = record of theme or table to be processed
anObj = object to be stored (attribute information only, no geometry)

RETURN: nothing

The given and returned variables should be declared where first called as:
Dim pmxDoc As IMxDocument
Dim theTheme As Variant
Dim aField, aRecord As Long
Dim anObj As Variant

LAYER (THEME) and TABLE RECORDS

5.3.8 Subroutine avSetValueG

This subroutine enables the programmer to store a shape in the SHAPE field of a specific row of a layer (theme). In using this subroutine, note the following:

1. Do not use this subroutine to store attribute information. Use this subroutine to store geometry in the SHAPE field only.
2. To store attribute information, use the subroutine avSetValue.
3. While in Avenue the same request may be used to write attribute information and geometry, there are two distinct Avenue requests because of there are two distinct interfaces in ArcObjects.
4. While the Avenue request operates on an FTab or VTab, the Avenue Wrap operates on a layer (theme) name.

The corresponding Avenue request is:
There is no corresponding Avenue request.

The call to this Avenue Wrap is:
Call **avSetValueG**(pmxDoc, theTheme, aField, aRecord, aShape)

avSetValueG

GIVEN: pmxDoc = the active view
theTheme = the theme or table to be processed
aField = field to be written to
aRecord = record of theme or table to be processed
aShape = shape to be stored (geometry only, no attribute information only)

RETURN: nothing

The given and returned variables should be declared where first called as:
Dim pmxDoc As IMxDocument
Dim theTheme As Variant
Dim aField, aRecord As Long
Dim aShape As IGeometry

5.4 Feature Handling Related Avenue Wraps

FEATURE HANDLING

5.4.1 Subroutine avFeatureInvalidate

This subroutine enables the programmer to redraw a feature.

The corresponding Avenue request is:

There is no corresponding Avenue request.

The call to this Avenue Wrap is:

Call **avFeatureInvalidate**(pmxDoc, theFeature)

avFeature Invalidate

GIVEN: pmxDoc = the active view
theFeature = the feature to be redrawn

RETURN: Nothing

The given and returned variables should be declared where first called as:

Dim pmxDoc As IMxDocument
Dim theFeature As IFeature

5.4.2 Subroutine avGetFeature

This subroutine enables the programmer to get the feature given a layer (theme) and an object ID.

The corresponding Avenue request is:

theFeature = aFTab.ReturnValue ("shape", theObjId)

The call to this Avenue Wrap is:

Call **avGetFeature**(pmxDoc, theTheme, theObjId, theFeature)

avGetFeature

GIVEN: pmxDoc = the active view
theTheme = the theme to be processed
theObjId = the object id of the desired feature

RETURN: theFeature = the feature

The given and returned variables should be declared where first called as:

Dim pmxDoc As IMxDocument
Dim theTheme As Variant
Dim theObjId As Long
Dim theFeature As IFeature

FEATURE HANDLING

5.4.3 Subroutine avGetFeatData

This subroutine enables the programmer to get the feature data of a given layer (theme) and object ID.

The corresponding Avenue request is:

There is no corresponding Avenue request.

avGetFeatData

The call to this Avenue Wrap is:

```
Call avGetFeatData(pmxDoc, theTheme, theObjId, _
                   theFeature, theShape, shapeType)
```

GIVEN:	pmxDoc	= the active view
	theTheme	= the theme to be processed
	theObjId	= the object id of the desired feature
RETURN:	theFeature	= the feature
	theShape	= the geometry of a feature
	shapeType	= the shape type of a feature

The given and returned variables should be declared where first called as:

```
Dim pmxDoc As IMxDocument
Dim theTheme As Variant
Dim theObjId As Long
Dim theFeature As IFeature
Dim theShape As IGeometry
Dim shapeType As esriGeometryType
```

5.4.4 Subroutine avGetGeometry

FEATURE HANDLING

This subroutine enables the programmer to get the geometry of a feature given its layer (theme) and object ID.

The corresponding Avenue request is:

There is no corresponding Avenue request.

The call to this Avenue Wrap is:

avGetGeometry

Call **avGetGeometry**(pmxDoc, theTheme, theObjId, theShape)

GIVEN: pmxDoc = the active view
theTheme = the theme to be processed
theObjId = the object id of the desired feature

RETURN: theShape = the geometry of a feature

The given and returned variables should be declared where first called as:

```
Dim pmxDoc As IMxDocument
Dim theTheme As Variant
Dim theObjId As Long
Dim theShape As IGeometry
```

5.4.5 Subroutine avUpdateAnno

This subroutine enables the programmer to apply a transformation to an existing annotation feature (a feature in a feature annotation layer). In using this subroutine, note the following:

1. The rotation angle is added to the existing angle of the annotation (positive value denotes a counter-clockwise rotation, while a negative value denotes a clockwise rotation).
2. A scale factor greater than 1.0 increases the size of the annotation, while a value less than 1.0 decreases the size.
3. The X scale factor is always used in the scaling process, the Scale method does not seem to work as it should on Annotation features when the X and Y scale factors are different.
4. The layer that the feature resides in must be in an editable state.

The corresponding Avenue request is:

There is no corresponding Avenue request.

FEATURE HANDLING

avUpdateAnno

The call to this Avenue Wrap is:

```
Call avUpdateAnno(pFeature, oldX, oldY, newX, newY, _
                  rotang, scaleX, scaleY, newFeature)
```

GIVEN:	pFeature	= the annotation feature to be modified
	oldX, oldY	= the coordinates of the feature's control point
	newX, newY	= the new coordinates of the feature's control point
	rotang	= the rotation angle in degrees to be added to the existing angle of the feature
	scaleX	= the X axis scale factor (greater than 0.0)
	scaleY	= the Y axis scale factor (greater than 0.0)
RETURN:	newFeature	= the new feature reflecting the transformation

The given and returned variables should be declared where first called as:

```
Dim pFeature As IFeature
Dim oldX, oldY, newX, newY, rotang, scaleX, scaleY As Double
Dim newFeature As IFeature
```

5.5 Calculating, Querying and Summarizing Layers (Themes) and Tables Related Avenue Wraps

CALCULATE QUERY and SUMMARIZE

5.5.1 Function avCalculate

This function enables the programmer to apply a calculation to a set of selected records for a specified field in a layer (theme) or table. If no records are selected, the entire layer (theme) or table is processed. Sample calculation strings are shown below. Note how the field names are handled depending upon the type of field being processed.

- Shapefile and Personal Geodatabase String field calculation:

```
aCalcString = """abcd"""
```

- Shapefile and Personal Geodatabase Numeric field calculation:

```
aCalcString = "([ID] - " + CStr(i) + ")"
```

The corresponding Avenue request is:

errFlag = aVTab.Calculate(aCalcString, aField)

The call to this Avenue Wrap is:

errFlag = **avCalculate**(pmxDoc, theTheme, aCalcString, aField)

avCalculate

GIVEN: pmxDoc = the active view
theTheme = name of theme or table to be processed
aCalcString = calculation string to be applied (see above)
aField = index value of the field to be populated. Index value is between 0 and n-1, where n is the total number of fields.

RETURN: errFlag = error flag as noted below
0 : no error
1 : theme or table not found
2 : error in performing calculation
3 : no records selected
4 : an edit session has not been started

The given and returned variables should be declared where first called as:

```
Dim pmxDoc As IMxDocument
Dim theTheme As Variant
Dim aCalcString As String
Dim aField As Long
Dim avCalculate As Integer
```

CALCULATE QUERY and SUMMARIZE

5.5.2 Subroutine avQuery

This subroutine enables the programmer to apply a query string to a layer (theme) or table. Sample query strings are shown below. Note how the field names are handled depending upon the type of field and the data source.

- Shapefile String field queries are case sensitive:
  `aQueryString = """PTCODE""" + " = 'BBBB'"`
- Personal Geodatabase String field queries are case insensitive:
  `aQueryString = "PTCODE = 'bbbb'"`
- Shapefile and Personal Geodatabase Numeric field query:
  `aQueryString = "SLN >= 10"`

The corresponding Avenue request is:

errFlag = aVTab.Query(aQueryString, selSet, setType)

The call to this Avenue Wrap is:

avQuery

Call **avQuery**(pmxDoc, theTheme, aQueryString, selSet, setType)

GIVEN:

pmxDoc	=	the active view
theTheme	=	name of theme or table to be processed
aQueryString	=	query string to be applied (see above)
selSet	=	theme selection set (see chapter 6 regarding selections)
setType	=	type of selection desired • "NEW" : new selection set • "ADD" : add to current selection set • "AND" : select from current selection set

RETURN: nothing — Performs the query. Use the avGetSelection (see Chapter 6) to get the selection set containing the results of the query.

The given and returned variables should be declared where first called as:

```
Dim pmxDoc As IMxDocument
Dim elmntTheme As Variant
Dim aQueryString As String
Dim selSet As ISelectionSet
Dim setType As String
```

5.5.3 Function avSummarize

CALCULATE QUERY and SUMMARIZE

This function enables the programmer to summarize a layer (theme) or table on a specified field. In using this function, note the following:

1. The table will be stored in the work space of the layer (theme) or table to be summarized. So do not specify a full path name, and do not include an extension such as .dbf. If an extension appears in the name, it will be removed without generating an error flag or message.
2. The type of summary (operation codes) to be performed on the items in the fieldList should be one of the following key words enclosed in double quotes:

   "Count" "Minimum" "Maximum" "Sum" "Average"
   "Variance" "StdDev" "Dissolve" (for use on the Shape field)
3. Since this routine passes NOTHING to theSumTable if an error is detected, make certain to check for this in the code that calls this function.
4. The number of items in the fieldList should be the same with that of the sumryList. If one of them is empty, so must be the other one.
5. If fieldList and sumryList are empty lists, or if they are passed in as NOTHING, the following default values will be used:
   - fieldList will contain two items each one being the value of aField.
   - The sumryList will contain two items, the first being the number of unique values within all rows of aField, and the second being the maximum unique value within all rows of aField.
6. If the table to be created exists on the disk, the routine will overwrite the existing table without asking or informing the user.

The corresponding Avenue request is:

```
theSumTable = aVTab.Summarize(aFileName, aType, aField,
                              fieldList, sumryList)
```

The call to this Avenue Wrap is:

avSummarize

```
Set theSumTable = avSummarize(pmxDoc, theTheme,
                              aFileName, aType, aField, fieldList,
                              sumryList)
```

GIVEN:

pmxDoc	=	the active view
theTheme	=	name of theme or table to be processed (see Note 1 above)
aFileName	=	string name of the output table theSumTable to be created (see Note 6 above)
aType	=	type of output table. Specify "dBase".

CALCULATE QUERY and SUMMARIZE

	aField	= field that the theme or table is summarized on
	fieldList	= additional fields to be summarized (see Note 4)
	sumryList	= operation codes to be performed on the items in the fieldList (see Notes 2 and 4)
RETURN:	theSumTable	= the object summary table, whose name is that of aFileName. If an error is detected during the processing, the keyword NOTHING will be returned and a message to that effect will be displayed. See Notes 3 and 6 above, as well as the example below.

The given and returned variables should be declared where first called as:

```
Dim pmxDoc As IMxDocument
Dim theTheme As Variant
Dim aFileName, aType, aField As String
Dim fieldList, sumryList As New Collection
Dim theSumTable As ITable
```

Example 1 For example purposes, let us assume that:

- The table to be summarized is called "SchoolZones", and contains the data shown in Table 5-1(A),
- We wish to summarize on the field ZONE to obtain:
- (a) a count of the unique zone identification values, (b) maximum area per unique zone, and (c) minimum perimeter per unique zone.
- The summary table that is to be created is to be called "Zones" and should be of dBase format.

The call to the Avenue Wrap would be:

```
Call CreateList(fieldList)
fieldList.Add("ZONE")
fieldList.Add("ZONE")
fieldList.Add("AREA")
fieldList.Add("PERIM")
Call CreateList(sumryList)
sumryList.Add("Count")
sumryList.Add("Maximum")
sumryList.Add("Maximum")
sumryList.Add("Minimum")
Set aSTable = avSummarize(pmxDoc, "SchoolZones", _
                    "Zones", "dBase", "ZONE", _
                    fieldList, sumryList)
```

Shown in Table 5-1(B) are the results of the above summation.

The sample code on the following pages illustrates how the table could be created, records added, populated and summarized programmatically.

Example 2 Now let us consider the same table of Example 1, but with both the fieldList and sumryList arguments passed in as empty lists. The call to the Avenue Wrap would then be:

```
Call CreateList(fieldList)
Call CreateList(sumryList)
Set aSTable = avSummarize(pmxDoc, "SchoolZones", _
                         "Zones", "dBase", "ZONE", _
                         fieldList, sumryList)
```

Shown in Table 5-1(C) are the results of the above summation.

Attributes of SchoolZones

OID	ID	ZONE	AREA	PERIM
0	0	C-2	15.349	3270.72
1	0	R-4	21.537	3874.33
2	0	A-2	18.968	3635.92
3	0	C-2	14.663	1023.03
4	0	C-2	17.318	3474.18
5	0	R-4	16.259	3366.28

Record: 1 Show: All Selected Records (0 out of 6 Selected.) Options

Table 5-1(A) Sample Table to Be Summarized

Attributes of Zones

OID	ZONE	Last_ZONE	Maximum_AREA	Minimum_PERIM
0	1	A-2	18.968	3635.92
1	3	C-2	17.318	1023.03
2	2	R-4	21.537	3366.28

Record: 1 Show: All Selected Records (0 out of 3 Selected.) Options

Table 5-1(B) Sample Table Summarized as per Example 1

Attributes of Zones

OID	ZONE	Last_ZONE
0	1	A-2
1	3	C-2
2	2	R-4

Record: 1 Show: All Selected Records (0 out of 3 Selected.) Options

Table 5-1(C) Sample Table Summarized as per Example 2

```
'
' ---
' ---VBA code that is associated with Example 1 illustrating how to
' ---create, add records, populate and summarize a table
' ---
'
Dim pMxApp As IMxApplication
Dim pmxDoc As IMxDocument
Dim pActiveView As IActiveView
Dim pMap As IMap
Dim sTblName, sTblPthName As String
Dim iok As Integer
Dim pTable As ITable
Dim irec As Long
Dim pFld1 As IFieldEdit
Dim pFld2 As IFieldEdit
Dim pFld3 As IFieldEdit
Dim fldList As New Collection
Dim theVTab As IFields
Dim col1, col2, col3 As Long
Dim sumTblName As String
Dim fieldList1 As New Collection
Dim sumryList2 As New Collection
Dim pSTable As ITable
'
' ---Get the active view
Call avGetActiveDoc(pMxApp, pmxDoc, pActiveView, pMap)
'
' ---Define the name of the table to be created
sTblName = "SchoolZones.dbf"
'
' ---Define the full pathname of the table
sTblPthName = "c:\temp\" + sTblName
'
' ---Delete the table if it exists
If (avFileExists(sTblPthName)) Then
   iok = avFileDelete(sTblPthName)
End If
'
' ---Create a dBase table
Set pTable = avVTabMakeNew(sTblPthName, "dbase")
'
' ---Make sure the table was created
If (Not pTable Is Nothing) Then
'
'    ---Add the table to the map, the .dbf extension will not
'    ---appear in the table of contents (TOC)
   iok = avAddDoc(pTable)
'
'    ---Add six records to the table
   irec = avAddRecord(pmxDoc, sTblName)
   irec = avAddRecord(pmxDoc, sTblName)
   irec = avAddRecord(pmxDoc, sTblName)
   irec = avAddRecord(pmxDoc, sTblName)
   irec = avAddRecord(pmxDoc, sTblName)
   irec = avAddRecord(pmxDoc, sTblName)
'
'    ---Create three fields to be added to the table
   Set pFld1 = avFieldMake("ZONE", "VCHAR", 3, 0)
   Set pFld2 = avFieldMake("AREA", "DOUBLE", 12, 4)
   Set pFld3 = avFieldMake("PERIM", "DOUBLE", 12, 4)
'
'    ---Add the fields to a list
   Call CreateList(fldList)
   fldList.Add pFld1
```

```
        fldList.Add pFld2
        fldList.Add pFld3
    '
    '   ---Add the field list to the table
        iok = avAddFields(pmxDoc, sTblName, fldList)
    '
    '   ---Get the attribute table
        Call avGetVTab(pmxDoc, sTblName, theVTab)
    '
    '   ---Make the table editable since when it is added to
    '   ---the map, it will not be editable
        Call avSetEditable(pmxDoc, sTblName, True)
    '
    '   ---Store the values for all six records that were added
        col1 = theVTab.FindField("ZONE")
        col2 = theVTab.FindField("AREA")
        col3 = theVTab.FindField("PERIM")
        Call avSetValue(pmxDoc, sTblName, col1, 0, "C-2")
        Call avSetValue(pmxDoc, sTblName, col2, 0, 15.349)
        Call avSetValue(pmxDoc, sTblName, col3, 0, 3270.72)
        Call avSetValue(pmxDoc, sTblName, col1, 1, "R-4")
        Call avSetValue(pmxDoc, sTblName, col2, 1, 21.537)
        Call avSetValue(pmxDoc, sTblName, col3, 1, 3874.33)
        Call avSetValue(pmxDoc, sTblName, col1, 2, "A-2")
        Call avSetValue(pmxDoc, sTblName, col2, 2, 18.968)
        Call avSetValue(pmxDoc, sTblName, col3, 2, 3635.92)
        Call avSetValue(pmxDoc, sTblName, col1, 3, "C-2")
        Call avSetValue(pmxDoc, sTblName, col2, 3, 14.663)
        Call avSetValue(pmxDoc, sTblName, col3, 3, 1023.03)
        Call avSetValue(pmxDoc, sTblName, col1, 4, "C-2")
        Call avSetValue(pmxDoc, sTblName, col2, 4, 17.318)
        Call avSetValue(pmxDoc, sTblName, col3, 4, 3474.18)
        Call avSetValue(pmxDoc, sTblName, col1, 5, "R-4")
        Call avSetValue(pmxDoc, sTblName, col2, 5, 16.259)
        Call avSetValue(pmxDoc, sTblName, col3, 5, 3366.28)
    '
    '   ---Commit the modifications to disk
        Call avSetEditable(pmxDoc, sTblName, False)
    '
    '   ---Define the name of the summary table to be created
        sumTblName = "Zones"
    '
    '   ---Define fields and operations to be used in summarization
        Call CreateList(fieldList1)
        fieldList1.Add ("ZONE")
        fieldList1.Add ("ZONE")
        fieldList1.Add ("AREA")
        fieldList1.Add ("PERIM")
        Call CreateList(sumryList2)
        sumryList2.Add ("Count")
        sumryList2.Add ("Maximum")
        sumryList2.Add ("Maximum")
        sumryList2.Add ("Minimum")
    '
    '   ---Summarize all records based upon the ZONE field
        Set pSTable = avSummarize(pmxDoc, sTblName, _
                                  sumTblName, "dBase", "ZONE", _
                                  fieldList1, sumryList2)
    '
    '   ---Check if the table summarized, if so add to map
        If (Not pSTable Is Nothing) Then
           iok = avAddDoc(pSTable)
        End If
    End If
```

SHAPEFILE and GEODATABASE

5.6 Shapefile and GeoDatabase Related Avenue Wraps

5.6.1 Function avOpenFeatClass

This function enables the programmer to open a dataset for editing purposes. A dataset may be a shapefile, raster image, tin, coverage or access database. The type of object returned is of IUknown type, however, depending upon the type of dataset (opmode) to be processed, the actual type of object returned will be for:

1. shapefiles, coverages and access databases, IFeatureClass, for
2. rasters, IRasterDataset and for
3. tins, ITin.

The corresponding Avenue request is:

There is no corresponding Avenue request.

The call to this Avenue Wrap is:

avOpenFeatClass

```
Set theObject = avOpenFeatClass(opmode, sDir, sName, _
                                aFCtype)
```

GIVEN: opmode = type of dataset to be opened. Specify

1 : shapefile 2 : raster 3 : tin

4 : coverage 5 : access database

sDir = directory location of the dataset

sName = name of the dataset (do not include any filename extension in the name)

aFCtype = feature class type (used only for coverages)

RETURN: theObject = dataset that has been opened. If the specified dataset cannot be found, or if found and it cannot be opened, due to permission rights or other reasons, then the keyword NOTHING is returned.

The given and returned variables should be declared where first called as:

```
Dim opmode As Integer
Dim sDir As String
Dim sName As String
Dim aFCtype As String
Dim theObject As IUnknown
```

SHAPEFILE and GEODATABASE

5.6.2 Function avOpenWorkspace

This function enables the programmer to open a workspace for processing. A workspace may be a shapefile, raster image, tin, coverage or access database.

The corresponding Avenue request is:

There is no corresponding Avenue request.

The call to this Avenue Wrap is:

Set theObject = **avOpenWorkspace**(opmode, sDir, sName)

avOpenWorkspace

GIVEN: opmode = type of workspace to be opened. Specify
1 : shapefile 2 : raster 3 : tin
4 : coverage 5 : access database
sDir = directory location of the workspace
sName = name of the workspace (do not include any filename extension in the name)

RETURN: theObject = workspace that has been opened. If the specified workspace cannot be found, or if found and it cannot be opened, due to permission rights or other reasons, then the keyword NOTHING is returned.

The given and returned variables should be declared where first called as:

```
Dim opmode As Integer
Dim sDir As String
Dim sName As String
Dim theObject As IWorkspace
```

5.6.3 Function CreateAccessDB

This subroutine enables the programmer to create a personal geodatabase by specifying a directory location and the name of the .mdb file to be created.

The corresponding Avenue request is:

There is no corresponding Avenue request.

The call to this Avenue Wrap is:

Set theObject = **CreateAccessDB**(sDir, sName, bOverWrite)

CreateAccessDB

SHAPEFILE and GEODATABASE

GIVEN: sDir = directory location of the workspace
sName = name of the workspace
bOverWrite = flag denoting whether the database should be overwritten if it exists (true = overwrite, false = do not overwrite)

RETURN: theObject = the workspace object representing the new personal geodatabase

The given and returned variables should be declared where first called as:

```
Dim sDir As String
Dim sName As String
Dim bOverWrite As Boolean
Dim theObject As IWorkspace
```

5.6.4 Function CreateAnnoClass

This subroutine enables the programmer to create an annotation feature class within a personal geodatabase.

The corresponding Avenue request is:

There is no corresponding Avenue request.

CreateAnnoClass

The call to this Avenue Wrap is:

```
Set theObject = CreateAnnoClass(pWorkspace, sName, _
                pFields, dRefScale, dUnits)
```

GIVEN: pWorkspace = workspace of the existing geodatabase
sName = name of the feature class to be created
pFields = attributes associated with the feature class
dRefScale = reference scale
dUnits = units of measure setting

RETURN: theObject = object representing the new annotation feature class in the existing geodatabase

The given and returned variables should be declared where first called as:

```
Dim pWorkspace As IWorkspace
Dim sName As String
Dim pFields As IFields
Dim dRefScale As Double
```

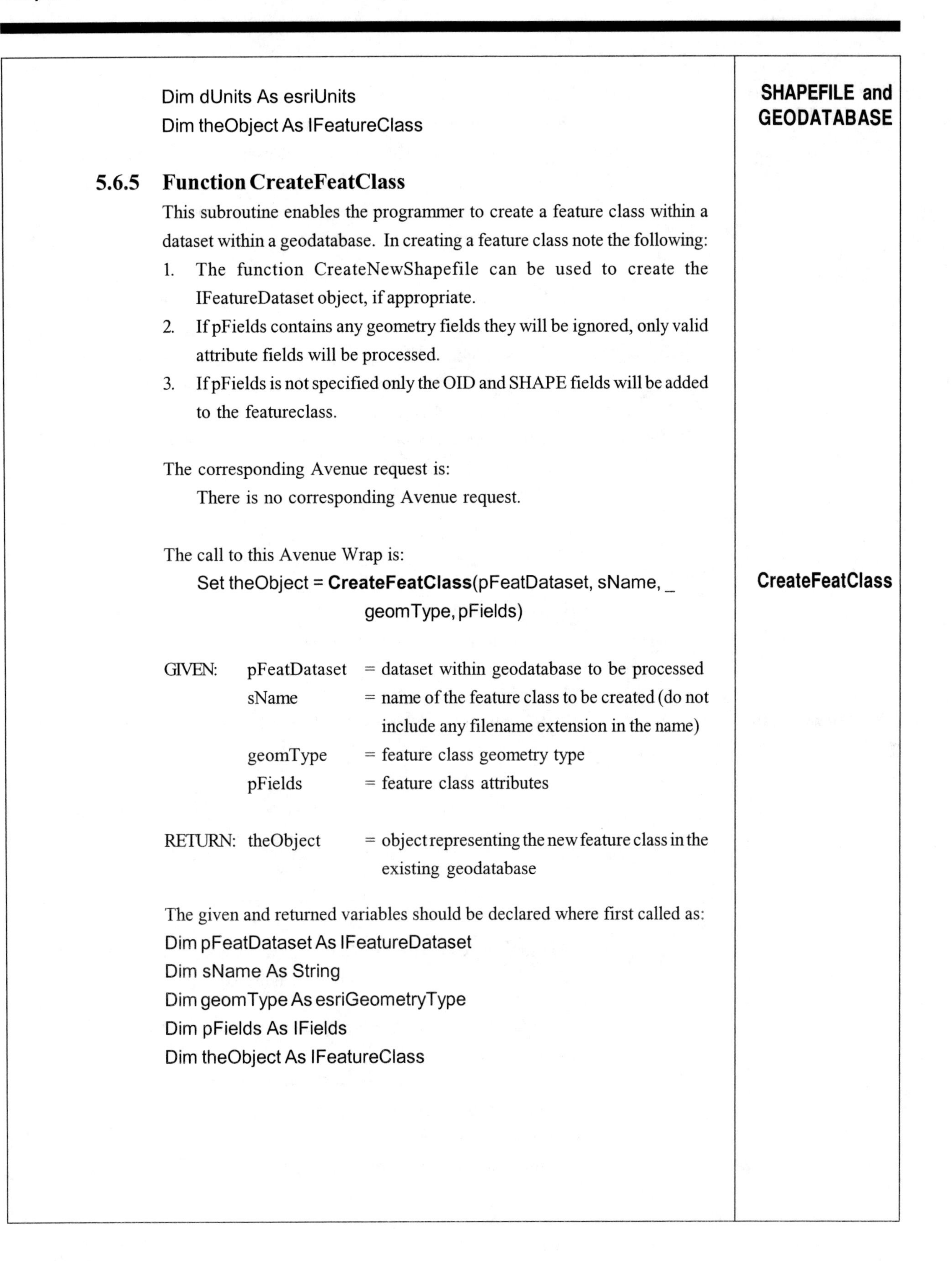

SHAPEFILE and GEODATABASE

```
Dim dUnits As esriUnits
Dim theObject As IFeatureClass
```

5.6.5 Function CreateFeatClass

This subroutine enables the programmer to create a feature class within a dataset within a geodatabase. In creating a feature class note the following:

1. The function CreateNewShapefile can be used to create the IFeatureDataset object, if appropriate.
2. If pFields contains any geometry fields they will be ignored, only valid attribute fields will be processed.
3. If pFields is not specified only the OID and SHAPE fields will be added to the featureclass.

The corresponding Avenue request is:

There is no corresponding Avenue request.

The call to this Avenue Wrap is:

CreateFeatClass

```
Set theObject = CreateFeatClass(pFeatDataset, sName, _
                                geomType, pFields)
```

GIVEN:	pFeatDataset	= dataset within geodatabase to be processed
	sName	= name of the feature class to be created (do not include any filename extension in the name)
	geomType	= feature class geometry type
	pFields	= feature class attributes
RETURN:	theObject	= object representing the new feature class in the existing geodatabase

The given and returned variables should be declared where first called as:

```
Dim pFeatDataset As IFeatureDataset
Dim sName As String
Dim geomType As esriGeometryType
Dim pFields As IFields
Dim theObject As IFeatureClass
```

SHAPEFILE and GEODATABASE

5.6.6 Function CreateNewGeoDB

This subroutine enables the programmer to create a personal geodatabase with a stand-alone annotation feature class using information that is specified by the user via a file dialog box. In using this function note the following:

1. Use CreateNewShapefile specifying the .mdb file name extension in the default filename to create a geodatabase that contains a feature class and not an annotation feature class.
2. The new annotation feature class is automatically added to the map once it has been created.
3. If an existing .mdb file is selected, the user can either abort the command (CANCEL), add to the .mdb file (NO) or overwrite the existing file (YES), see Figure 5-1(D).
4. When an existing .mdb file is appended the root name of the default filename is used as the name of the new annotation class.
5. When an existing .mdb file is to be overwritten, if the file exists in the map the function will not delete the file but will inform the user and abort the function.

The corresponding Avenue request is:

There is no corresponding Avenue request.

CreateNewGeoDB

The call to this Avenue Wrap is:

```
Set theObject = CreateNewGeoDB(pFieldsI, geomType, _
                        defName, aTitle)
```

GIVEN:
- pFieldsI = attributes to be stored in the new geodatabase
- geomType = feature class geometry type
- defName = default filename
- aTitle = file dialog message box title

RETURN:
- theRect = object representing the new annotation feature class in the existing geodatabase

The given and returned variables should be declared where first called as:

```
Dim pFieldsI As esriCore.IFields
Dim geomType As esriCore.esriGeometryType
Dim defName As String
Dim aTitle As String
Dim theObject As esriCore.IFeatureClass
```

5.6.7 Function CreateNewShapeFile

SHAPEFILE and GEODATABASE

This function enables the programmer to create a new shapefile or a personal geodatabase using information that is specified by the user via a file dialog box, see Figure 5-1(A). In using this function, note the following:

1. If the pFieldsI argument is set to NOTHING, a default shape field with a default spatial reference will be assigned, and one attribute called ID will be added to the shapefile.
2. The geometry type (geomType) should be specified as:
   - esriGeometryPoint,
   - esriGeometryPolyline, or
   - esriGeometryPolygon.
3. The default name of the shapefile that is specified (defName) will appear in the file dialog box.
4. If the defName argument contains the .shp filename extension, the dataset type that will be created will be a shapefile. If the .mdb filename extension is found, the type of dataset created will be a personal geodatabase. If no filename extension is given both types will appear in the list of available types and the user can pick the desired type, see Figure 5-1(C).
5. The new shapefile or geodatabase is automatically added to the map once it has been created.
6. If an existing .shp file is selected, the user can either abort the command (NO), or overwrite the existing file (YES), see Figure 5-1(B).
7. If an existing .mdb file is selected, the user can either abort the command (CANCEL), add to the .mdb file (NO) or overwrite the existing file (YES), see Figure 5-1(D).
8. When an existing .mdb file is appended the root name of the default filename is used as the name of the new feature class that is created.
9. When an existing .mdb file is to be overwritten, if the file exists in the map the function will not delete the file but will inform the user and abort the function.

The corresponding Avenue request is:

There is no corresponding Avenue request.

The call to this Avenue Wrap is:

CreateNewShapeFile

```
Set NewShapeFile = CreateNewShapeFile(pFieldsI, _
                          geomType, defName, aTitle)
```

SHAPEFILE and GEODATABASE

GIVEN: pFieldsI = attributes to be stored in the new shapefile
geomType = shapefile geometry type
defName = default filename
aTitle = file dialog message box title

RETURN: NewShapeFile = feature class that is created

The given and returned variables should be declared where first called as:

```
Dim pFieldsI As esriCore.IFields
Dim geomType As esriCore.esriGeometryType
Dim defName As String
Dim aTitle As String
Dim NewShapeFile As esriCore.IFeatureClass
```

5.6.8 Function CreateShapeFile

This function enables the programmer to create a new shapefile using information explicitly defined in the calling arguments (no user interaction). In using this function, note the following:

1. The name of the shapefile to be created (strName) can or can not contain the .shp extension. If it does, it will be stripped off.
2. The geometry type (geomType) should be specified as:
   - esriGeometryPoint,
   - esriGeometryPolyline, or
   - esriGeometryPolygon.
3. The pFields argument is optional (can be omitted from the argument list). If it is not specified, a default shape field with a default spatial reference will be assigned, and one attribute called ID will be added to the shapefile.
4. The pCLSID argument is optional (can be omitted from the argument list). If it is specified, the pFields argument must also be specified.

The corresponding Avenue request is:

There is no corresponding Avenue request.

CreateShapeFile

The call to this Avenue Wrap is:

```
Set NewShapeFile = CreateShapeFile(featWorkspace, _
                   strName, geomType, pFields, pCLSID)
```

SHAPEFILE and GEODATABASE

GIVEN: featWorkspace = directory location
strName = shapefile name
geomType = shapefile geometry type
pFields = shapefile attributes
pCLSID = geometry type subclass

RETURN: NewShapeFile = feature class that is created

The given and returned variables should be declared where first called as:

Dim featWorkspace As esriCore.IFeatureWorkspace
Dim strName As String
Dim geomType As esriCore.esriGeometryType
Dim pfields As esriCore.IFields
Dim pCLSID As esriCore.UID
Dim NewShapefile As esriCore.IFeatureClass

```
'
' ---
' ---Sample illustrating how to create a new shapefile that
' ---has a default spatial reference and three attributes
' ---using a name that the user enters in a file dialog box.
' ---The shapefile is to contain Polyline features and will
' ---be added to the map once it has been created.
' ---
'
  Dim pMxApp As esriCore.IMxApplication
  Dim pmxDoc As esriCore.IMxDocument
  Dim pActiveView As esriCore.IActiveView
  Dim pMap As esriCore.IMap
  Dim aDefName As String
  Dim pFieldsEdit As esriCore.IFieldsEdit
  Dim pFieldEdit As esriCore.IFieldEdit
  Dim pSR As esriCore.ISpatialReference
  Dim pGeomDef As esriCore.IGeometryDef
  Dim pGeomDefEdit As esriCore.IGeometryDefEdit
  Dim aMessage As String
  Dim pNShapeFile As esriCore.IFeatureClass
  Dim aMsg, aTitle2 As String
  Dim theTheme As Variant
  Dim theFTab As esriCore.IFields
  Dim pFeatureClass As esriCore.IFeatureClass
  Dim aLayer As esriCore.IFeatureLayer
'
' ---Get the active view
  Call avGetActiveDoc(pMxApp, pmxDoc, pActiveView, pMap)
'
```

SHAPEFILE and GEODATABASE

```
'    ---Define the default shapefile name (since there is no
'    ---extension specified in the name, the Save as type: drop
'    ---down list will contain both Shapefile and Personal
'    ---Geodatabases)
     aDefName = "L_poly"
'
'    ---Check if the shapefile is in the map, we can not
'    ---create a shapefile if it exists in the map
     If (avFindDoc(aDefName) <> -1) Then
'       ---Remove the shapefile from the map, does not
'       ---delete it from the hard drive (disk)
        Call avRemoveDoc(aDefName)
     End If
'
'    ---Create the required shapefile attributes
'
'    ---Define the object ID field
     Set pFieldsEdit = New esriCore.Fields
     Set pFieldEdit = New esriCore.Field
     With pFieldEdit
         .name = "OID"
         .Type = esriCore.esriFieldTypeOID
         .aliasName = "Object ID"
        .IsNullable = False
     End With
     pFieldsEdit.AddField pFieldEdit
'
'    ---Assign the default spatial reference
     Set pSR = New esriCore.UnknownCoordinateSystem
     pSR.SetDomain -9999999999#, 9999999999#, _
                   -9999999999#, 9999999999#
     pSR.SetFalseOriginAndUnits 0, 0, 100000#
'
'    ---Define geometry type for shape field to be Polyline
     Set pGeomDef = New esriCore.GeometryDef
     Set pGeomDefEdit = pGeomDef
     With pGeomDefEdit
         .GeometryType = esriCore.esriGeometryPolyline
         .GridCount = 1
         .GridSize(0) = 10
         .AvgNumPoints = 2
         .HasM = False
         .HasZ = False
         Set .SpatialReference = pSR
     End With
'
```

```
'  ---Define the Shape Field
   Set pFieldEdit = New esriCore.Field
   With pFieldEdit
       .name = "Shape"
       .Type = esriCore.esriFieldTypeGeometry
       .IsNullable = True
       .Editable = True
       .aliasName = "Shape"
       Set .GeometryDef = pGeomDef
   End With
   pFieldsEdit.AddField pFieldEdit
'
'  ---Add the desired attributes into the attribute list
'  ---In this example we will add an integer attribute, a
'  ---double attribute and a string attribute using arbitrary
'  ---field names and sizes
'
'  ---Map Number
   Set pFieldEdit = New esriCore.Field
   pFieldEdit.name = "MAP"
   pFieldEdit.Type = esriCore.esriFieldTypeInteger
   pFieldEdit.DomainFixed = False
   pFieldEdit.Editable = True
   pFieldEdit.IsNullable = False
   pFieldEdit.Precision = 8
   pFieldsEdit.AddField pFieldEdit
'
'  ---Line Length
   Set pFieldEdit = New esriCore.Field
   With pFieldEdit
       .name = "LEN"
       .Editable = True
       .IsNullable = False
       .Precision = 14
       .Scale = 4
       .Type = esriCore.esriFieldTypeDouble
   End With
   pFieldsEdit.AddField pFieldEdit
'
'  ---Description associated with the polyline
   Set pFieldEdit = New esriCore.Field
   pFieldEdit.name = "LINE_DESC"
   pFieldEdit.Type = esriCore.esriFieldTypeString
   pFieldEdit.Editable = True
   pFieldEdit.IsNullable = False
   pFieldEdit.Precision = 40
   pFieldsEdit.AddField pFieldEdit
'
'  ---Define the file dialog message box title
   aMessage = "Enter the name of the Shapefile " + _
              "to contain Lines"
'
```

SHAPEFILE and GEODATABASE

SHAPEFILE and GEODATABASE

```
'
'   ---Prompt the user to specify the shapefile name
    Set pNShapeFile = CreateNewShapefile(pFieldsEdit, _
                          esriCore.esriGeometryPolyline, _
                          aDefName, aMessage)
'
'   ---Check if the command has been canceled (aborted)
    If (ugerror = 1) Then
       Exit Sub
    End If
'
'   ---Check if any problems were detected
    If pNShapeFile Is Nothing Then
'
'      ---Inform user of the problem
       aMsg = "Error creating Shapefile, check permissions."
       aTitle2 = "Create Shapefile"
       Call avMsgBoxWarning(aMsg, aTitle2)
       Exit Sub
'
'   ---Shapefile created properly
    Else
'
'      ---Get the name of the shapefile
       theTheme = pNShapeFile.aliasName
    End If
'
'   ---Get the attribute table for the theme
    Call avGetFTab(pmxDoc, theTheme, _
                   theFTab, pFeatureClass, aLayer)
'
```

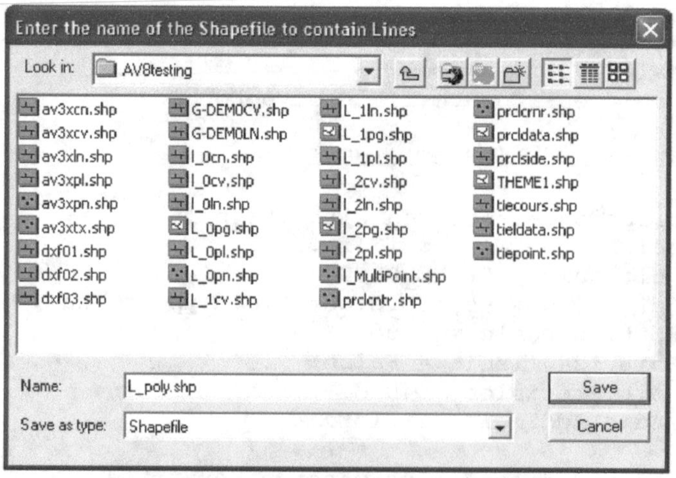

Figure 5-1(A)
CreateNewShapeFile File Dialog Box
when default filename contains the .shp extension

SHAPEFILE and GEODATABASE

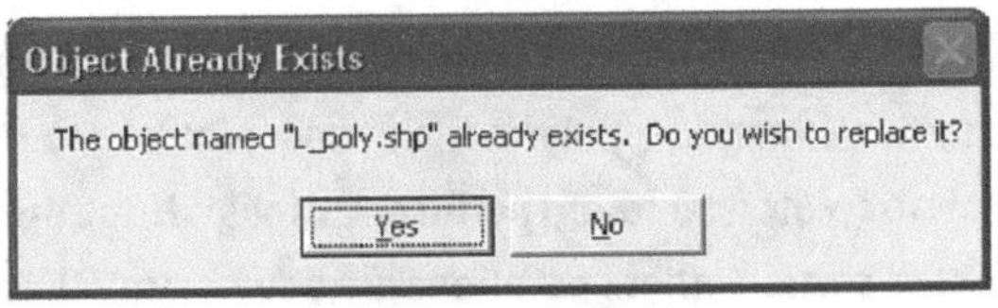

Object Already Exists

The object named "L_poly.shp" already exists. Do you wish to replace it?

Yes No

Figure 5-1(B)
Okay to Overwrite Existing Shapefile Query

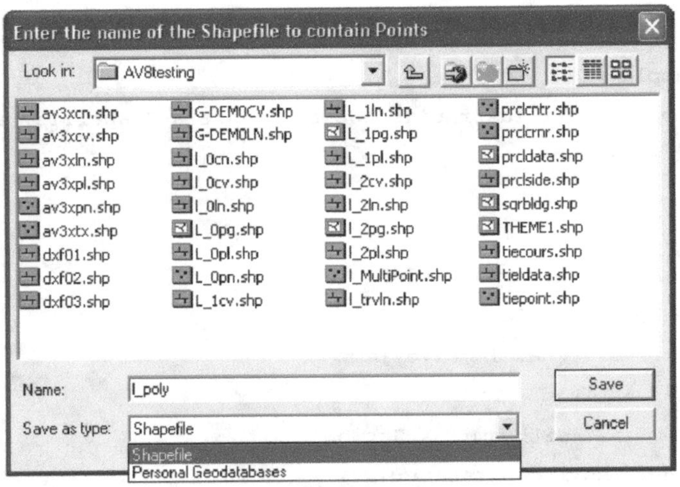

Enter the name of the Shapefile to contain Points

Look in: AV8testing

av3xcn.shp G-DEMOCV.shp L_1ln.shp prclcntr.shp
av3xcv.shp G-DEMOLN.shp L_1pg.shp prclcrnr.shp
av3xln.shp l_0cn.shp L_1pl.shp prcldata.shp
av3xpl.shp l_0cv.shp l_2cv.shp prclside.shp
av3xpn.shp l_0ln.shp l_2ln.shp sqrbldg.shp
av3xtx.shp L_0pg.shp l_2pg.shp THEME1.shp
dxf01.shp L_0pl.shp l_2pl.shp tiecours.shp
dxf02.shp L_0pn.shp l_MultiPoint.shp tieldata.shp
dxf03.shp L_1cv.shp l_trvln.shp tiepoint.shp

Name: l_poly Save

Save as type: Shapefile Cancel

Shapefile
Personal Geodatabases

Figure 5-1(C)
CreateNewShapeFile File Dialog Box when default filename contains no extension

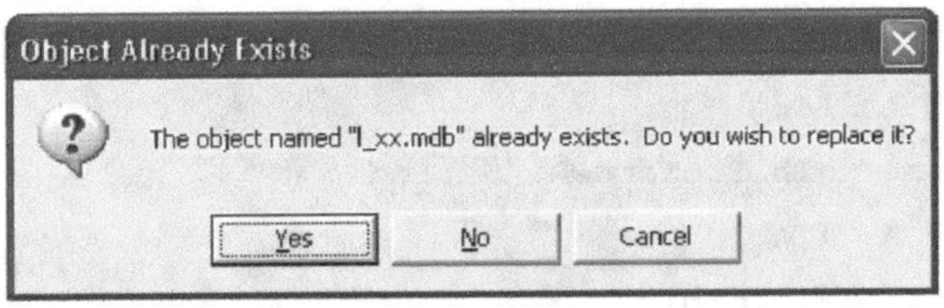

Object Already Exists

The object named "l_xx.mdb" already exists. Do you wish to replace it?

Yes No Cancel

Figure 5-1(D)
Okay to Overwrite (Yes) or Append (No) an Existing Personal GeoDatabase Query

SAMPLE CODE

5.7 Sample Code

The sample code below contains two examples, (a) one that illustrates how to create a shapefile, in this example a polyline, and add a feature to it, and (b) another that illustrates how to create a table and perform various editing operations. In the course of these samples, certain other operations are demonstrated, some of which are used strictly for illustration purposes. Note that the various Avenue Wraps that are called below have been highlighted in bold font. Some of these Avenue Wraps are discussed in detail in other chapters.

```
'
' ---
' ---Example #1
' ---Sample code illustrating how to create a Shapefile, and
' ---add a feature to it.
' ---
'

Dim pMxApp As IMxApplication
Dim pmxDoc As IMxDocument
Dim pActiveView As IActiveView
Dim pMap As IMap
Dim sThmName, sPthName As String
Dim PTheme As IFeatureLayer
Dim aIndex As Long
Dim iok As Integer
Dim iRec As Long
Dim theFTab As IFields
Dim pFeatCls As IFeatureClass
Dim pLayer As IFeatureLayer
Dim pLineX As IPolyline
Dim aField As Long
Dim sTblName, sTblPthName As String
Dim pTable As ITable
Dim pFld1 As IFieldEdit
Dim pFld2 As IFieldEdit
Dim pFld3 As IFieldEdit
Dim fldList As New Collection
Dim theVTab As IFields
Dim col, nRec As Long
```

SAMPLE CODE

```
    Dim sel As ISelectionSet
    Dim aCalcString, aQueryString As String
    Dim sumTblName As String
    Dim fieldList1 As New Collection
    Dim sumryList2 As New Collection
    Dim pSTable As ITable
'
'   ---Get the active view                                      <<<------
    Call avGetActiveDoc(pMxApp, pmxDoc, pActiveView, pMap)
'
'   ---Define the name of the shapefile to be created
    sThmName = "L_poly.shp"
'
'   ---Define the full pathname of the shapefile
    sPthName = "c:\temp\" + sThmName
'
'   ---Create a polyline shapefile
    Set PTheme = avFTabMakeNew(sPthName, "POLYLINE")
'
'   ---Make sure the shapefile was actually created.
'   ---It is possible that, due to certain restrictions that may have
'   ---been imposed on the operating system by its administrator,
'   ---the shapefile may not have been created. In addition, if
'   ---the shapefile exists, it will not be created.
'
'   ---Handle the case when the shapefile is to be created
    Else
       MsgBox "Shapefile: " + sThmName + " created"
'
'      ---Add the shapefile to the map
       iok = avAddDoc(PTheme)
       MsgBox "Shapefile: " + sThmName + " added to TOC"
'
'      ---Make the shapefile editable
       Call avSetEditable(pmxDoc, sThmName, True)
'
'      ---Start an operation that will be added to the Undo list
       Call avStartOperation
'
```

SAMPLE CODE

```
'     ---Add a record to the shapefile, this is a new feature that
'     ---has been added
      iRec = avAddRecord(pmxDoc, sThmName)
'
'     ---Get the attribute table
      Call avGetFTab(pmxDoc, sThmName, theFTab, pFeatCls, pLayer)
'
'     ---Create a line that will represent the geometry of a new
'     ---feature in the shapefile
      Set pLineX = avPolyline2Pt(20000#, 20000#, 30000#, 25000#)
'
'     ---Store the geometry for the new feature in the shape field
'     ---of the layer
      aField = theFTab.FindField("SHAPE")
      Call avSetValueG(pmxDoc, sThmName, aField, iRec, pLineX)
'
'     ---Redraw the theme to refresh the display
      Call avThemeInvalidate(pmxDoc, sThmName, True)
'
'     ---Stop the editing operation so that the operation consists
'     ---only of adding a single feature.
'     ---Note that the editor will remain in an edit state so that
'     ---the Undo capabilities can be utilized, if so desired
      Call avStopOperation("Add Feature")
      MsgBox "Feature added to map"
   End If
'
'  ---
'  ---Example #2
'  ---Sample illustrating how to create a dBase Table, and perform
'  ---various table editing operations.
'  ---
'
'  ---Define the name of the table to be created
   sTblName = "table1.dbf"
'
'  ---Define the full pathname of the table
   sTblPthName = "c:\temp\" + sTblName
'
```

SAMPLE CODE

```
'  ---Create a dBase table
   Set pTable = avVTabMakeNew(sTblPthName, "dbase")
'
'  ---Make sure the table was actually created.  It is possible that
'  ---the table was not created:
'  ---(a) due to certain restrictions that may have been imposed on
'  ---    the operating system by its administrator, or
'  ---(b) because the table may exist on the disk.
   If (pTable Is Nothing) Then
'     ---The table was not created.  Display an error message
      MsgBox "Error in creating table: " + sTblName
'
'     ---Check if the table exists in the disk.  If so remove it.
      If (avFileExists(sTblPthName)) Then
         MsgBox "Table: " + sTblPthName + " exists"
'        ---Check if the table exists in the map.  If so, remove it.
         aIndex = avFindDoc(sTblName)
         If (aIndex <> -1) Then
'           ---Remove the table from the map
            Call avRemoveDoc(sTblName)
            MsgBox "Table: " + sTblPthName + " removed from TOC"
         End If
'        ---Delete the table from disk
         iok = avDeleteDS(sTblPthName)
         If (iok = 0) Then
            MsgBox "Table: " + sTblPthName + " deleted " + CStr(iok)
         Else
            MsgBox "Error deleting table"
         End If
      Else
         MsgBox "Table: " + sTblName + " does not exist" + _
                Chr(13) + "and could not create the table"
      End If
'
'  ---Handle the case when the table is created
   Else
      MsgBox "Table: " + sTblName + " created"
'
```

SAMPLE CODE

```
'       ---Add the table to the map
        iok = avAddDoc(pTable)
        MsgBox "Table: " + sTblName + " added to TOC"
'
'       ---Add three records to the table
        iRec = avAddRecord(pmxDoc, sTblName)
        iRec = avAddRecord(pmxDoc, sTblName)
        iRec = avAddRecord(pmxDoc, sTblName)
        MsgBox "3 records added to " + sTblName
'
'       ---Create three fields that will be added to the table
        Set pFld1 = avFieldMake("StringF", "vchar", 20, 0)
        Set pFld2 = avFieldMake("DoubleF", "double", 12, 4)
        Set pFld3 = avFieldMake("LongF", "long", 10, 0)
'
'       ---Add the fields to a collection
        Call CreateList(fldList)
        fldList.Add pFld1
        fldList.Add pFld2
        fldList.Add pFld3
'
'       ---Add the fields collection to the table
        iok = avAddFields(pmxDoc, sTblName, fldList)
'
'       ---Get the attribute table
        Call avGetVTab(pmxDoc, sTblName, theVTab)
'
'       ---Check to see whether the table is editable or not.
'       ---If not, make it so.
        If (Not avIsEditable(sTblName)) Then
           MsgBox "Table: " + sTblName + " is not editable"
'
'          ---Make the table editable
           Call avSetEditable(pmxDoc, sTblName, True)
           If (avIsEditable(sTblName)) Then
              MsgBox "Table: " + sTblName + " is now editable"
           End If
'
```

SAMPLE CODE

```
'       ---Store a value in the table, under a specific field,
'       ---for all three records that were added
        col = theVTab.FindField("StringF")
        Call avSetValue(pmxDoc, sTblName, col, 0, "test string")
        Call avSetValue(pmxDoc, sTblName, col, 1, "second string")
        Call avSetValue(pmxDoc, sTblName, col, 2, "third string")
'
'       ---Store a value in a different field for a specific record
        col = theVTab.FindField("DoubleF")
        Call avSetValue(pmxDoc, sTblName, col, 1, 14.3456)
'
'       ---Commit the modifications to the disk
        Call avSetEditable(pmxDoc, sTblName, False)
'
'       ---Determine the number of records in the table
        nRec = avGetNumRecords(pmxDoc, sTblName)
        MsgBox "Number of records in " + sTblName + " = " + _
                                                   CStr(nRec)
'
'       ---Select all of the records in the table
        Call avSetAll(pmxDoc, sTblName, sel)
        MsgBox CStr(sel.Count) + " records selected (all)"
'
'       ---Clear the selection
        Call avClearSelection(pmxDoc, sTblName)
        Call avGetSelection(pmxDoc, sTblName, sel)
        MsgBox CStr(sel.Count) + " records selected (none)"
'
'       ---Select the second and third records in the table
        Call avBitmapSet(pmxDoc, sTblName, 1)
        Call avBitmapSet(pmxDoc, sTblName, 2)
        Call avGetSelection(pmxDoc, sTblName, sel)
        MsgBox CStr(sel.Count) + " records selected "
'
'       ---Clear the second record from the selection
        Call avGetSelectionClear(pmxDoc, sTblName, 1)
        MsgBox "1 selected record deselected"
'
```

SAMPLE CODE

```
'        ---Start editing on the table
         Call avSetEditable(pmxDoc, sTblName, True)
'
'        ---Add 16 records to the table
         iRec = avAddRecord(pmxDoc, sTblName)
         iRec = avAddRecord(pmxDoc, sTblName)
         iRec = avAddRecord(pmxDoc, sTblName)
         iRec = avAddRecord(pmxDoc, sTblName)
         iRec = avAddRecord(pmxDoc, sTblName)
         iRec = avAddRecord(pmxDoc, sTblName)
         iRec = avAddRecord(pmxDoc, sTblName)
         iRec = avAddRecord(pmxDoc, sTblName)
         iRec = avAddRecord(pmxDoc, sTblName)
         iRec = avAddRecord(pmxDoc, sTblName)
         iRec = avAddRecord(pmxDoc, sTblName)
         iRec = avAddRecord(pmxDoc, sTblName)
         iRec = avAddRecord(pmxDoc, sTblName)
         iRec = avAddRecord(pmxDoc, sTblName)
         iRec = avAddRecord(pmxDoc, sTblName)
         iRec = avAddRecord(pmxDoc, sTblName)
         MsgBox "multiple records added"
'
'        ---Clear the selection set for the table
         Call avClearSelection(pmxDoc, sTblName)
'
'        ---Select two records
         Call avBitmapSet(pmxDoc, sTblName, 0)
         Call avBitmapSet(pmxDoc, sTblName, 1)
         MsgBox "two records selected"
'
'        ---Delete the selected records in the table
         Call avRemoveRecord(pmxDoc, sTblName, -1)
         MsgBox "two records deleted"
'
'        ---Stop editing on the table
         Call avSetEditable(pmxDoc, sTblName, False)
      End If
   End If
'
```

SAMPLE CODE

```
'   ---Get the attribute table
    Call avGetVTab(pmxDoc, sTblName, theVTab)
'
'   ---Make sure the table exists
    If (Not theVTab Is Nothing) Then
'
'      ---Make the table editable
       Call avSetEditable(pmxDoc, sTblName, True)
'
'      ---Build an arbitrary calculation string
       col = theVTab.FindField("LongF")
       nRec = avGetNumRecords(pmxDoc, sTblName)
       aCalcString = "([DoubleF] - " + CStr(nRec) + ")"
'
'      ---Apply a Calculation to two selected records
       Call avClearSelection(pmxDoc, sTblName)
       Call avBitmapSet(pmxDoc, sTblName, 0)
       Call avBitmapSet(pmxDoc, sTblName, 1)
       iok = avCalculate(pmxDoc, sTblName, aCalcString, col)
       MsgBox "2 records applied a calculation"
'
'      ---Apply a new Calculation to one selected record
       Call avClearSelection(pmxDoc, sTblName)
       Call avBitmapSet(pmxDoc, sTblName, 2)
       aCalcString = "([DoubleF] - 10)"
       iok = avCalculate(pmxDoc, sTblName, aCalcString, col)
       MsgBox "1 record applied a calculation"
'
'      ---Stop the editor
       Call avSetEditableTheme(pmxDoc, Null, Null)
'
'      ---Apply a Query to the table
       aQueryString = "LongF = 0.0"
       Call avQuery(pmxDoc, sTblName, aQueryString, sel, "NEW")
       Call avGetSelection(pmxDoc, sTblName, sel)
       MsgBox CStr(sel.Count) + " records selected"
'
'      ---Check if the Summary table exists in the TOC
       aIndex = avFindDoc("sumTable")
```

SAMPLE CODE

```
      If (aIndex <> -1) Then
         Call avRemoveDoc("sumTable")
         MsgBox "sumTable removed from TOC"
      End If
      If (avFileExists("sumTable.dbf")) Then
         Call avFileDelete("sumTable.dbf")
         MsgBox "sumTable deleted from disk"
      End If
'
'     ---Define the name of the summary table to be created
      sumTblName = "sumTable"
'
'     ---Summarize the selected records in the table based upon the
'     ---LongF field.
'     ---The default operation codes will be used.
'     ---That is why fieldList1 and sumryList2 are empty colections.
      Call CreateList(fieldList1)
      Call CreateList(sumryList2)
      Set pSTable = avSummarize(pmxDoc, sTblName, _
                               sumTblName, "dBase", UCase("LongF"), _
                               fieldList1, sumryList2)
'
'     ---Check if the table could not be summarized
      If (pSTable Is Nothing) Then
         MsgBox "Error in summarizing " + sTblName
'     ---Handle case when the table was summarized without error
      Else
         iok = avAddDoc(pSTable)
         MsgBox "Summary table: " + sumTblName + " added to TOC"
      End If
'
'  ---Handle case when table does not exist
   Else
      MsgBox "Table: " + sTblName + " does not exist"
   End If
'
```

CHAPTER 6

Feature Selection Avenue Wraps

This chapter contains Avenue Wraps pertaining to the handling of themes and their tables including (a) theme and table selections and retrieval, (b) creation and clearing of selection sets, (c) querying, editing and summarizing tables, and (d) performing calculations on table cells. These Avenue Wraps include the following:

The source listing of each of the above Avenue Wraps may be found in Appendix D of this publication.

6.1 Making Selections Related Avenue Wraps

MAKING SELECTIONS

6.1.1 Subroutine avBitmapSet

This subroutine enables the programmer to add a record to the selected set of a layer (theme) or table.

The corresponding Avenue request is:

```
aBitMap.Set(theRcrd)
```

The call to this Avenue Wrap is:

Call **avBitmapSet**(pmxDoc, theTheme, theRcrd)

avBitmapSet

GIVEN: pmxDoc = the active view
theTheme = the theme to be processed
theRcrd = the record to be added to the selection

RETURN: nothing Adds the record to the selection set

The given and returned variables should be declared where first called as:

```
Dim pmxDoc As IMxDocument
Dim theTheme As Variant
Dim theRcrd As Long
```

6.1.2 Subroutine avGetSelection

This subroutine enables the programmer to get the selected set for a layer (theme) or Table.

The corresponding Avenue request is:

```
psTableSel = aVTab.GetSelection
```

The call to this Avenue Wrap is:

Call **avGetSelection**(pmxDoc, theTheme, psTableSel)

avGetSelection

GIVEN: pmxDoc = the active view
theTheme = the theme to be processed

RETURN: psTableSel = the selection set for the theme

The given and returned variables should be declared where first called as:

```
Dim pmxDoc As IMxDocument
```

MAKING SELECTIONS

Dim theTheme As Variant
Dim psTableSel As ISelectionSet

6.1.3 Subroutine avGetSelectionIDs

This subroutine enables the programmer to get a collection (list) of object identification numbers (OIDs) for a selection set.

The corresponding Avenue request is:

There is no corresponding Avenue request.

The call to this Avenue Wrap is:

avGetSelectionIDs

Call **avGetSelectionIDs**(psTableSel, selRecsList)

GIVEN: psTableSel = the selection set for a theme

RETURN: selRecsList = the list of OIDs for the selection set

The given and returned variables should be declared where first called as:

Dim psTableSel As ISelectionSet
Dim selRecsList as New Collection

6.1.4 Subroutine avGetSelFeatures

This subroutine enables the programmer to preserve the record numbers of selected features for those layers (themes) having selected features, depending upon a specified numeric selection mode. Shown below are the selection modes and the type of themes to be preserved, the record numbers of which are to be saved. Reference is made to the example at the end of this section.

=0 preserve all selected features of a theme in themeList regardless of type

=1 preserve only point features from the selected set of a theme in themeList

=2 preserve only polyline features from the selected set of a theme in themeList

=3 preserve only polygon features from the selected set of a theme in themeList

=10 preserve all selected features in a theme, including themes in themeList that do not have selected features.

The corresponding Avenue request is:

There is no corresponding Avenue request.

MAKING SELECTIONS

avGetSelFeatures

The call to this Avenue Wrap is:

```
Call avGetSelFeatures(pmxDoc, themeList, selMode, _
                      selThmList, selRecList)
```

GIVEN: pmxDoc = the active view
themeList = list of layers (themes) to be processed (themes from which object identification numbers will be saved, depending upon value of selMode)
selMode = mode of selection, see above mode listing

RETURN: selThmList = list of themes with selected features
selRecList = list of selected features record numbers

The given and returned variables should be declared where first called as:

```
Dim pmxDoc As IMxDocument
Dim themeList As New Collection
Dim selMode As Integer
Dim selThmList As New Collection
Dim selRecList As New Collection
```

THEME1 (points)	THEME2 (polylines)	THEME3 (points)	THEME4 (points)
0 AA	0 HHH	0 OOO	0 VVV
1 BBB	1 IIIIII	1 PPP	1 WW
2 CCC	2 JJJJJ	2 QQ	2 XXX
3 DDD	3 KKK	3 RRR	3 YYY
4 EEE	4 LLL	4 SSS	4 ZZZ
5 FFF	5 MM	5 TTT	
6 GGG	6 NNN	6 UUU	

Figure 6-1 List of Themes for avGetSelFeatures

Example 1 Consider the three layers (themes) of Figure 6-1, and the selected records (shaded) in each one of them.

```
Call CreateList(themeList)
themeList.Add(Theme1)
themeList.Add(Theme2)
```

MAKING SELECTIONS

```
themeList.Add(Theme3)
themeList.Add(Theme4)
Call avGetSelFeatures(pmxDoc, themeList, 1, _
                      selThmList, selRecList)
```

In this example the two returned lists will contain the following data:

selThmList: THEME1 / 2 / THEME4 / 4

selRecList: 5 / 6 / 1 / 2 / 3 / 4

Example 2 Consider the same themeList, but a different call.

```
Call avGetSelFeatures(pmxDoc, themeList, 0, _
                      selThmList, selRecList)
```

In this example the two returned lists will contain the following data:

selThmList: THEME1 / 2 / THEME2 / 3 / THEME4

selRecList: 5 / 6 / 1 / 2 / 3 / 1 / 2 / 3 / 4

Example 3 Consider the same themeList, but a different call.

```
Call avGetSelFeatures(pmxDoc, themeList, 10, _
                      selThmList, selRecList)
```

In this example the two returned lists will contain the following data:

selThmList: THEME1 / 2 / THEME2 / 3 / THEME3 / 0 / THEME4

selRecList: 5 / 6 / 1 / 2 / 3 / 1 / 2 / 3 / 4

6.1.5 Subroutine avInvSelFeatures

This subroutine enables the programmer to redraw, for a specified set of layers (themes), the selected features within the specified layers (themes). In using this subroutine, note that the composition of the argument, selThmList, is the same as that of the previously discussed avGetSelFeatures Avenue Wrap.

The corresponding Avenue request is:

There is no corresponding Avenue request.

avInvSelFeatures

The call to this Avenue Wrap is:

Call **avInvSelFeatures**(pmxDoc, selThmList)

GIVEN: pmxDoc = the active view

selThmList = list of themes with selected features

MAKING SELECTIONS

RETURN: nothing

The given and returned variables should be declared where first called as:

```
Dim pmxDoc As IMxDocument
Dim selThmList As New Collection
```

6.1.6 Subroutine avSelectByFTab

This subroutine enables the programmer to select features in a layer (theme) based upon the selected features in another layer (theme).

The corresponding Avenue request is:

```
anFTab.SelectByFTab (anotherFTab, aRelType, aDistance,
                     aSelType)
```

The call to this Avenue Wrap is:

avSelectByFTab

Call **avSelectByFTab**(pmxDoc, elmntTheme, seltrTheme,_
selMod, selTol, setType)

GIVEN:

pmxDoc	=	the active view
elmntTheme	=	name of theme to be processed
seltrTheme	=	selector theme to be used
selMod	=	selection mode of operation: "INTERSECTS", or "ISWITHINDISTANCEOF"
selTol	=	selection distance tolerance
setType	=	type of selection desired: "NEW" : new selection set, or "ADD" : add to selection set, or "AND" : select from selection set

RETURN: nothing

The given and returned variables should be declared where first called as:

```
Dim pmxDoc As IMxDocument
Dim elmntTheme As Variant
Dim seltrTheme As Variant
Dim selMod As String
Dim selTol As Double
Dim setType As String
```

MAKING SELECTIONS

6.1.7 Subroutine avSelectByPoint

This subroutine enables the programmer to select features in a layer (theme) based upon a point. In using this subroutine, note that:

- The Avenue request operates on an FTab, while this Avenue Wrap operates on a layer (theme) name.
- This Avenue Wrap is not a substitute for the Avenue request anFTheme.SelectByPoint (aPoint, aSelType)

The corresponding Avenue request is:

```
aFTab.SelectByPoint (thePoint, selTol, setType)
```

avSelectByPoint

The call to this Avenue Wrap is:

```
Call avSelectByPoint(pmxDoc, elmntTheme, thePoint, _
                     selTol, setType)
```

GIVEN:

GIVEN:	pmxDoc	= the active view
	elmntTheme	= name of theme to be processed
	thePoint	= selector theme to be used
	selTol	= selection distance tolerance
	setType	= type of selection desired "NEW" : new selection set, or "ADD" : add to selection set, or "AND" : select from selection set

RETURN: nothing

The given and returned variables should be declared where first called as:

```
Dim pmxDoc As IMxDocument
Dim elmntTheme As Variant
Dim thePoint As IPoint
Dim selTol As Double
Dim setType As String
```

6.1.8 Subroutine avSelectByPolygon

This subroutine enables the programmer to select features in a layer (theme) based upon a polygon.

- The Avenue request operates on an FTab, while this Avenue Wrap operates on a layer (theme) name.
- This Avenue Wrap is not a substitute for the Avenue request anFTheme.SelectByPolygon (aPolygon, aSelType)

MAKING SELECTIONS

The corresponding Avenue request is:

anFTab.SelectByPolygon (aPolygon, aSelType)

avSelectBy Polygon

The call to this Avenue Wrap is:

Call **avSelectByPolygon**(pmxDoc, elmntTheme, theGeom, _
setType)

GIVEN: pmxDoc = the active view
elmntTheme = name of theme to be processed
theGeom = geometry to be used
setType = type of selection desired
"NEW" : new selection set, or
"ADD" : add to selection set, or
"AND" : select from selection set

RETURN: nothing

The given and returned variables should be declared where first called as:

Dim pmxDoc As IMxDocument
Dim elmntTheme As Variant
Dim theGeom As IGeometry
Dim setType As String

6.1.9 Subroutine avSetActive

This subroutine enables the programmer to make a layer (theme) selectable or not.

The corresponding Avenue request is:

aTheme.SetActive (sStatus)

avSetActive

The call to this Avenue Wrap is:

Call **avSetActive**(pmxDoc, theTheme, sStatus)

GIVEN: pmxDoc = the active view
theTheme = theme to be processed
sStatus = selectable status (True = selectable)
(False = not selectable)

RETURN: nothing

MAKING SELECTIONS

The given and returned variables should be declared where first called as:

```
Dim pmxDoc As IMxDocument
Dim theTheme As Variant
Dim sStatus As Boolean
```

6.1.10 Subroutine avSetAll

This subroutine enables the programmer to select all of the features or rows in a layer (theme) or table.

The corresponding Avenue request is:

```
aBitMap.SetAll
```

avSetAll

The call to this Avenue Wrap is:

Call **avSetAll**(pmxDoc, theTheme, psTableSel)

GIVEN:	pmxDoc	= the active view
	theTheme	= the theme to be processed
RETURN:	psTableSel	= the selection set for the theme or table with all features or rows selected

The given and returned variables should be declared where first called as:

```
Dim pmxDoc As IMxDocument
Dim theTheme As Variant
Dim psTableSel As ISelectionSet
```

6.1.11 Subroutine avSetSelection

This subroutine enables the programmer to set the selected set for a layer (theme) or table.

The corresponding Avenue request is:

```
aVTab.SetSelection (aBitmap)
```

avSetSelection

The call to this Avenue Wrap is:

Call **avSetSelection**(pmxDoc, theTheme, psTableSel)

GIVEN:	pmxDoc	= the active view
	theTheme	= the theme to be processed
	psTableSel	= the selection set for the theme
RETURN:	nothing	

MAKING SELECTIONS

The given and returned variables should be declared where first called as:
Dim pmxDoc As IMxDocument
Dim theTheme As Variant
Dim psTableSel As ISelectionSet

6.1.12 Subroutine avSetSelFeatures

This subroutine enables the programmer to add to the selections, for a specified set of layers (themes), a specified set of features. In using this subroutine, note that the composition of each of the given arguments, selThmList and selRecList, is the same as that of the previously discussed avGetSelFeatures Avenue Wrap.

The corresponding Avenue request is:
There is no corresponding Avenue request.

The call to this Avenue Wrap is:
Call **avSetSelFeatures**(pmxDoc, selThmList, selRecList)

avSetSelFeatures

GIVEN: pmxDoc = the active view
selThmList = list of themes with selected features
selRecList = list of selected features record numbers

RETURN: nothing

The given and returned variables should be declared where first called as:
Dim pmxDoc As IMxDocument
Dim selThmList As New Collection
Dim selRecList As New Collection

6.1.13 Subroutine avUpdateSelection

This subroutine enables the programmer to update the attribute table of a layer (theme) to reflect the current selection set for the layer (theme).

The corresponding Avenue request is:
aVTab.UpdateSelection

The call to this Avenue Wrap is:
Call **avUpdateSelection**(pmxDoc, theTheme)

avUpdateSelection

MAKING SELECTIONS

GIVEN: pmxDoc = the active view
theTheme = theme to be processed

RETURN: nothing

The given and returned variables should be declared where first called as:

```
Dim pmxDoc As IMxDocument
Dim theTheme As Variant
```

6.1.14 Subroutine avXORSelection

This subroutine enables the programmer to perform an exclusive XOR on the selection set of a layer (theme) or table. When using this subroutine, note that:

1. The current selection set for the layer (theme) or table is used in the XORing with the set that is passed in (orgSelSet). If the layer (theme) or table has no selection set the command will select all features (rows) in the layer (theme) or table and use this set in the XOR.
2. In essence, if we have certain rows of a table selected and represented by the given argument orgSelSet, this subroutine will deselect these rows, and will select the rows that were not selected originally, placing them in the xorSelSet.

The corresponding Avenue request is:

```
xorSelSet.XOr (orgSelSet)
```

avXORSelection

The call to this Avenue Wrap is:

```
Call avXORSelection(pmxDoc, theTheme, orgSelSet, _
                    xorSelSet)
```

GIVEN: pmxDoc = the active view
theTheme = theme to be processed
orgSelSet = the selection set to be XORed

RETURN: xorSelSet = the selection set after XOR was performed

The given and returned variables should be declared where first called as:

```
Dim pmxDoc As IMxDocument
Dim theTheme As Variant
Dim orgSelSet
Dim xorSelSet
```

6.2 Clearing and Removing Selections Related Avenue Wraps

CLEARING AND REMOVING FROM SELECTIONS

6.2.1 Subroutine avBitmapClear

This subroutine enables the programmer to remove a record from the selected set for a layer (theme) or Table. The difference between this subroutine and the avGetSelectionClear is in the given argument list.

The corresponding Avenue request is:

```
psTableSet.Clear (theRcrd)
```

The call to this Avenue Wrap is:

Call **avBitmapClear**(psTableSel, theRcrd)

avBitmapClear

GIVEN: psTableSel = selection set for a theme
theRcrd = record to be removed from the selection

RETURN: nothing The record is removed

The given and returned variables should be declared where first called as:

```
Dim psTableSel As ISelectionSet
Dim theRcrd As Long
```

6.2.2 Subroutine avClearSelection

This subroutine enables the programmer to clear the selection (deselect all selected features or rows) from a layer (theme) or table.

The corresponding Avenue request is:

```
theTheme.ClearSelection
```

The call to this Avenue Wrap is:

Call **avClearSelection**(pmxDoc, theTheme)

avClearSelection

GIVEN: pmxDoc = the active view
theTheme = the layer (theme) or table to be processed. If NULL, all selected features in all themes will be deselected.

RETURN: nothing

CLEARING AND REMOVING FROM SELECTIONS

The given and returned variables should be declared where first called as:
Dim pmxDoc As IMxDocument
Dim theTheme As Variant

6.2.3 Subroutine avGetSelectionClear

This subroutine enables the programmer to remove a record from the selected set for a layer (theme) or Table. Note that this subroutine removes the record from the selected set and not from the database. The difference between this subroutine and the avClear is in the given argument list.

The corresponding Avenue request is:
There is no corresponding Avenue request.

avGetSelection Clear

The call to this Avenue Wrap is:
Call **avGetSelectionClear**(pmxDoc, theTheme, theRcrd)

GIVEN:	pmxDoc	= the active view
	theTheme	= the layer (theme) or table to be processed
	theRcrd	= the record to be removed from the selection
RETURN:	nothing	The record is removed from the selection set

The given and returned variables should be declared where first called as:
Dim pmxDoc As IMxDocument
Dim theTheme As Variant
Dim theRcrd As Long

6.2.4 Subroutine avRemoveRecord

This subroutine enables the programmer to delete a record from, or the selected features in, a layer (theme) or table. In using this subroutine, note the following:

1. The records to be deleted, are removed from the database. Hence the layer (theme) or table must have been made editable prior to calling this subroutine.
2. The given argument theRcrd controls what is to be deleted. If it is:
   =0 The first feature or row will be deleted, regardless whether that feature or row is selected or not.
   >0 The specified number plus 1 denotes the feature or row to be deleted, regardless whether that feature or row is selected or not.
   =-1 Any features or rows that are selected will be deleted. If there are no features or rows selected, nothing is deleted.

CLEARING AND REMOVING FROM SELECTIONS

3. The Avenue counterpart request pertains only to the case in which the given argument, theRcrd, is greater than or equal to 0.

The corresponding Avenue request is:

aVTab.RemoveRecord (theRcrd)

The call to this Avenue Wrap is:

Call **avRemoveRecord**(pmxDoc, theTheme, theRcrd)

avRemoveRecord

GIVEN: pmxDoc = the active view
theTheme = the layer (theme) or table to be processed
theRcrd = mode of deletion. If it is
>=0 : record of feature (row) for deletion
=-1 : delete selected features (rows) in the theTheme

RETURN: nothing The record is removed

The given and returned variables should be declared where first called as:

Dim pmxDoc As IMxDocument
Dim theTheme As Variant
Dim theRcrd As Long

CHAPTER 7

MESSAGE AND MENU BOXES AVENUE WRAPS

This chapter is comprised of various Avenue Wraps that help (a) display information to the user in the status bar or in message boxes, (b) present the user with message boxes containing questions for action or selection, and (c) present the user with message boxes with provision for key entering input of data, and/or selection from predefined lists. As a general note to all message box Avenue Wraps, the programmer should remember that all arguments that may be returned from a message box are either of the variant or string type. Hence, if numbers are to be returned, the programmer will need to convert the string item, which is passed back by the message box, into a numeric item in a manner described in Chapter 2.

These Avenue Wraps include the following:

The source listing of each of the above Avenue Wraps may be found in Appendix D of this publication.

INFORMATION WARNING MESSAGES

7.1 Information and Warning Messages Related Avenue Wraps

7.1.1 Subroutine avMsgBoxInfo

avMsgBoxInfo

This subroutine enables the programmer to display an information message box. A sample message box is shown in Figure 7-1, and the given arguments are identified therein.

The corresponding Avenue request is:

```
MsgBox.Info (aMsg, aHeading)
```

The call to this Avenue Wrap is:

Call **avMsgBoxInfo**(aMsg, aHeading)

GIVEN: aMsg = the message to be displayed
aHeading = message box caption heading (title)

RETURN: nothing

The given and returned variables should be declared where first called as:

```
Dim aMsg, aHeading As Variant
```

aHeading ⟶ Example of avMsgBoxInfo Avenue Wrap
aMsg ⟶ Processing completed. Press OK to continue
OK

Figure 7-1 avMsgBoxInfo Message Box

7.1.2 Subroutine avMsgBoxList

This subroutine enables the programmer to display a list of strings, for information purposes only, within a message box. A sample message box is shown in Figure 7-2, and the given and returned arguments are identified therein. An example for calling this subroutine is also presented. Although the user is able to click on the displayed list and highlight the selected item, the selected item will not be returned. What is returned is the value of the selected button, vbOK (1), or vbCancel (2).

INFORMATION WARNING MESSAGES

avMsgBoxList

The corresponding Avenue request is:

ians = MsgBox.List (aList, aMsg, aHeading)

The call to this Avenue Wrap is:

Call **avMsgBoxList**(aList, aMsg, aHeading, ians)

GIVEN:	aList	= the collection of strings to be displayed
	aMsg	= the message to be displayed
	aHeading	= message box caption heading (title)
RETURN:	ians	= the button that was selected (vbOK, or vbCancel)

The given and returned variables should be declared where first called as:

Dim aList As New Collection
Dim aMsg, aHeading As Variant
Dim ians As Integer

Presented below is a sample code for testing this subroutine.

```
'
' ---Example of avMsgBoxList Avenue Wrap
Dim aList As New Collection
Dim aMsg, aHed As Variant
Dim ians As Integer
aList.Add ("Item Number 1:")
aList.Add ("Item Number 2:")
aList.Add ("Item Number 3:")
aList.Add ("1.0")
aList.Add ("2.0")
aList.Add ("AA")
aList.Add ("BB")
aList.Add ("CC")
aMsg = "Select item from list:"
aHed = "Example of avMsgBoxList Avenue Wrap"
Call avMsgBoxList(aList, aMsg, aHed, ians)
If (ians = vbCancel) Then
   MsgBox "User Cancelled"
Else
   MsgBox "OK button was selected"
End If
```

INFORMATION
WARNING
MESSAGES

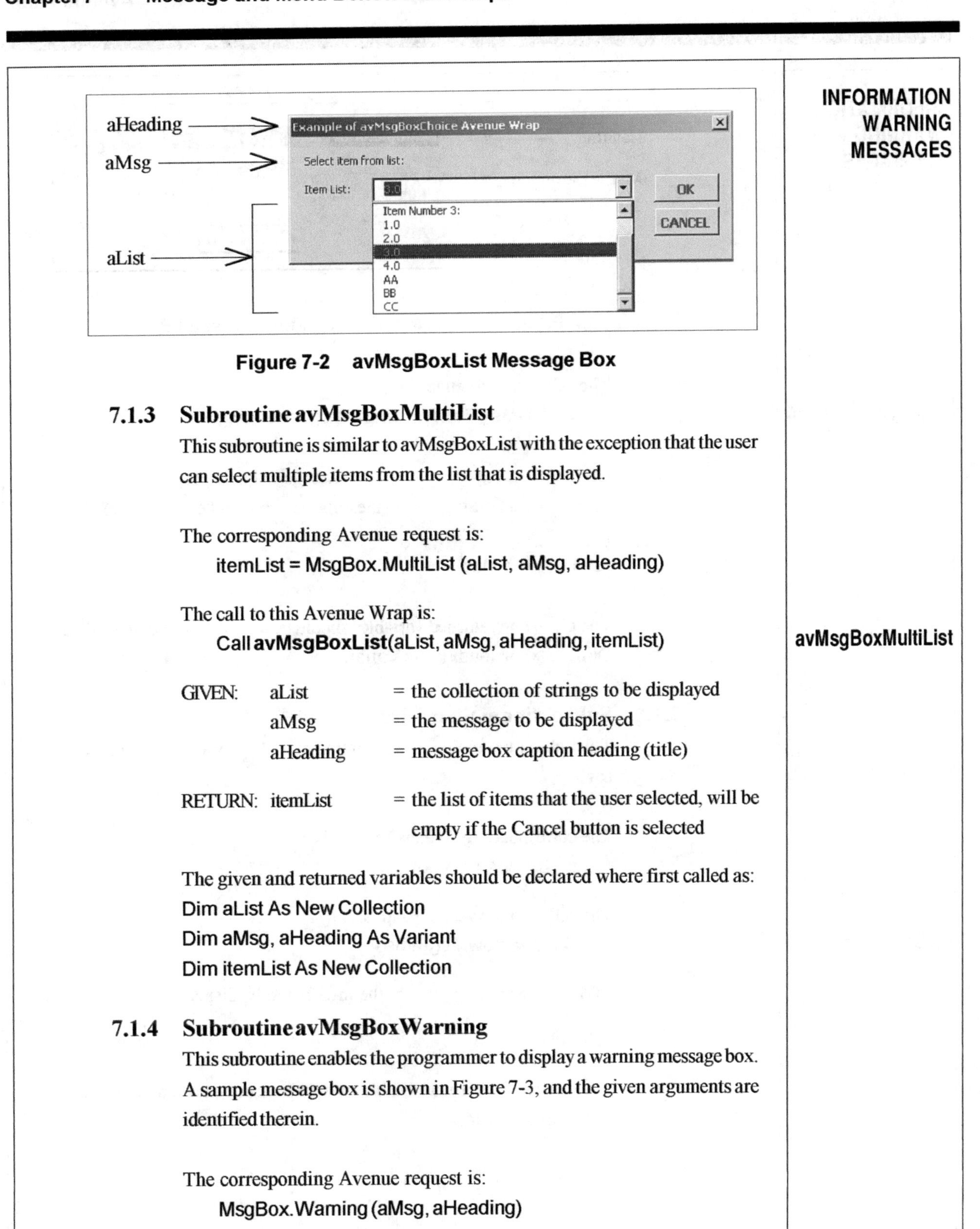

Figure 7-2 avMsgBoxList Message Box

7.1.3 Subroutine avMsgBoxMultiList

This subroutine is similar to avMsgBoxList with the exception that the user can select multiple items from the list that is displayed.

The corresponding Avenue request is:

```
itemList = MsgBox.MultiList (aList, aMsg, aHeading)
```

The call to this Avenue Wrap is:

avMsgBoxMultiList

Call **avMsgBoxList**(aList, aMsg, aHeading, itemList)

GIVEN:	aList	= the collection of strings to be displayed
	aMsg	= the message to be displayed
	aHeading	= message box caption heading (title)
RETURN:	itemList	= the list of items that the user selected, will be empty if the Cancel button is selected

The given and returned variables should be declared where first called as:

```
Dim aList As New Collection
Dim aMsg, aHeading As Variant
Dim itemList As New Collection
```

7.1.4 Subroutine avMsgBoxWarning

This subroutine enables the programmer to display a warning message box. A sample message box is shown in Figure 7-3, and the given arguments are identified therein.

The corresponding Avenue request is:

```
MsgBox.Warning (aMsg, aHeading)
```

INFORMATION WARNING MESSAGES

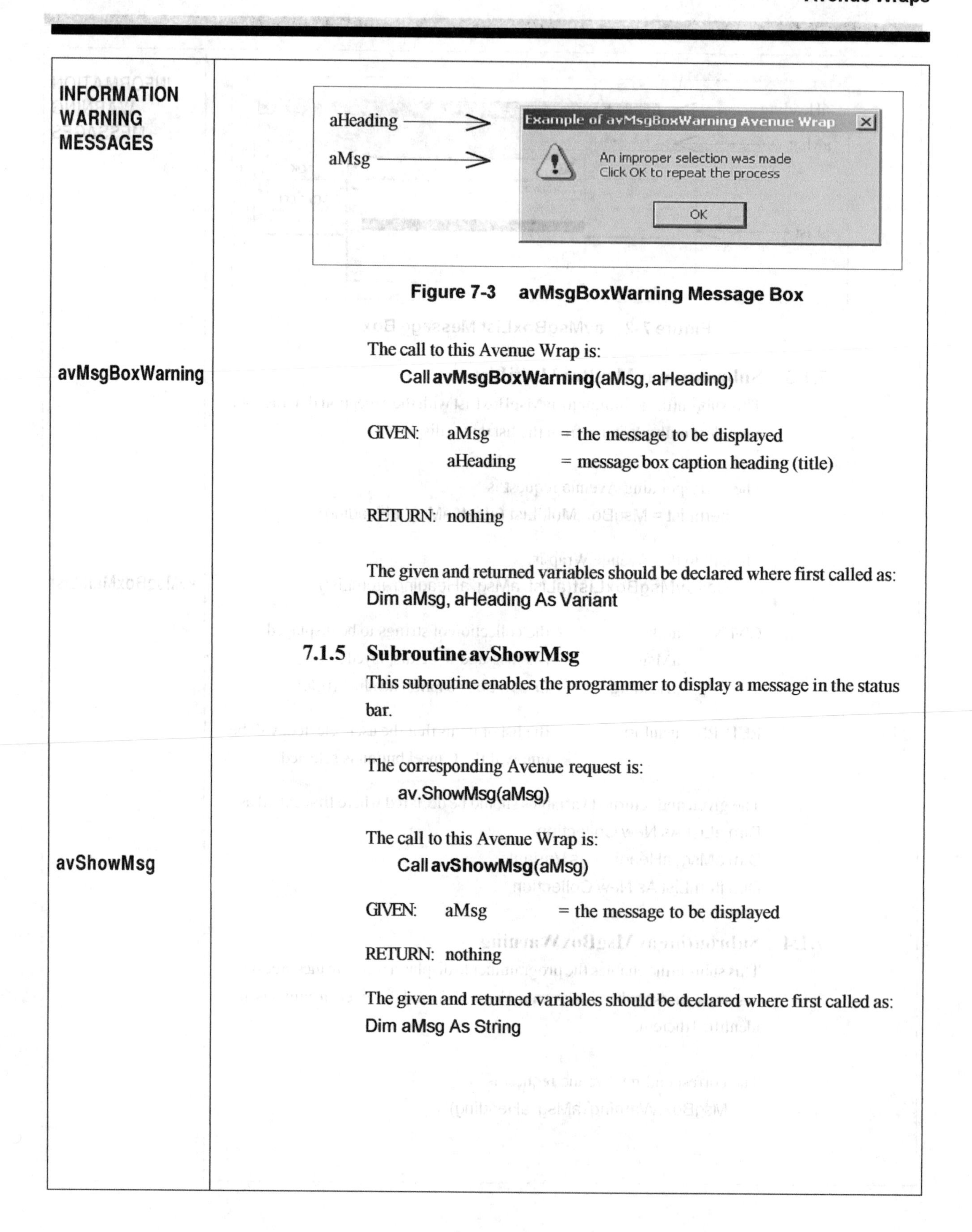

Figure 7-3 avMsgBoxWarning Message Box

avMsgBoxWarning

The call to this Avenue Wrap is:

Call **avMsgBoxWarning**(aMsg, aHeading)

GIVEN: aMsg = the message to be displayed
aHeading = message box caption heading (title)

RETURN: nothing

The given and returned variables should be declared where first called as:

Dim aMsg, aHeading As Variant

7.1.5 Subroutine avShowMsg

This subroutine enables the programmer to display a message in the status bar.

The corresponding Avenue request is:

av.ShowMsg(aMsg)

avShowMsg

The call to this Avenue Wrap is:

Call **avShowMsg**(aMsg)

GIVEN: aMsg = the message to be displayed

RETURN: nothing

The given and returned variables should be declared where first called as:

Dim aMsg As String

DATA INPUT MESSAGE BOXES

7.2 Data Input Message Boxes Related Avenue Wraps

7.2.1 Subroutine avMsgBoxInput

This subroutine enables the programmer to display a single line input message box. A sample message box is shown in Figure 7-4, and the given arguments are identified therein.

The corresponding Avenue request is:

MsgBox.Input (aMsg, aHeading, aDefault)

avMsgBoxInput

The call to this Avenue Wrap is:

Call **avMsgBoxInput**(aMsg, aHeading, aDefault, ians)

GIVEN:	aMsg	= the message to be displayed
	aHeading	= message box caption heading (title)
	aDefault	= the default button value a value
RETURN:	ians	= the response from the user. If the user Cancels the command, ians will be equal to vbCancel.

The given and returned variables should be declared where first called as:

Dim aMsg, aHeading, aDefault, ians As Variant

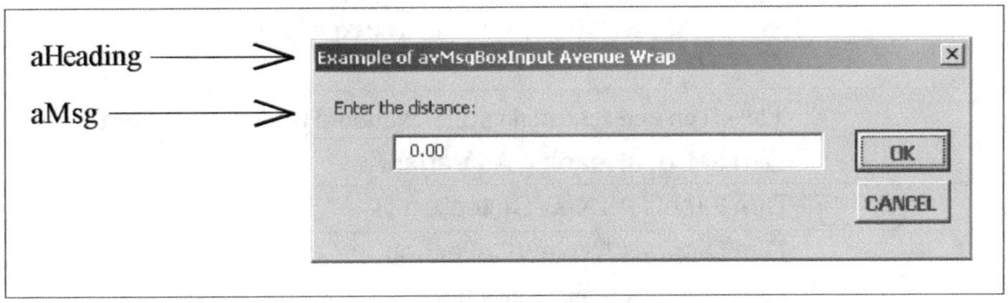

aHeading → Example of avMsgBoxInput Avenue Wrap

aMsg → Enter the distance:

0.00

OK

CANCEL

Figure 7-4 avMsgBoxInput Message Box

7.2.2 Subroutine avMsgBoxMultiInput

This subroutine enables the programmer to display a multi-line input message box (dialog box) containing labels and corresponding data fields with default values. A sample message box is shown in Figure 7-5, and the given arguments are identified therein. In using this subroutine, note the following:

1. Although there is no limit to the number of items that can be prompted for (labels and data fields with default values) in the dialog box, the

DATA INPUT MESSAGE BOXES

difficulty exists in that the dialog box may exceed the visible area of the screen, if there are too many data items displayed. It is recommend that avMsgBoxMultiInput2 be used when more than 12 to 15 items are to be displayed.

2. If the dialog box contains numeric values, the string items contained in the returned list will need to be converted into numbers.

The corresponding Avenue request is:

aList = MsgBox.MultiInput (aMsg, aHeading, Labels, Defaults)

avMsgBox MultiInput

The call to this Avenue Wrap is:

Call **avMsgBoxMultiInput**(aMsg, aHeading, Labels, Defaults,_
aList)

GIVEN:	aMsg	= instruction message to be displayed
	aHeading	= message box caption heading (title)
	Labels	= list of labels for each of the items that the user is prompted to key enter a value for
	Defaults	= list of default values for each of the items that the user is prompted to key enter a value for
RETURN:	aList	= list of responses for each of the items that were displayed. If the user Cancels the command: aList will be an empty list, that is, aList.Count will be equal to 0.

The given and returned variables should be declared where first called as:

Dim aMsg, aHeading As Variant
Dim Labels As New Collection
Dim Defaults As New Collection
Dim aList As New Collection

Presented below is a sample code for testing this subroutine.

```
'
'   ---Example of avMsgBoxMultiInput Avenue Wrap
Dim Labels As New Collection
Dim Defaults As New Collection
Dim aMsg, aHed As Variant
Dim aList As New Collection
Labels.Add ("Distance:")
```

DATA INPUT MESSAGE BOXES

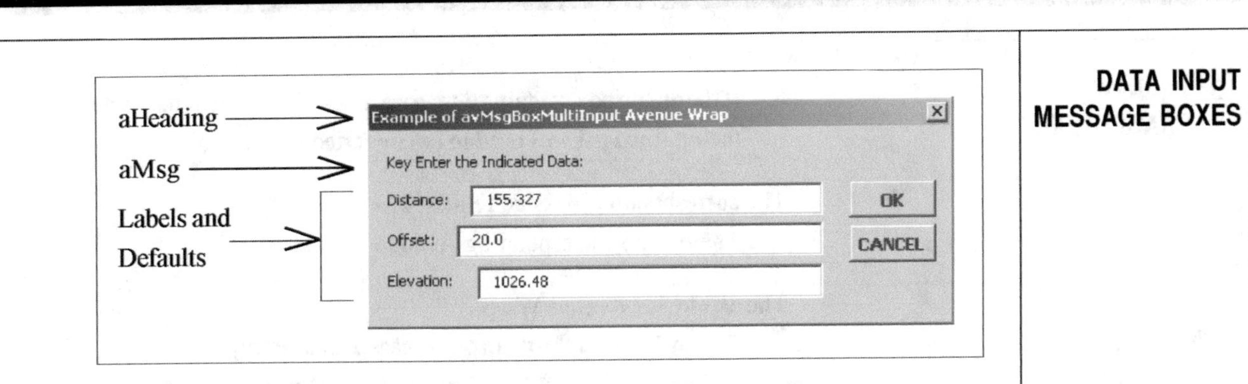

Figure 7-5 avMsgBoxMultiInput Message Box

```
Labels.Add ("Offset:")
Labels.Add ("Elevation:")
Defaults.Add ("10.0")
Defaults.Add ("20.0")
Defaults.Add ("30.0")
aMsg = "Key Enter the Indicated Data:"
aHed = "Example of avMsgBoxMultiInput Avenue Wrap"
Call avMsgBoxMultiInput(aMsg, aHed, labels, _
                        defaults, aList)
```

Note that the default values of the data fields were programmed to be 10.0, 20.0 and 30.0, but were changed by the user to be as shown in Figure 7-5.

7.2.3 Subroutine avMsgBoxMultiInput2

This subroutine is similar to avMsgBoxMultiInput, except that it is used when each of the labels and defaults lists exceeds 10. This subroutine operates as follows:

1. Upon initial display of the message box, see Figure 7-6(A), the first ten items of the label and default list are shown. The user may accept any of the default values, or replace them by key entering new values.
2. To display the subsequent 10 items of said lists, see Figure 7-6(B), and make changes to their default values, the OK button is pressed.
3. This process may continue until the last set of labels and defaults is displayed, see Figure 7-6(C), and any changes to the default values are made.
4. Pressing the OK button on the message box during the display of the last set of items terminates the command.
5. Pressing the BACK button during the display of any items beyond the first 10 causes the display of the preceding 10 items.

DATA INPUT MESSAGE BOXES

6. If the dialog box contains numeric values, the string items contained in the returned list will need to be converted into numbers.

The corresponding Avenue request is:
There is no corresponding Avenue request.

avMsgBox Multilnput2

The call to this Avenue Wrap is:
Call **avMsgBoxMultiInput2**(aMsg, aHeading,_
Labels, Defaults, aList)

GIVEN:
- aMsg = instruction message to be displayed
- aHeading = message box caption heading (title)
- Labels = list of labels for each of the items that the user is prompted to key enter a value
- Defaults = list of default values for each of the items that the user is prompted to key enter a value

RETURN:
- aList = list of responses for each of the items that were displayed. If the user Cancels the command: aList will be an empty list, that is, aList.Count will equal 0.

The given and returned variables should be declared where first called as:

```
Dim aMsg, aHeading As Variant
Dim Labels As New Collection
Dim Defaults As New Collection
Dim aList As New Collection
```

Presented below is a sample code for testing this subroutine.

```
'
'   ---Example of avMsgBoxMultiInput Avenue Wrap
Dim aMsg, aHed As Variant
Dim i As Long
Dim LabList As New Collection
Dim DefList As New Collection
Dim aList As New Collection
aMsg = "Key Enter the Indicated Data:"
aHed = "Example of avMsgBoxMultiInput2 Avenue Wrap"
For i = 1 To 25
    LabList.Add ("Item Number " + CStr(i) + ":")
    DefList.Add (CStr(i) + ".0")
Next
Call avMsgBoxMultiInput2(aMsg, aHed, LabList, _
                            DefList, aList)
```

DATA INPUT MESSAGE BOXES

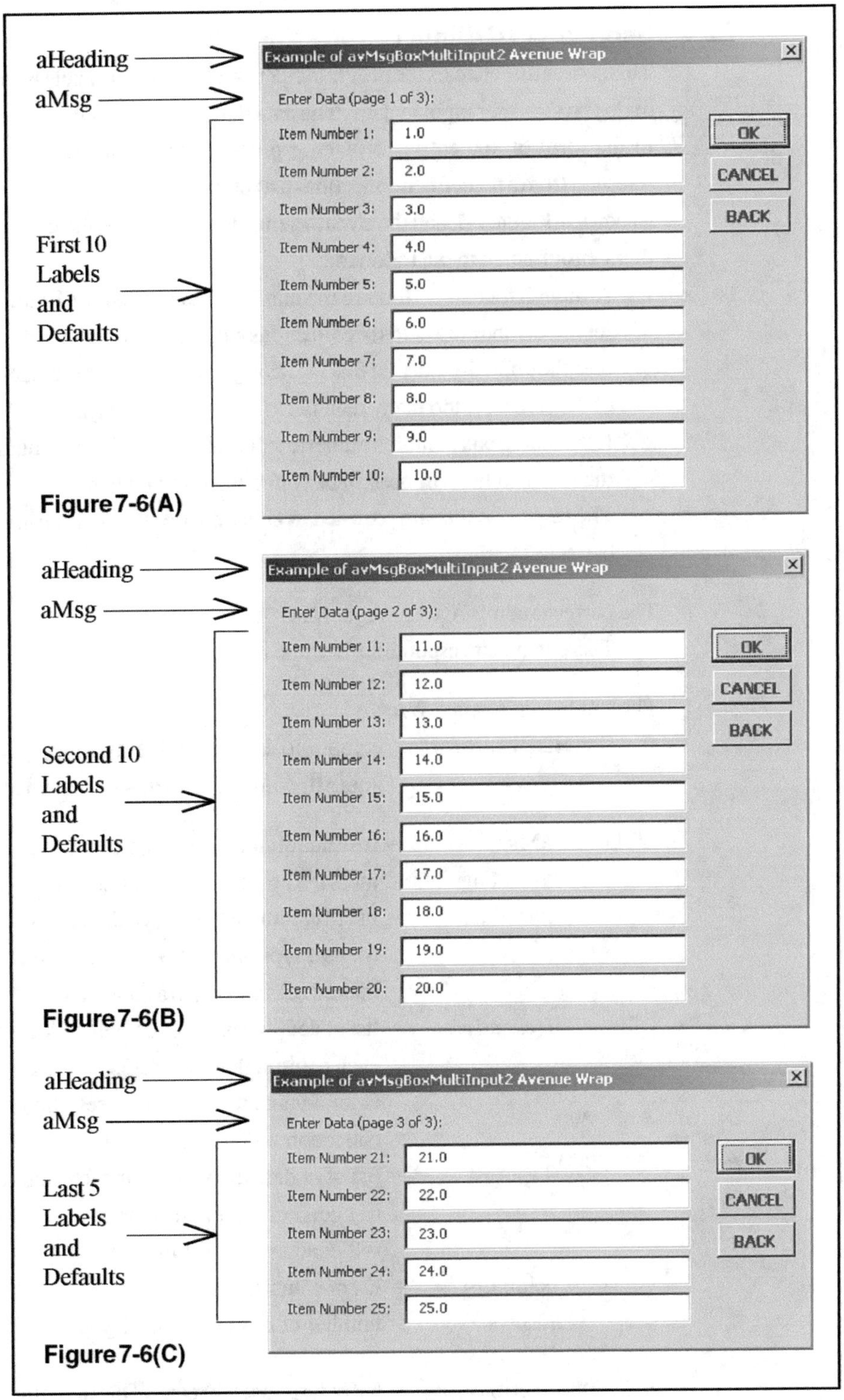

aHeading → Example of avMsgBoxMultiInput2 Avenue Wrap

aMsg → Enter Data (page 1 of 3):

First 10 Labels and Defaults →

Item Number 1: 1.0
Item Number 2: 2.0
Item Number 3: 3.0
Item Number 4: 4.0
Item Number 5: 5.0
Item Number 6: 6.0
Item Number 7: 7.0
Item Number 8: 8.0
Item Number 9: 9.0
Item Number 10: 10.0

OK CANCEL BACK

Figure 7-6(A)

aHeading → Example of avMsgBoxMultiInput2 Avenue Wrap

aMsg → Enter Data (page 2 of 3):

Second 10 Labels and Defaults →

Item Number 11: 11.0
Item Number 12: 12.0
Item Number 13: 13.0
Item Number 14: 14.0
Item Number 15: 15.0
Item Number 16: 16.0
Item Number 17: 17.0
Item Number 18: 18.0
Item Number 19: 19.0
Item Number 20: 20.0

OK CANCEL BACK

Figure 7-6(B)

aHeading → Example of avMsgBoxMultiInput2 Avenue Wrap

aMsg → Enter Data (page 3 of 3):

Last 5 Labels and Defaults →

Item Number 21: 21.0
Item Number 22: 22.0
Item Number 23: 23.0
Item Number 24: 24.0
Item Number 25: 25.0

OK CANCEL BACK

Figure 7-6(C)

Figure 7-6 avMsgBoxMultiInput2 Message Box

DATA INPUT MESSAGE BOXES

7.2.4 Subroutine HDBbuild

This subroutine enables the programmer to build a custom-made horizontal dialog box for user input of data. The data that is entered by the user may be in the form of key entries replacing preassigned default values, and/or selection from a predefined collection (list) of strings. A sample message box is shown in Figure 7-7, and the given arguments are identified therein. In using this subroutine, note the following:

1. Although there is no limit to the number of items that can be prompted for (labels and data fields or collections with default values) in the dialog box, the difficulty exists in that the dialog box may exceed the visible area of the screen, if too many data items (rows) are displayed.
2. If the dialog box contains numeric values, the string items contained in the returned list will need to be converted into numbers.
3. The responses that are returned in UserInfo are stored on a row by row basis, left to right.

The corresponding Avenue request is:

There is no corresponding Avenue request.

The call to this Avenue Wrap is:

```
Call HDBbuild(aMsg, aHeading, LabelList, DefaultInfo,_
              TypeList, colmnList, nRows, UserInfo)
```

GIVEN:

GIVEN:	aMsg	= instruction message to be displayed
	aHeading	= message box caption heading (title)
	LabelList	= list of column labels for each of the items that the user is prompted to key enter a value for, or make a selection from a collection
	DefaultInfo	= list of default values or default collections for each column that the user is prompted to key enter a value for, or make a selection from a collection
	TypeList	= list of of data item types for each column = 1 denotes a data line item = 2 denotes a collection item
	colmnList	= list of column widths
	nRows	= number of rows to be displayed
RETURN:	UserInfo	= list of responses for each of the items that were displayed. If the user Cancels the command:

DATA INPUT MESSAGE BOXES

aList will be an empty list, that is, aList.Count will be equal to 0.

The given and returned variables should be declared where first called as:

Dim aMsg, aHeading As Variant
Dim LabelList As New Collection
Dim DefaultInfo As New Collection
Dim TypeList As New Collection
Dim colmnList As New Collection
Dim nRows As Integer
Dim UserInfo As New Collection

Example The example below is to build a horizontal dialog box comprised of seven columns. All of the columns except the fifth and sixth contain a collection, from which, the user is to select one of the predefined default strings. The fifth and sixth columns are to be of data line entry type where the user can key enter a desired response. Reference is made to Figure 7-7.

```
'
Dim nRows As Integer
Dim annFields(4) As Variant
Dim aFFList As New Collection
Dim aStrng As String
Dim ii, nItems As Long
Dim aFFListV() As Variant
Dim aQuadList(4) As Variant
Dim aSTList(4) As Variant
Dim txtsiz As String
Dim aFSList(32) As Variant
Dim aMsg, aHeading As Variant
Dim LabelList As New Collection
Dim colmnList As New Collection
Dim defltList As New Collection
Dim aTypeList As New Collection
Dim UserInfo As New Collection
Dim annAttrib As New Collection
Dim annFonts As New Collection
Dim annStyle As New Collection
Dim annSize As New Collection
Dim annPrefix As New Collection
Dim annSuffix As New Collection
Dim annQuad As New Collection
Dim jj As Long
Dim aField, aFont, aStyle, aSize As Variant
Dim aPrefix, aSuffix, aQuad As Variant
'
```

DATA INPUT MESSAGE BOXES

```
'   ---Define the number of rows to be displayed
    nRows = 4
'
'   ---Build an array of the attributes to be annotated
    annFields(1) = "PNT"
    annFields(2) = "X"
    annFields(3) = "Y"
    annFields(4) = "Z"
'
'   ---Get a list of Windows fonts that are available
    Call avGetWinFonts(aFFList)
'
'   ---Check if the Arial font is present
    aStrng = " "
    For ii = 1 To aFFList.Count
        aStrng = aFFList.Item(ii)
        If (UCase(aStrng) = "ARIAL") Then
           Exit For
        End If
    Next
'
'   ---When the Arial font is present add it at the top
'   ---of list as the default
    If (aStrng <> " ") Then
       aFFList.Add aStrng, before:=1
    End If
'
'   ---Build an array of the available Windows fonts
    nItems = aFFList.Count
    ReDim aFFListV(nItems)
    For ii = 1 To nItems
        aStrng = aFFList.Item(ii)
        aFFListV(ii) = aStrng
    Next ii
'
'   ---Build a list of quadrants where the annotation
'   ---can appear
    aQuadList(1) = "E"
    aQuadList(2) = "N"
    aQuadList(3) = "S"
    aQuadList(4) = "W"
'
'   ---Build a list of font styles
    aSTList(1) = "Normal"
    aSTList(2) = "Italic"
    aSTList(3) = "Bold Italic"
    aSTList(4) = "Bold"
'
```

DATA INPUT MESSAGE BOXES

```
'     ---Define the default font size
      txtsiz = 10
'
'     ---Build a list of available font sizes with the
'     ---default font size being the first in the list
      aFSList(1) = txtsiz
      aFSList(2) = "2"
      aFSList(3) = "3"
      aFSList(4) = "4"
      aFSList(5) = "5"
      aFSList(6) = "6"
      aFSList(7) = "7"
      aFSList(8) = "8"
      aFSList(9) = "9"
      aFSList(10) = "10"
      aFSList(11) = "11"
      aFSList(12) = "12"
      aFSList(13) = "13"
      aFSList(14) = "14"
      aFSList(15) = "15"
      aFSList(16) = "16"
      aFSList(17) = "17"
      aFSList(18) = "18"
      aFSList(19) = "19"
      aFSList(20) = "20"
      aFSList(21) = "21"
      aFSList(22) = "22"
      aFSList(23) = "23"
      aFSList(24) = "24"
      aFSList(25) = "30"
      aFSList(26) = "36"
      aFSList(27) = "40"
      aFSList(28) = "48"
      aFSList(29) = "60"
      aFSList(30) = "72"
      aFSList(31) = "96"
      aFSList(32) = "127"
'
'     ---Define the dialog box prompt
      aMsg = "Enter annotation parameters:"
'
'     ---Define the dialog box title
      aHeading = "Annotate Text or Attribute"
'
'     ---Build the list of column labels
      Call CreateList(LabelList)
      LabelList.Add ("Attribute")
      LabelList.Add ("Font")
      LabelList.Add ("Style")
```

DATA INPUT MESSAGE BOXES

```
        LabelList.Add ("Size")
        LabelList.Add ("Prefix")
        LabelList.Add ("Suffix")
        LabelList.Add ("Quadrant")
'
'       ---Build the list of column widths in pixels
        Call CreateList(colmnList)
        colmnList.Add (100)
        colmnList.Add (130)
        colmnList.Add (80)
        colmnList.Add (45)
        colmnList.Add (80)
        colmnList.Add (80)
        colmnList.Add (50)
'
'       ---For each column that is to appear in the dialog box
'       ---build the default values and data item type lists,
'       ---where 1 = data line, 2 = choice box
        Call CreateList(defltList)
        Call CreateList(aTypeList)
        defltList.Add annFields
        aTypeList.Add (2)
        defltList.Add aFFListV
        aTypeList.Add (2)
        defltList.Add aSTList
        aTypeList.Add (2)
        defltList.Add aFSList
        aTypeList.Add (2)
        defltList.Add (" ")
        aTypeList.Add (1)
        defltList.Add (" ")
        aTypeList.Add (1)
        defltList.Add aQuadList
        aTypeList.Add (2)
'
'       ---Build the horizontal dialog box enabling the user
'       ---to specify the annotation attributes
        Call HDBbuild(aMsg, aHeading, LabelList, _
                      defltList, aTypeList, colmnList, nRows, _
                      UserInfo)
'
'       ---Make sure the display is current
        Call avGetDisplayFlush
'
'       ---Check if the user wishes to abort the command
        nItems = UserInfo.Count
        If (nItems <= 0) Then
           Exit Sub
        End If
```

DATA INPUT MESSAGE BOXES

```
'
'     ---Define the end loop value accounting for the fact
'     ---that the user responses are stored on a row by row
'     ---basis with the columns processed left to right
      nItems = nItems / colmnList.Count
'
'     ---Build lists to be used in annotating the attributes
'     ---The 7 collections that are populated contain the
'     ---responses on a row by row basis for each column
'     ---that appeared in the dialog box
      jj = 0
      For ii = 1 To nItems
          jj = jj + 1
          aField = UserInfo.Item(jj)
          jj = jj + 1
          aFont = UserInfo.Item(jj)
          jj = jj + 1
          aStyle = UserInfo.Item(jj)
          jj = jj + 1
          aSize = UserInfo.Item(jj)
          jj = jj + 1
          aPrefix = UserInfo.Item(jj)
          jj = jj + 1
          aSuffix = UserInfo.Item(jj)
          jj = jj + 1
          aQuad = UserInfo.Item(jj)
          annAttrib.Add (aField)         ' Attribute column
          annFonts.Add (aFont)           ' Font column
          annStyle.Add (aStyle)          ' Style column
          annSize.Add (aSize)            ' Size column
          annPrefix.Add (aPrefix)        ' Prefix column
          annSuffix.Add (aSuffix)        ' Suffix column
          annQuad.Add (aQuad)            ' Quadrant column
      Next
```

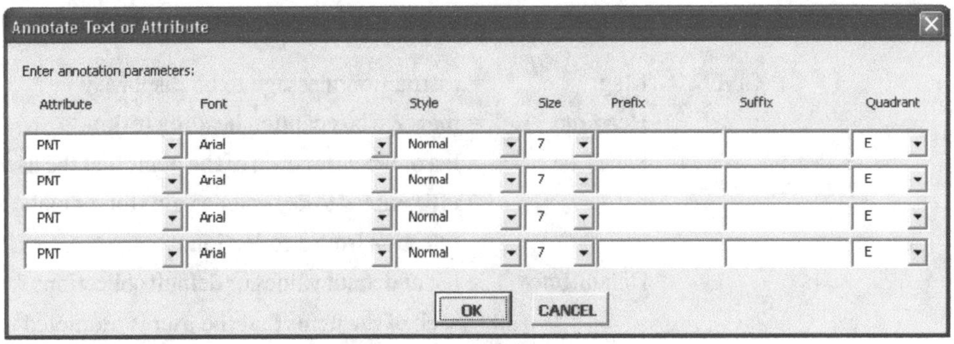

Annotate Text or Attribute

Enter annotation parameters:

Attribute	Font	Style	Size	Prefix	Suffix	Quadrant
PNT	Arial	Normal	7			E
PNT	Arial	Normal	7			E
PNT	Arial	Normal	7			E
PNT	Arial	Normal	7			E

OK CANCEL

Figure 7-7 HDBbuild Message Box

DATA INPUT MESSAGE BOXES

7.2.5 Subroutine VDBbuild

This subroutine enables the programmer to build a custom-made vertical dialog box for user input of data. The data that is entered by the user may be in the form of key entries replacing preassigned default values, and/or selection from a predefined collection (list) of strings. Sample message boxes are shown in Figure 7-8(A), 7-8(B) and 7-8(C), and the given arguments are identified therein. In using this subroutine, note the following:

1. Although there is no limit to the number of items that can be prompted for (labels and data fields or collections with default values) in the dialog box, the difficulty exists in that the dialog box may exceed the visible area of the screen, if too many data items are displayed. It is recommended that VDBbuild2 be used when more than 12 to 15 items are to be displayed.
2. If the dialog box contains numeric values, the string items contained in the returned list will need to be converted into numbers.
3. In addition to providing the ability to create a combination of data entry and choice selection items in the same dialog box, VDBbuild is used as the source subroutine by other message box Avenue Wraps.
4. If the dialog box to be built with VDBbuild is to contain only data line entry items, then this subroutine is similar to the avMsgBoxMultiInput.
5. If the dialog box to be built with VDBbuild is to contain a mixture of data line key entry and selection from one or more collection items, then the programmer should note the comments of the example below regarding the given arguments DefaultInfo and TypeList.

The corresponding Avenue request is:

There is no corresponding Avenue request.

VDBbuild

The call to this Avenue Wrap is:

Call **VDBbuild**(Msg, Heading, LabelList, DefaultInfo, TypeList, _
UserInfo)

GIVEN:

Msg	=	instruction message to be displayed
Heading	=	message box caption heading (title)
LabelList	=	list of labels for each of the items that the user is prompted to key enter a value for, or make a selection from a collection
DefaultInfo	=	list of default values or default collections for each of the items that the user is prompted to key enter a value for, or make a selection from a collection

DATA INPUT MESSAGE BOXES

TypeList = type of data items to be displayed. If it is
= 1 denotes a data line item
= 2 denotes a collection item
= 3 denotes a list box item

RETURN: UserInfo = list of responses for each of the items that were displayed. If the user Cancels the command: aList will be an empty list, that is, aList.Count will be equal to 0.

The given and returned variables should be declared where first called as:

```
Dim Msg, Heading As Variant
Dim LabelList As New Collection
Dim DefaultInfo As New Collection
Dim TypeList As New Collection
Dim UserInfo As New Collection
```

Example The example below is to build a dialog box of three items, the first and third of which are to be key entry type and the second one is to contain a collection, from which, the user is to select one of the predefined default strings. Reference is made to Figures 7-8(A) and 7-8(B). The reader should note that:

1. There is nothing different about the LabelList given argument. Each label carries a description for each input item.
2. Since the first and third items are to be key entries, the corresponding DefaultInfo(1) and DefaultInfo(3) items contain a single default string value ("1" and "3" respectively).
3. Since the second item is to contain a list of strings, from which the user is to select one, the array DataArray is assigned to DefaultInfo(2).
4. For this example, the array DataArray is to contain 16 strings, as such, the array is declared as shown in the code below.
5. Since the first and third items are to be key entries, and the second item is to contain a collection of strings, the corresponding TypeList(1) and TypeList(3) items have been assigned the value of 1 (data line), while the TypeList(2) item has been assigned the value 2 (collection item).

```
'
'   ---Example of VDBbuild Avenue Wrap
    Dim aMsg, aHeading As Variant
    Dim DataArray(16)
```

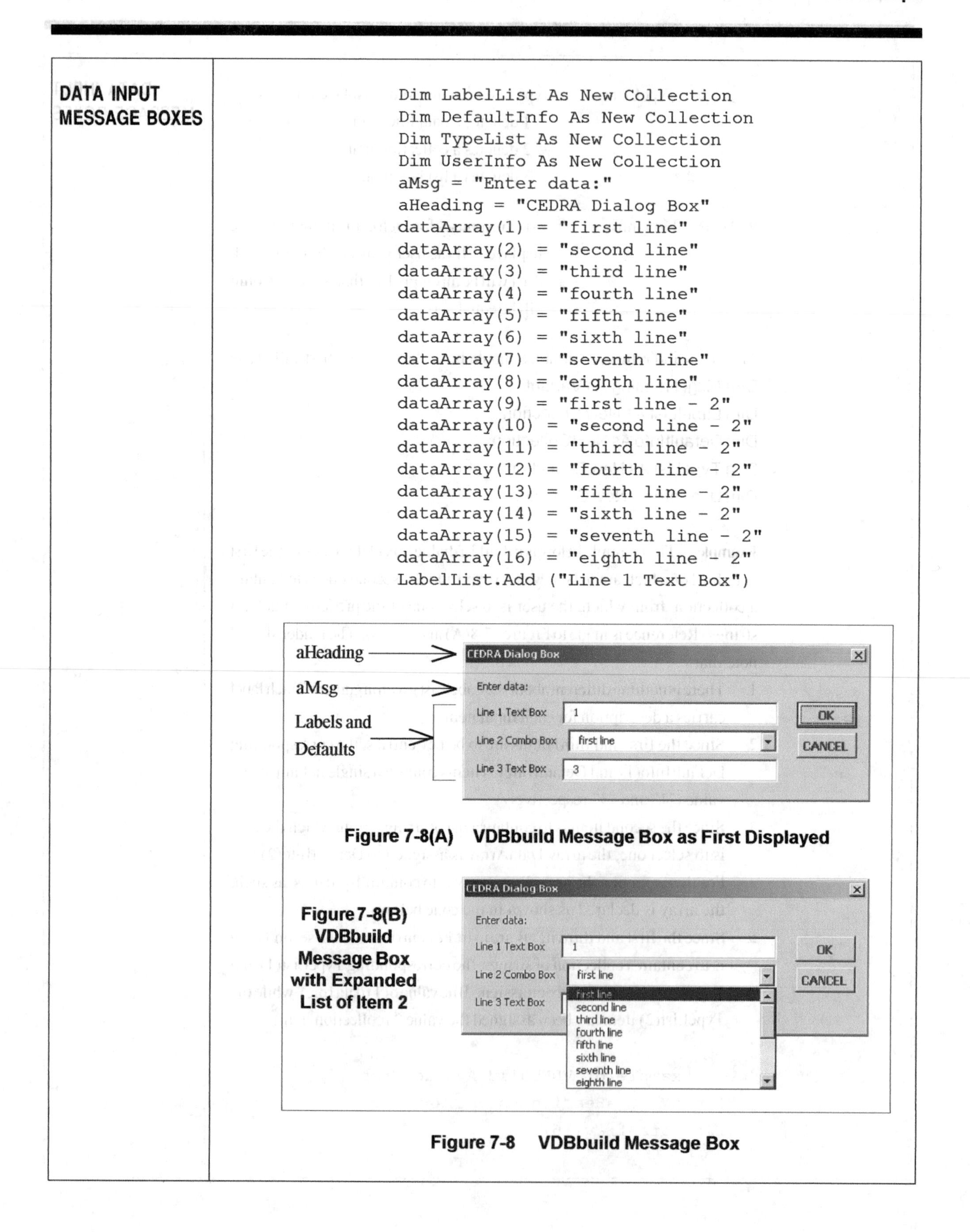

DATA INPUT MESSAGE BOXES

```
Dim LabelList As New Collection
Dim DefaultInfo As New Collection
Dim TypeList As New Collection
Dim UserInfo As New Collection
aMsg = "Enter data:"
aHeading = "CEDRA Dialog Box"
dataArray(1) = "first line"
dataArray(2) = "second line"
dataArray(3) = "third line"
dataArray(4) = "fourth line"
dataArray(5) = "fifth line"
dataArray(6) = "sixth line"
dataArray(7) = "seventh line"
dataArray(8) = "eighth line"
dataArray(9) = "first line - 2"
dataArray(10) = "second line - 2"
dataArray(11) = "third line - 2"
dataArray(12) = "fourth line - 2"
dataArray(13) = "fifth line - 2"
dataArray(14) = "sixth line - 2"
dataArray(15) = "seventh line - 2"
dataArray(16) = "eighth line - 2"
LabelList.Add ("Line 1 Text Box")
```


Figure 7-8(A) VDBbuild Message Box as First Displayed

Figure 7-8(B) VDBbuild Message Box with Expanded List of Item 2

Figure 7-8 VDBbuild Message Box

DATA INPUT MESSAGE BOXES

```
LabelList.Add ("Line 2 Combo Box")
LabelList.Add ("Line 3 Text Box")
DefaultInfo.Add ("1")
DefaultInfo.Add (dataArray)
DefaultInfo.Add ("3")
TypeList.Add (1)
TypeList.Add (2)
TypeList.Add (1)
Call VDBbuild(aMsg, aHeading, LabelList, _
              DefaultInfo, TypeList, UserInfo)
```

Example The example below is to build a dialog box that contains a single item, which is a list box, see Figure 7-8(C). When a list box item is to be displayed the dialog box should only contain this control.

```
'
    Dim aStrngList(4) As Variant
    Dim entryList As New Collection
    Dim defaultInfo As New Collection
    Dim typeList As New Collection
    Dim aMsg, Heading As Variant
    Dim userInfo As New Collection
'
'   ---Build an array containing the items
    aStrngList(1) = "Item number 1"
    aStrngList(2) = "Item number 2"
    aStrngList(3) = "Item number 3"
    aStrngList(4) = "Item number 4"
'
'   ---Initialize dialog box lists for the
'   ---label, default value and data item type
    Call CreateList(entryList)
    Call CreateList(defaultInfo)
    Call CreateList(typeList)
'
'   ---Define the label for the data item
    entryList.Add ("Item List:")
'
'   ---Define default value for the data item
    defaultInfo.Add (aStrngList)
'
'   ---Define the type of data item
'   ---1: TextBox, 2: ComboBox
'   ---3: ListBox with Multiselect
    typeList.Add (3)
'
'   ---Define the dialog box prompt
    aMsg = "Select the item(s) from the list:"
'
```

```
'    ---Define the message box title
     Heading = "ListBox Control Test"
'
'    ---Display the items in a list box control
     Call VDBbuild(aMsg, Heading, entryList, _
                   defaultInfo, typeList, _
                   userInfo)
```

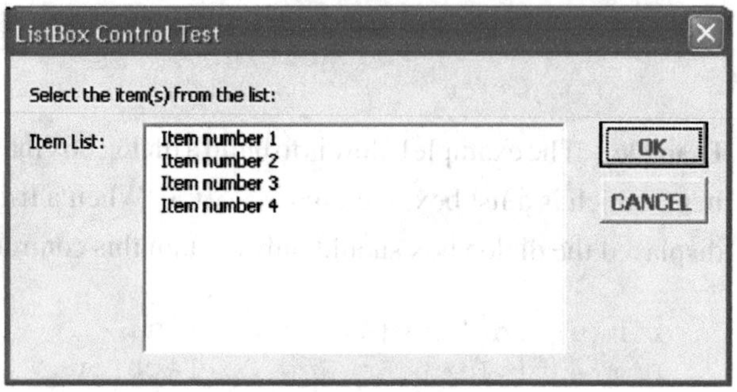

ListBox Control Test
Select the item(s) from the list:
Item List: Item number 1 Item number 2 Item number 3 Item number 4
OK
CANCEL

Figure 7-8(C)
VDBbuild Message Box with ListBox Control

VDBbuild2

7.2.6 Subroutine VDBbuild2

This subroutine enables the programmer to build a custom-made vertical dialog box for user input of data, which is to contain more than 10 combination items of data line entries and collections, with the ability to display a BACK button so that the user may advance from page to page. Its composition is identical to that of the subroutine VDBbuild, and its operation of the OK and BACK buttons is identical that of the subroutine avMsgBoxMultiInput2.

Shown in Figure 7-9 is the result if VDBbuild2 was used, in the previous example, rather than VDBbuild. Note that Figure 7-9 was produced in a Windows XP environment, while Figure 7-8 was produced in a Windows 2000 environment. That is why the look of the dialog boxes are slightly different.

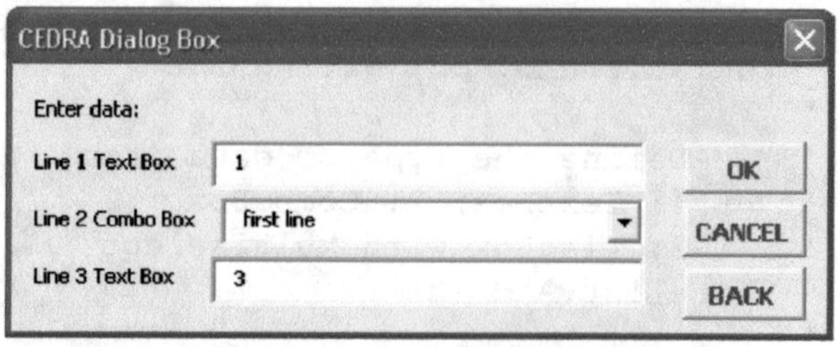

CEDRA Dialog Box
Enter data:
Line 1 Text Box 1
Line 2 Combo Box first line
Line 3 Text Box 3
OK
CANCEL
BACK

Figure 7-9 VDBbuild2 Message Box

7.3 Selection Message Boxes Related Avenue Wraps

SELECTION MESSAGE BOXES

7.3.1 Subroutine avMsgBoxChoice

This subroutine enables the programmer to display a choice message box, that contains a list of string items, from which the user selects one string that will be passed back. This subroutine is similar to the avMsgBoxList subroutine. The only difference is in the returned value. Whereas the return argument, ians, in the avMsgBoxList subroutine contains the value of the button clicked (vbOK, or vbCancel), in the avMsgBoxChoice subroutine it contains the string that was selected from the message box display. The message box of avMsgBoxChoice looks exactly like that of Figure 7-2.

The corresponding Avenue request is:

```
ians = MsgBox.Choice (aList, aMsg, aHeading)
```

The call to this Avenue Wrap is:

Call **avMsgBoxChoice**(aList, aMsg, aHeading, ians)

avMsgBoxChoice

GIVEN:	aList	= the list of items to be displayed
	aMsg	= the message to be displayed
	aHeading	= message box caption
RETURN:	ians	= the item selected by the user. If the user Cancels the command, ians will be equal to vbCancel.

The given and returned variables should be declared where first called as:

```
Dim aList As New Collection
Dim aMsg, aHeading As Variant
Dim ians As Variant
```

7.3.2 Subroutine avMsgBoxYesNo

This subroutine enables the programmer to display an option string in a message box, which the user may accept or reject by clicking the Yes or No button respectively. See Figure 7-10.

The corresponding Avenue request is:

```
ians = MsgBox.YesNo (aMsg, aHeading, aDefault)
```

The call to this Avenue Wrap is:

Call **avMsgBoxYesNo**(aMsg, aHeading, aDefault, ians)

avMsgBoxYesNo

SELECTION MESSAGE BOXES

GIVEN: aMsg = the message to be displayed
aHeading = message box caption
aDefault = the default button setting
true= YES, false = NO

RETURN: ians = selected button (vbYes or vbNo)

The given and returned variables should be declared where first called as:

```
Dim aMsg, aHeading As Variant
Dim aDefault As Boolean
Dim ians As Integer
```

aHeading → Example of avMsgBoxYesNo Avenue Wrap

aMsg → OK to accept the previous results ?

Yes No

Figure 7-10 avMsgBoxYesNo Message Box

7.3.3 Subroutine avMsgYesNoCancel

avMsgBox YesNoCancel

This subroutine enables the programmer to display an option string in a message box, that the user may accept, reject or cancel the display of the box by clicking the Yes, No or Cancel button respectively. See Figure 7-11.

The corresponding Avenue request is:

```
ians = MsgBox.YesNoCancel (aMsg, aHeading , aDefault)
```

The call to this Avenue Wrap is:

Call **avMsgBoxYesNo**(aMsg, aHeading , aDefault, ians)

GIVEN: aMsg = the message to be displayed
aHeading = message box caption
aDefault = the default button setting
true= YES, false = NO

RETURN: ians = selected button (vbYes, vbNo or vbCancel)

PROJECT RELATED COMMANDS

aHeading → Example of avMsgBoxYesNoCancel Avenue Wrap

aMsg → OK to accept the previous results ?

Yes No Cancel

Figure 7-11 avMsgBoxYesNoCancel Message Box

The given and returned variables should be declared where first called as:

```
Dim aMsg, aHeading As Variant
Dim aDefault As Boolean
Dim ians As Integer
```

CHAPTER 8

GEOMETRIC ROUTINES AVENUE WRAPS

This chapter contains Avenue Wraps that enable the programmer to create and retrieve geometric features such points, lines and polygons, and to intersect, merge or union two shapes. These Avenue Wraps include the following:

The source listing of each of the above Avenue Wraps may be found in Appendix D of this publication.

GENERAL GEOMETRY

8.1 General Geometric Avenue Wraps

8.1.1 Subroutine avaClassMake

This subroutine enables the programmer to create a point, polyline or polygon type geometry object given a collection of points (shapeList) and the desired object type (aClass). In using this subroutine, note the following:

1. The given argument aClass specifies the desired object type, and its numeric value indicates the following object type to be created:

11 for PolyLineM	31 for PolyLineM and PolyLineZ
12 for PolyLineZ	32 for PolygonM and PolygonZ
13 for PolygonM	33 for PointM and PointZ
14 for PolygonZ	34 for MultiPointM and MultiPointZ
15 for PointM	41 for PolyLine
16 for PointZ	42 for Polygon
17 for MultiPointM	43 for Point
18 for MultiPointZ	44 for MultiPoint

2. The given argument shapeList is a collection comprised of the following items: nParts / nPoints / xPt / yPt / zPt / mPt / idPt

   where: nParts / nPoints / idPt are declared as long integer numbers denoting the number of parts, number of points in a part and identification number of a point,

   and: xPt / yPt / zPt / mPt are declared as double precision floating numbers denoting the x, y and z coordinates, and the measure of a point.

3. As an example of the composition of the shapeList collection, consider a multi-point, polyline or polygon comprised of three parts, the first having three points, the second two points and the third two points. The

Part 1			Part 2		Part 3	
3 (Number of Parts)						
3 (Number of Points in Part 1)			**2** (Points in Part 2)		**2** (Points in Part 3)	
x11	**x12**	**x13**	**x21**	**x22**	**x31**	**x32**
y11	**y12**	**y13**	**y21**	**y22**	**y31**	**y32**
z11	**z12**	**z13**	**z21**	**z22**	**z31**	**z32**
m11	**m12**	**m13**	**m21**	**m22**	**m31**	**m32**
idPt11	**idPt12**	**idPt13**	**idPt21**	**idPt22**	**idPt31**	**idPt32**

Figure 8-1 Composition of Sample shapeList Collection - avaClassMake

GENERAL GEOMETRY

contents of shapeList would then be as shown in Figure 8-1, with the first suffix indicating the part, and the second indicating the point number.

The corresponding Avenue request is:
There is no corresponding Avenue request.

avaClassMake

The call to this Avenue Wrap is:
Call **avaClassMake**(aClass, shapeList, theFeat)

GIVEN: aClass = the type of special feature. See Note 1 above.
shapeList = the list of points comprising the feature

RETURN: theFeat = the special feature

The given and returned variables should be declared where first called as:
Dim aClass As Integer
Dim shapeList As New Collection
Dim theFeat As IPoint, IMultiPoint, IPolyline, or IPolygon

8.1.2 Subroutine avAsList

This subroutine enables the programmer to create a collection of points that comprise a polyline or polygon from the **IFeature** interface. The composition of the resultant collection (shapeList) is shown in Figure 8-2, which represents a collection of collections of points, each of last said collections representing a part of the polyline or polygon feature (theFeature).

The corresponding Avenue request is:
shapeList = theFeature.AsList

avAsList

The call to this Avenue Wrap is:
Call **avAsList**(theFeature, shapeList)

GIVEN: theFeature = feature to be processed

RETURN: shapeList = the shape's list of list of points

The given and returned variables should be declared where first called as:
Dim theFeature As IFeature
Dim shapeList As New Collection

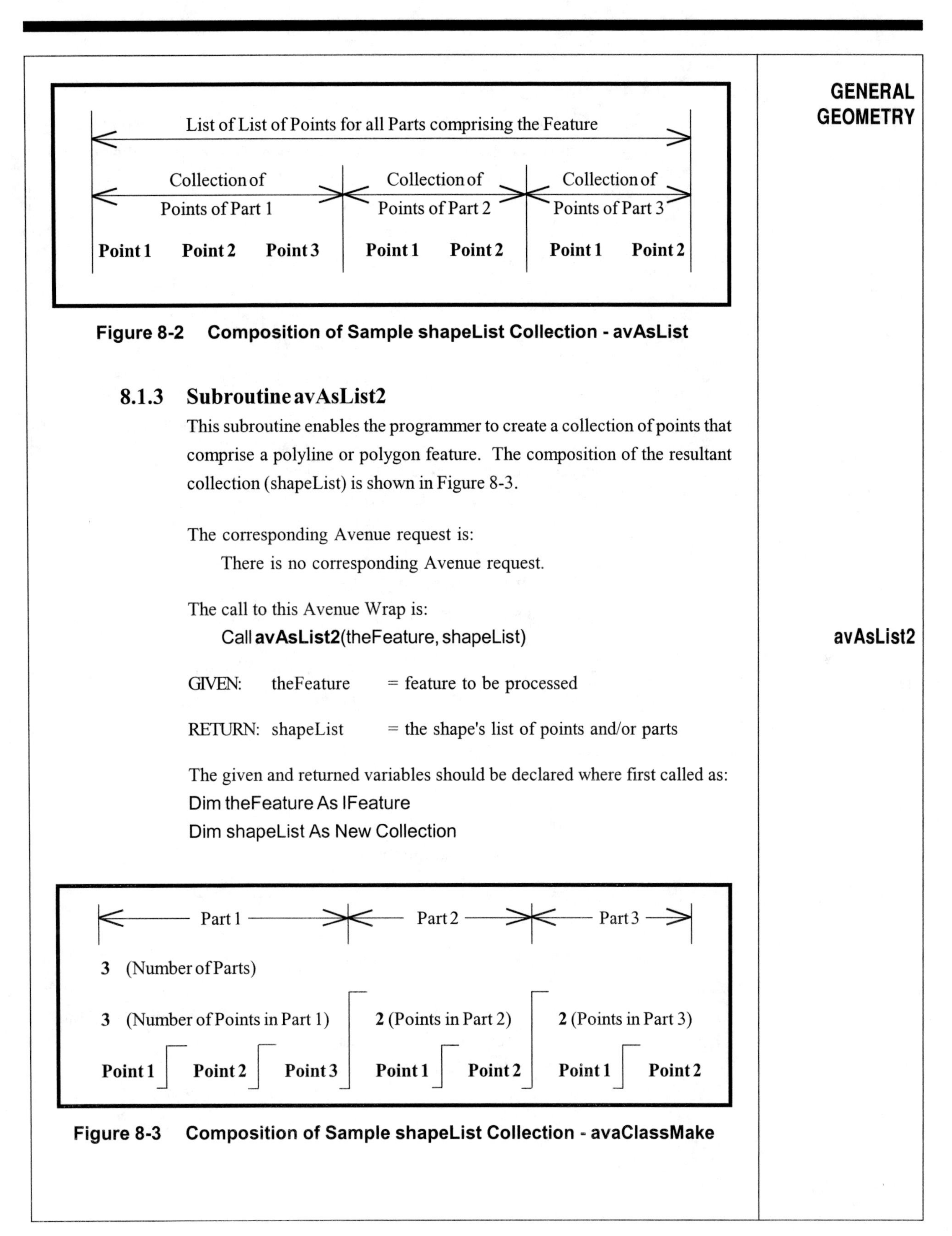

GENERAL GEOMETRY

Figure 8-2 Composition of Sample shapeList Collection - avAsList

8.1.3 Subroutine avAsList2

This subroutine enables the programmer to create a collection of points that comprise a polyline or polygon feature. The composition of the resultant collection (shapeList) is shown in Figure 8-3.

The corresponding Avenue request is:

There is no corresponding Avenue request.

The call to this Avenue Wrap is:

Call **avAsList2**(theFeature, shapeList)

avAsList2

GIVEN: theFeature = feature to be processed

RETURN: shapeList = the shape's list of points and/or parts

The given and returned variables should be declared where first called as:

Dim theFeature As IFeature
Dim shapeList As New Collection

Figure 8-3 Composition of Sample shapeList Collection - avaClassMake

GENERAL GEOMETRY

8.1.4 Subroutine avAsList3

This subroutine enables the programmer to create a collection of points that comprise a polyline or polygon geometry object. The composition of the resultant collection (shapeList) is shown in Figure 8-3.

The corresponding Avenue request is:
    There is no corresponding Avenue request.

avAsList3

The call to this Avenue Wrap is:
    Call **avAsList3**(theGeometry, shapeList)

GIVEN: theGeometry = feature to be processed

RETURN: shapeList = the shape's list of points and/or parts

The given and returned variables should be declared where first called as:

```
Dim theGeometry As IGeometry
Dim shapeList As New Collection
```

8.1.5 Subroutine avPlAsList

This subroutine enables the programmer to create a collection of points from a geometry object. The composition of the resultant collection (shapeList) is shown in Figure 8-2. This routine can be used for polyline, polygon and multi-point features, but it can not be used for point features.

The corresponding Avenue request is:
    shapeList = pFeatureGeom.AsList

avPlAsList

The call to this Avenue Wrap is:
    Call **avPlAsList**(pFeatureGeom, shapeList)

GIVEN: pFeatureGeom = geometry object to be processed

RETURN: shapeList = list of list of points comprising the geometry

The given and returned variables should be declared where first called as:

```
Dim pFeatureGeom As IGeometry
Dim shapeList As New Collection
```

GENERAL GEOMETRY

8.1.6 Subroutine avPlAsList2

This subroutine enables the programmer to create a collection of points from a polyline geometry object. The composition of the resultant collection (shapeList) is shown in Figure 8-1. This routine can be used for polyline, polygon and multi-point features, but it can not be used for point features.

The corresponding Avenue request is:

There is no corresponding Avenue request.

The call to this Avenue Wrap is:

Call **avPlAsList2**(theLine, shapeList)

avPlAsList2

GIVEN: theLine = geometry object to be processed

RETURN: shapeList = list of points comprising the polyline

The given and returned variables should be declared where first called as:

Dim theLine As IGeometry
Dim shapeList As New Collection

8.1.7 Subroutine avPlFindVertex

This subroutine enables the programmer to find (a) the point of a multi-point, (b) the endpoint of a line, or (c) the vertex of a polyline or polygon point list, as shown in Figure 8-1, nearest to a given location.

The corresponding Avenue request is:

There is no corresponding Avenue request.

The call to this Avenue Wrap is:

Call **avPlFindVertex**(ipmode, elmntList, X, Y, thePart, thePt)

avPLFindVertex

GIVEN: ipmode = the mode of operation
0 : find the vertex close to a location
elmntList = list of points comprising the feature
X, Y = coordinates of the location nearest which a vertex is to be returned

RETURN: thePart = the part of the polyline. Part numbers begin at zero (0), not one (1)

GENERAL GEOMETRY

thePt = the sequential point number (starting at 1) in the part (thePart) nearest to the given location (X,Y).

The given and returned variables should be declared where first called as:

```
Dim ipmode As Integer
Dim elmntList As New Collection
Dim X, Y As Double
Dim thePart, thePt As Long
```

8.1.8 Subroutine avPlGet3Pt

This subroutine enables the programmer to get the coordinates of three points from a specific part in a specified collection of a feature point. The composition of the given point collection (shapeList) is as shown in Figure 8.1. The Avenue Wrap avPlAsList2 may be used to extract this collection if it is not known by any other means.

The corresponding Avenue request is:

There is no corresponding Avenue request.

avPLGet3Pt

The call to this Avenue Wrap is:

Call **avPlGet3Pt**(shapeList, thePart, X1, Y1, XM, YM, X2, Y2)

GIVEN: shapeList = list of points comprising the feature

thePart = the part of the polyline. Part numbers begin at zero (0), not one (1)

RETURN: X1,Y1 = start point coordinates of part

XM,YM = mid point coordinates of part

X2,Y2 = end point coordinates of part

The given and returned variables should be declared where first called as:

```
Dim shapeList As New Collection
Dim thePart As Long
Dim X1, Y1, XM, YM, X2, Y2 As Double
```

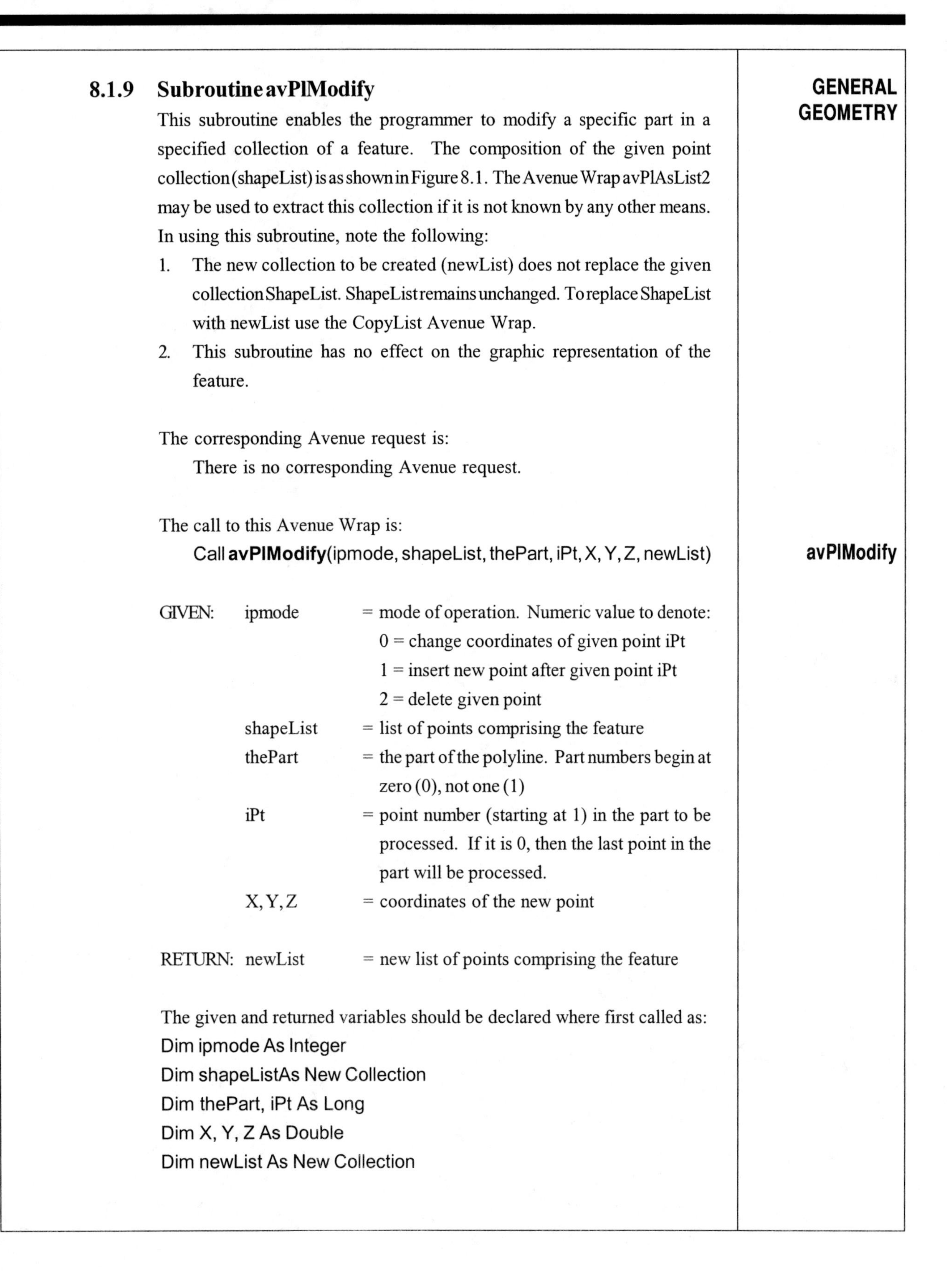

GENERAL GEOMETRY

8.1.9 Subroutine avPlModify

This subroutine enables the programmer to modify a specific part in a specified collection of a feature. The composition of the given point collection (shapeList) is as shown in Figure 8.1. The Avenue Wrap avPlAsList2 may be used to extract this collection if it is not known by any other means. In using this subroutine, note the following:

1. The new collection to be created (newList) does not replace the given collection ShapeList. ShapeList remains unchanged. To replace ShapeList with newList use the CopyList Avenue Wrap.
2. This subroutine has no effect on the graphic representation of the feature.

The corresponding Avenue request is:

There is no corresponding Avenue request.

The call to this Avenue Wrap is:

Call **avPlModify**(ipmode, shapeList, thePart, iPt, X, Y, Z, newList)

avPlModify

GIVEN:

GIVEN:	ipmode	= mode of operation. Numeric value to denote: 0 = change coordinates of given point iPt 1 = insert new point after given point iPt 2 = delete given point
	shapeList	= list of points comprising the feature
	thePart	= the part of the polyline. Part numbers begin at zero (0), not one (1)
	iPt	= point number (starting at 1) in the part to be processed. If it is 0, then the last point in the part will be processed.
	X, Y, Z	= coordinates of the new point
RETURN:	newList	= new list of points comprising the feature

The given and returned variables should be declared where first called as:

```
Dim ipmode As Integer
Dim shapeListAs New Collection
Dim thePart, iPt As Long
Dim X, Y, Z As Double
Dim newList As New Collection
```

8.2 Geometric Feature Creation Avenue Wraps

GEOMETRIC FEATURE CREATION

The routines in this section create the geometric attributes that comprise the indicated geometric feature only. They do not create the graphic representation of the feature.

8.2.1 Function avAsPolygon

This function enables the programmer to change an IUnknown polygon interface into an IGeometry polygon interface.

The corresponding Avenue request is:

There is no corresponding Avenue request.

The call to this Avenue Wrap is:

Set thePolygon = **avAsPolygon**(pInput)

avAsPolygon

GIVEN: pInput = the IUnknown polygon interface to be converted

RETURN: thePolygon = the IGeometry polygon interface

The given and returned variables should be declared where first called as:

Dim pInput As IUnknown
Dim thePolygon As IGeometry

8.2.2 Function avCircleMakeXY

This function enables the programmer to create a circle given the coordinates of its center and its radius.

The corresponding Avenue request is:

theCircle = Circle.Make (aPoint, aRadius)

accepts as input an object (aPoint) rather than the X and Y coordinates of the center point

The call to this Avenue Wrap is:

Set theCircle = **avCircleMakeXY**(xPt, yPt, rad)

avCircleMakeXY

GIVEN: xPt, yPt = X and Y coordinates of the circle's center
rad = radius of the circle

RETURN: theCircle = the curve feature

GEOMETRIC FEATURE CREATION

The given and returned variables should be declared where first called as:

```
Dim xPt, yPt, rad As Double
Dim theCircle As ICurve
```

8.2.3 Function avPointMake

This function enables the programmer to create a point given its X and Y coordinates.

The corresponding Avenue request is:

```
thePoint = Point.Make (xPt, yPt)
```

avPointMake

The call to this Avenue Wrap is:

Set thePoint = **avPointMake**(xPt, yPt)

GIVEN: xPt = X coordinate of point
yPt = Y coordinate of point

RETURN: thePoint = the point feature

The given and returned variables should be declared where first called as:

```
Dim xPt, yPt As Double
Dim thePoint As IPoint
```

8.2.4 Function avPolygonMake

This function enables the programmer to create a polygon object from a given collection of points, which collection is composed as per Figure 8-2. The last point of said collection may or may not be a repetition of the first point. If it is not, the function will force a closure to the first point.

The corresponding Avenue request is:

```
thePolygon = Polygon.Make(shapeList)
```

avPolygonMake

The call to this Avenue Wrap is:

Set thePolygon = **avPolygonMake**(shapeList)

GIVEN: shapeList = the list of list of points comprising the polygon

RETURN: thePolygon = the polygon object feature

GEOMETRIC FEATURE CREATION

The given and returned variables should be declared where first called as:
Dim shapeList As New Collection
Dim thePolygon As IPolygon

8.2.5 Function avPolygonMake2

This function enables the programmer to create a polygon object from a given collection of points, which collection is composed as per Figure 8-1. The last point of said collection may or may not be a repetition of the first point. If it is not, the function will force a closure to the first point.

The corresponding Avenue request is:
There is no corresponding Avenue request.

The call to this Avenue Wrap is:
Set thePolygon = **avPolygonMake2**(shapeList)

avPolygonMake2

GIVEN: shapeList = the list of points comprising the polygon

RETURN: thePolygon = the polygon object feature

The given and returned variables should be declared where first called as:
Dim shapeList As New Collection
Dim thePolygon As IPolygon

8.2.6 Function avPolylineMake

This function enables the programmer to create a polyline object from a given collection of points, which collection is composed as per Figure 8-2.

The corresponding Avenue request is:
theLine = Polyline.Make(shapeList)

The call to this Avenue Wrap is:
Set theLine = **avPolylineMake**(shapeList)

avPolylineMake

GIVEN: shapeList = the list of points comprising the polygon

RETURN: theLine = the polyline feature

The given and returned variables should be declared where first called as:
Dim shapeList As New Collection
Dim theLine As IPolyline

GEOMETRIC FEATURE CREATION

8.2.7 Function avPolylineMake2

This function enables the programmer to create a polyline object from a given collection of points, which collection is composed as per Figure 8-1.

The corresponding Avenue request is:

There is no corresponding Avenue request.

The call to this Avenue Wrap is:

avPolylineMake2

```
Set theLine = avPolylineMake2(shapeList)
```

GIVEN: shapeList = the list of points comprising the polygon

RETURN: theLine = the polyline feature

The given and returned variables should be declared where first called as:

```
Dim shapeList As New Collection
Dim theLine As IPolyline
```

8.2.8 Function avPolyline2Pt

This function enables the programmer to create a polyline given the X and Y coordinates of two points.

The corresponding Avenue request is:

```
theLine = Polyline.Make({{X1, Y1, X2, Y2}})
```

The call to this Avenue Wrap is:

avPolyline2Pt

```
Set theLine = avPolyline2Pt(X1, Y1, X2, Y2)
```

GIVEN: X1, Y1 = X and Y coordinate of the start point
X2, Y2 = X and Y coordinate of the end point

RETURN: theLine = the polyline feature

The given and returned variables should be declared where first called as:

```
Dim X1, Y1, X2, Y2 As Double
Dim theLine As IPolyline
```

GEOMETRIC FEATURE CREATION

8.2.9 Function avRectMakeXY

This function enables the programmer to create a rectangle given the X and Y coordinates of two opposite corners.

The corresponding Avenue request is:

```
theRect = Rect.MakeXY(X1, Y1, X2, Y2)
```

The call to this Avenue Wrap is:

Set theRect = **avRectMakeXY**(X1, Y1, X2, Y2)

avRectMakeXY

GIVEN:	X1,Y1	= X and Y coordinate of the start point of a diagonal
	X2,Y2	= X and Y coordinate of the end point of a diagonal
RETURN:	theRect	= the rectangle (polygon) feature

The given and returned variables should be declared where first called as:

```
Dim X1, Y1, X2, Y2 As Double
Dim theRect As IPolygon
```

GEOMETRIC ATTRIBUTES

8.3 Geometric Attributes Avenue Wraps

8.3.1 Function avReturnArea

This function enables the programmer to get the area of an IGeometry object. Note that if an invalid geometry is specified, the function, avReturnArea, will return zero.

The corresponding Avenue request is:

theArea = theGeom.ReturnArea

avReturnArea

The call to this Avenue Wrap is:

theArea = **avReturnArea**(theGeom)

GIVEN: theGeom = the geometry to be processed

RETURN: theArea = the area of the geometry

The given and returned variables should be declared where first called as:

```
Dim theGeom As IGeometry
Dim theArea As Double
```

8.3.2 Function avReturnCenter

This function enables the programmer to get a point object representing the center of an IGeometry object. Note that if an invalid geometry is specified, the function, avReturnCenter, will return NOTHING.

The corresponding Avenue request is:

theCenter = theGeom.ReturnCenter

avReturnCenter

The call to this Avenue Wrap is:

Set theCenter = **avReturnCenter**(theGeom)

GIVEN: theGeom = the geometry to be processed

RETURN: theCenter = the centroid of the geometry

The given and returned variables should be declared where first called as:

```
Dim theGeom As IGeometry
Dim theCenter As IPoint
```

GEOMETRIC ATTRIBUTES

8.3.3 Function avReturnLength

This function enables the programmer to get the length of an IGeometry object (length of a line, perimeter of a polygon or circumference of a circle). When using this function, note the following:

1. For multi-part features, avReturnLength will return the total length, which includes all parts.
2. If an invalid geometry is specified the function, avReturnLength, will return zero.

The corresponding Avenue request is:

theLength = theGeom.ReturnLength

The call to this Avenue Wrap is:

theLength = **avReturnLength**(theGeom)

avReturnLength

GIVEN: theGeom = the geometry to be processed

RETURN: theLength = the length as described above

The given and returned variables should be declared where first called as:

Dim theGeom As IGeometry
Dim theLength As Double

GEOMETRIC EDITING

8.4 Geometric Editing Avenue Wraps

Reference is made to the functions avReturnIntersection, avReturnMerged and avReturnUnion presented below. The general operation of and differences between these three functions are identified below. In perusing them, refer to Figure 8-3.

- All three operate on a given pair of similar geometry feature types of multipoint, polyline or polygon.
- avReturnIntersection returns only those points, line segments or polygon areas that are common to both given features.
- avReturnMerged returns only those points, line segments or polygon areas that are not common to both given features.
- avReturnUnion returns all points, line segments or polygons except those that are duplicate to both given features.

8.4.1 Function avClean

This function enables the programmer to verify and enforce the correctness of a shape. In general, this means that duplicate points, vertices and line segments are removed from the shape.

The corresponding Avenue request is:

```
CleanShape = aShape1.Clean
```

avClean

The call to this Avenue Wrap is:

Set CleanShape = **avClean**(aShape1)

GIVEN:	aShape1	= shape to be cleaned
RETURN:	CleanShape	= new shape reflecting the cleaning

The given and returned variables should be declared where first called as:

```
Dim aShape1 As IGeometry
Dim CleanShape As IGeometry
```

8.4.2 Function avIntersects

This function enables the programmer to check whether two shapes intersect with each other.

The corresponding Avenue request is:

```
anIntersect = aShape1.Intersects(aShape2)
```

GEOMETRIC EDITING

The call to this Avenue Wrap is:

anIntersect = **avIntersects**(aShape1, aShape2)

avIntersects

GIVEN:	aShape1	= base shape
	aShape2	= second shape to be intersected with the base shape
RETURN:	anIntersect	= intersection flag of the input objects. If: true = shapes intersect, false = they do not

The given and returned variables should be declared where first called as:

```
Dim aShape1 As IGeometry
Dim aShape2 As IGeometry
Dim anIntersect As Boolean
```

8.4.3 Function avReturnIntersection

This function enables the programmer to intersect two shapes to form a new shape. Refer to the commentary at the beginning of this section and to Figures 8-3(A) and 8-3(C) regarding the given shapes and returned shape of this Avenue Wrap.

The corresponding Avenue request is:

NewShape = aShape1.ReturnIntersection(aShape2)

The call to this Avenue Wrap is:

Set NewShape = **avReturnIntersection**(aShape1, aShape2)

avReturn Intersection

GIVEN:	aShape1	= base shape
	aShape2	= second shape to be merged with the base shape
RETURN:	NewShape	= new shape reflecting the intersection

The given and returned variables should be declared where first called as:

```
Dim aShape1 As IGeometry
Dim aShape2 As IGeometry
Dim NewShape As IGeometry
```

GEOMETRIC EDITING

8.4.4 Function avReturnMerged

This function enables the programmer to merge two shapes together to form a new shape. Refer to the commentary at the beginning of this section and to Figures 8-3(A) and 8-3(D) regarding the given shapes and returned shape of this Avenue Wrap.

The corresponding Avenue request is:

```
NewShape = aShape1.ReturnMerged(aShape2)
```

avReturnMerged

The call to this Avenue Wrap is:

Set NewShape = **avReturnMerged**(aShape1, aShape2)

GIVEN:	aShape1	= base shape
	aShape2	= second shape to be merged with the base shape
RETURN:	NewShape	= new shape reflecting the merging

The given and returned variables should be declared where first called as:

```
Dim aShape1 As IGeometry
Dim aShape2 As IGeometry
Dim NewShape As IGeometry
```

8.4.5 Function avReturnUnion

This function enables the programmer to union two shapes together to form a new shape. Refer to the commentary at the beginning of this section and to Figures 8-3(A) and 8-3(B) regarding the given shapes and returned shape of this Avenue Wrap.

The corresponding Avenue request is:

```
NewShape = aShape1.ReturnUnion(aShape2)
```

avReturnUnion

The call to this Avenue Wrap is:

Set NewShape = **avReturnUnion**(aShape1, aShape2)

GIVEN:	aShape1	= base shape
	aShape2	= second shape to be united with the base shape
RETURN:	NewShape	= new shape reflecting the union

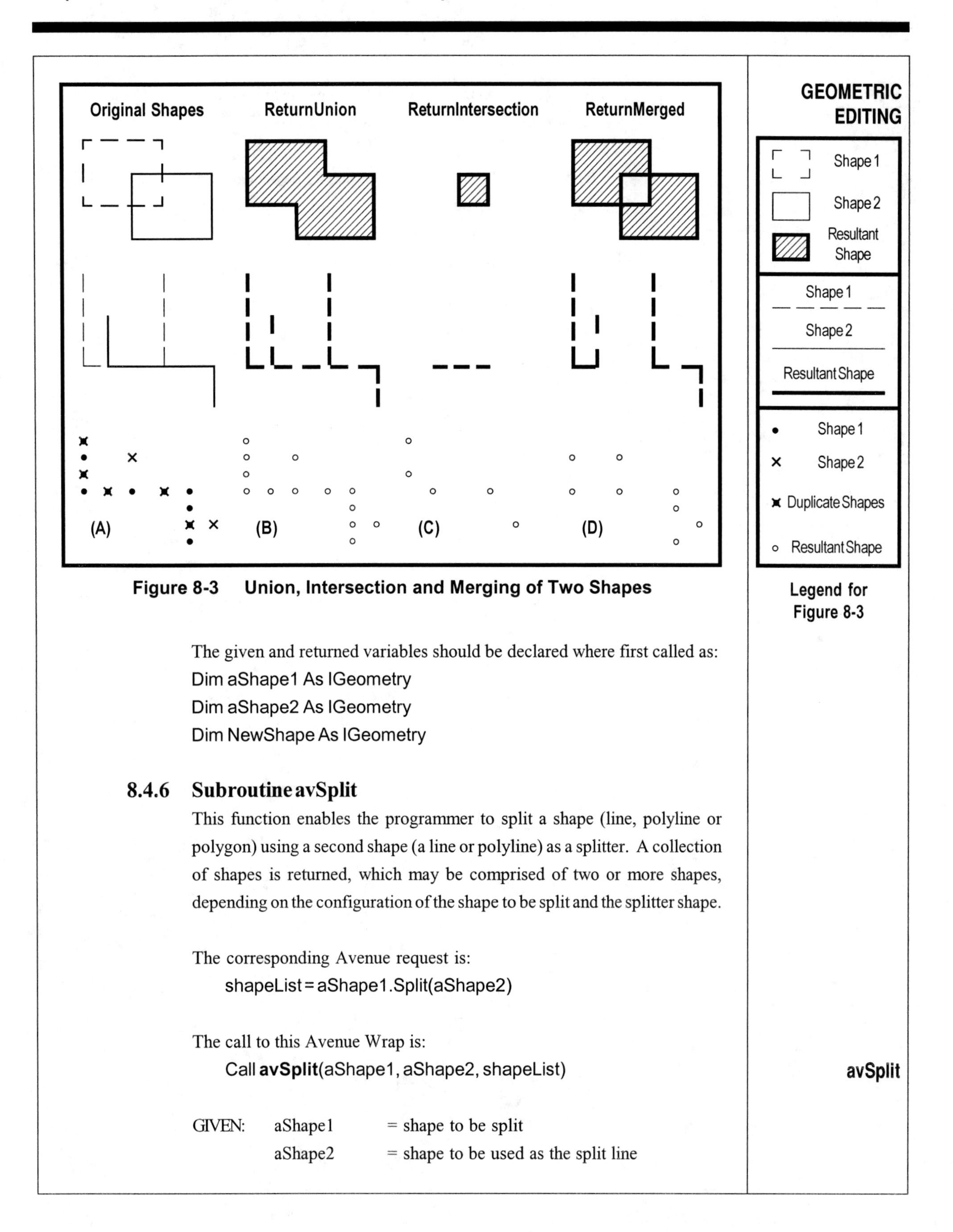

Figure 8-3 Union, Intersection and Merging of Two Shapes

Legend for Figure 8-3

The given and returned variables should be declared where first called as:

```
Dim aShape1 As IGeometry
Dim aShape2 As IGeometry
Dim NewShape As IGeometry
```

8.4.6 Subroutine avSplit

This function enables the programmer to split a shape (line, polyline or polygon) using a second shape (a line or polyline) as a splitter. A collection of shapes is returned, which may be comprised of two or more shapes, depending on the configuration of the shape to be split and the splitter shape.

The corresponding Avenue request is:

```
shapeList = aShape1.Split(aShape2)
```

The call to this Avenue Wrap is:

Call **avSplit**(aShape1, aShape2, shapeList)

avSplit

GIVEN: aShape1 = shape to be split
aShape2 = shape to be used as the split line

GEOMETRIC EDITING

RETURN: shapeList = list of new shapes created as a result of the splitting process

The given and returned variables should be declared where first called as:

```
Dim aShape1 As IGeometry
Dim aShape2 As IGeometry
Dim shapeList As New Collection
```

SAMPLE CODE

8.5 Sample Code for Shape Editing

The example below demonstrates the use of the shape editing Avenue Wraps presented in the previous section. Four sample tests are carried out: (a) splitting of a polygon, (b) merging of two polygons, (c) intersection of two polygons, and (d) union of two polygons. To use the sample code below, do the following:

➢ 1 Create a module with the ArcMap VBA Editor and load or key enter the sample code below.

➢ 2 Go back to ArcMap and, using conventional ArcMap functionality, create seven polygons and and one polyline. In drawing these features, intersect the polyline with the first polygon that is drawn, and draw the other six polygons as three pairs of overlapping polygons.

➢ 3 In drawing the polygons and polyline, following the drawing order shown below:
(a) the polygon which is to be split by a polyline,
(b) the polyline to be used in splitting the polygon,
(c) the two polygons to be merged,
(d) the two polygons to be intersected, and
(e) the two polygons to be United.

➢ 4 Go back to the ArcMap Editor and execute the module with the sample code by clicking at the ▶ tool. If less then seven features were selected, a message will be displayed to this effect and the program will terminate, in which case go back to Step 3 above. If more than seven polygons are selected, only the first seven will be considered. The order of how the features are processed is based upon the order in which they were created. That is why the order of feature creation is important. Upon completion of each of the four tests, a message will be displayed and the resultant shape of the operation that was performed will be highlighted. At the end of the fourth pass, the program will terminate.

➢ 5 If desired, go back to Step 2 above, and repeat the test by modifying the figures that were drawn, and observe the results.

SAMPLE CODE

```
'
' ---
' ---Sample code illustrating how to perform various shape
' ---editing operations.
' ---This sample requires that seven polygon features
' ---and one polyline feature be selected prior to
' ---executing this macro.
' ---The first selected polygon and the selected polyline
' ---features will be used in a split operation.
' ---The remaining selected polygons will be used to
' ---demonstrate the merging, intersecting and uniting
' ---operations.
' ----
'
```

Declaration Statements

```
Dim pMxApp As IMxApplication
Dim pmxDoc As IMxDocument
Dim pActiveView As IActiveView
Dim pMap As IMap
Dim selPG As ISelectionSet
Dim selPL As ISelectionSet
Dim selPGlist As New Collection
Dim selPLlist As New Collection
Dim iOpr As Integer
Dim pFeatPG As IFeature
Dim pFeatPL As IFeature
Dim pGeomPG As IGeometry
Dim pGeomPL As IGeometry
Dim theOpr As String
Dim pFeatPG1 As IFeature
Dim pFeatPG2 As IFeature
Dim pGeomPG1 As IGeometry
Dim pGeomPG2 As IGeometry
Dim shapeList As New Collection
Dim mergedPoly As IGeometry
Dim intrsPoly As IGeometry
Dim unionPoly As IGeometry
Dim i As Long
Dim pg As IGeometry
Dim pCurGraLyr1 As IGraphicsLayer
Dim graPT As IElement
Dim pSymbol As ISymbol
Dim iIntrs As Boolean
'
```

Get the Document and the Polygon Selections

```
' ---Get the active view
Call avGetActiveDoc(pMxApp, pmxDoc, pActiveView, pMap)
'
' ---Get the selected polygons from the theme L_0pg
Call avGetSelection(pmxDoc, "L_0pg", selPG)
```

SAMPLE CODE

```
'
' ---Get the selected polyline from the theme L_0pl
  Call avGetSelection(pmxDoc, "L_0pl", selPL)
'
' ---Get the OIDs for the selection sets
  Call avGetSelectionIDs(selPG, selPGlist)
  Call avGetSelectionIDs(selPL, selPLlist)
'
' ---Note that at least seven polygons and one polyline
' ---must have been selected prior to invoking this
' ---subroutine
  If ((selPGlist.Count > 6) And (selPLlist.Count > 0)) Then
'
'    ---Perform 4 shape editing operations
'    ---Loop 1 splits a polygon (selected polygon #1) using
'    ---the selected polyline as a splitter.
'    ---Loop 2 merges two polygons (selected polygons #2 and
'    ---#3)
'    ---Loop 3 intersects two polygons (selected polygons #4
'    ---and #5)
'    ---Loop 4 unites two polygons (selected polygons #6 and
'    ---#7)
'
     For iOpr = 1 To 4
'
'        ---Get the first selected polygon and the polyline
         If (iOpr = 1) Then
            Call avGetFeature(pmxDoc, "L_0pg", _
                              selPGlist.Item(1), pFeatPG)
            Call avGetFeature(pmxDoc, "L_0pl", _
                              selPLlist.Item(1), pFeatPL)
'           ---Get the geometries of the selected features
            Set pGeomPG = pFeatPG.Shape
            Set pGeomPL = pFeatPL.Shape
'           ---Define the editing operation
            theOpr = "Split"
'
'        ---The other editing operations require two
'        ---polygons, so get the next selected polygons
         Else
            If (iOpr = 2) Then
               Call avGetFeature(pmxDoc, "L_0pg", _
                                 selPGlist.Item(2), pFeatPG1)
               Call avGetFeature(pmxDoc, "L_0pg", _
                                 selPGlist.Item(3), pFeatPG2)
'              ---Define the editing operation
               theOpr = "Merge"
            End If
```

Get the OIDs associated with the selections

Get the Features from the Selected Collections for Splitting

Get the Features from the Selected Collections for Merging

SAMPLE CODE	
Get the Features from the Selected Collections for Intersecting	
Get the Features from the Selected Collections for Unioning	
Loop1 Split a Polygon	
Loop 2 Merge Two Polygons	
Loop 3 Intersect Two Polygons	
Loop 4 Union Two Polygons	

```
            If (iOpr = 3) Then
               Call avGetFeature(pmxDoc, "L_0pg", _
                                 selPGlist.Item(4), pFeatPG1)
               Call avGetFeature(pmxDoc, "L_0pg", _
                                 selPGlist.Item(5), pFeatPG2)
'              ---Define the editing operation
               theOpr = "Intersect"
            End If
            If (iOpr = 4) Then
               Call avGetFeature(pmxDoc, "L_0pg", _
                                 selPGlist.Item(6), pFeatPG1)
               Call avGetFeature(pmxDoc, "L_0pg", _
                                 selPGlist.Item(7), pFeatPG2)
'              ---Define the editing operation
               theOpr = "Union"
            End If
'           ---Get the geometries of the selected features
            Set pGeomPG1 = pFeatPG1.Shape
            Set pGeomPG2 = pFeatPG2.Shape
         End If
'
'        ---Split the polygon using the polyline
         If (iOpr = 1) Then
            Call avSplit(pGeomPG, pGeomPL, shapeList)
         End If
'
'        ---Merge two polygons together (will create a
'        ---hole if the polygons overlap)
         If (iOpr = 2) Then
            Set mergedPoly = avReturnMerged(pGeomPG1, _
                                            pGeomPG2)
            Call CreateList(shapeList)
            shapeList.Add mergedPoly
         End If
'
'        ---Intersect two polygons (returns an empty shape
'        ---if the polygons do not intersect)
         If (iOpr = 3) Then
            Set intrsPoly = avReturnIntersection(pGeomPG1, _
                                                 pGeomPG2)
            Call CreateList(shapeList)
            shapeList.Add intrsPoly
         End If
'
'        ---Union two polygons
         If (iOpr = 4) Then
            Set unionPoly = avReturnUnion(pGeomPG1, _
                                          pGeomPG2)
```

SAMPLE CODE

```
            Call CreateList(shapeList)
            shapeList.Add unionPoly
         End If
'
'        ---Check if any new polygons were created.  If so
'        ---cycle thru them, and display each new polygon
'        ---in red to indicat the shape of the new polygon.
'        ---In addition, display in a message box the area
'        ---of the polygon and the operation that was
'        ---performed.
         If (shapeList.Count > 0) Then
            For i = 1 To shapeList.Count
'               ---Grab a polygon from the list
                Set pg = shapeList.Item(i)
'               ---Set the current active graphics layers as
'               ---the basic graphics layer
                Call avSetGraphicsLayer(Null, pCurGraLyr1)
'               ---Create a graphic polygon using the new
'               ---polygon as its shape and assign a red
'               ---fill to it
                Set graPT = avGraphicShapeMake("FILL", pg)
                Set pSymbol = avSymbolMake("FILL")
                Call avSymbolSetColor("FILL", pSymbol, _
                                      "RED")
                Call avGraphicSetSymbol("FILL", graPT, _
                                        pSymbol)
'               ---Add the graphic to the display
                Call avViewAddGraphic(graPT)
'               ---Display in a message box the area of the
'               ---new polygon and the operation that was
'               ---performed
                If (iOpr <> 3) Then
                   MsgBox theOpr + " operation" + Chr(13) + _
                        "Polygon " + CStr(i) + " Area = " + _
                        CStr(avReturnArea(pg))
                Else
'                  ---Determine if the polygons intersect
                   iIntrs = avIntersects(pGeomPG1, pGeomPG2)
'                  ---In addition to the usual information,
'                  ---inform user whether the polygons
'                  ---intersect each other or not
                   MsgBox theOpr + " operation" + Chr(13) + _
                       "Polygon " + CStr(i) + " Area = " + _
                       CStr(avReturnArea(pg)) + Chr(13) + _
                       "Intersection = " + CStr(iIntrs)
                End If
'               ---Get rid of the graphic
                Call avRemoveGraphic(graPT)
            Next
```

Inform user as to the results of the editing process

SAMPLE CODE

```
'           ---Handle the case when and operation does not
'           ---produce new polygons
            Else
               MsgBox theOpr + " produced no new shapes"
            End If
        Next
'
'   ---Handle the case when not enough features selected for
'   ---the various editing operations to be performed
    Else
        MsgBox "Not enough features selected"
    End If
'
```

CHAPTER 9

USER DOCUMENT INTERACTION

This chapter presents instructions on how to create a tool bar and place therein various tools. The code presented, herein, pertains to the establishment of five different types of tools, which enable the programmer to interactively extract from the user: a point, a polyline, a circle, a polygon or a rectangle. Using these examples, the programmer may then develop other custom tools. Examples 9.2.2, 9.2.3, 9.2.4 and 9.2.5, presuppose that the subroutine MyCommand, is to act upon the geometry that is established by the user. The name of this subroutine can be changed, if desired.

9.1 Creating a Tool Bar and Tool

The instructions below illustrate how to create a tool bar and place therein a tool icon. These instructions will be common to all of the examples presented subsequently.

➤ 1 Invoke ArcMap and click at the Tools menu at the top of the screen and then at the Customize sub-menu (see Figure 9-1), at which time the Customize window of Figure 9-2(A) is displayed.

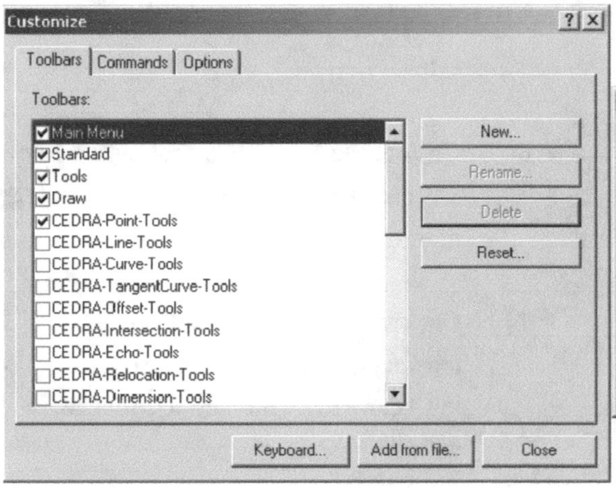

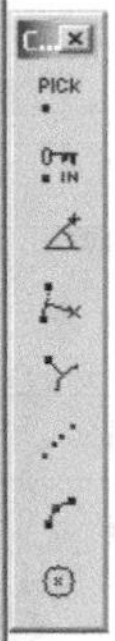

Figure 9-1

Figure 9-2(A) Customize Window

9-2 (B)

➢ 2 If the Toolbars tab is not active, click at it to make it active as shown in Figure 9-2(A). At the left side of the Customize window, one will note squares, of which the top five have been checked. Note that upon execution of Step 2 above, only the top four squares will be checked.

At this time the programmer has the option to either create a new tool bar, or to display an existing tool bar, and in either case, add a new tool. At this time we will consider the case of adding a new tool to an existing toolbar.

Add tool to an existing tool bar

➢ 3 Check the fifth square (CEDRA-Point-Tools or any other desired square) at which time the corresponding tool bar will be displayed somewhere within the Customize window. Drag it by its top off said window, perhaps to the location shown in Figure 9-2(B).

➢ 4 Click at the Commands tab of the Customize window. Two sub-windows, Categories and Commands, are now displayed within the Customize window as shown in Figure 9-3.

➢ 5 Scroll down on the Categories (left) sub-window and click at the UIControls category.

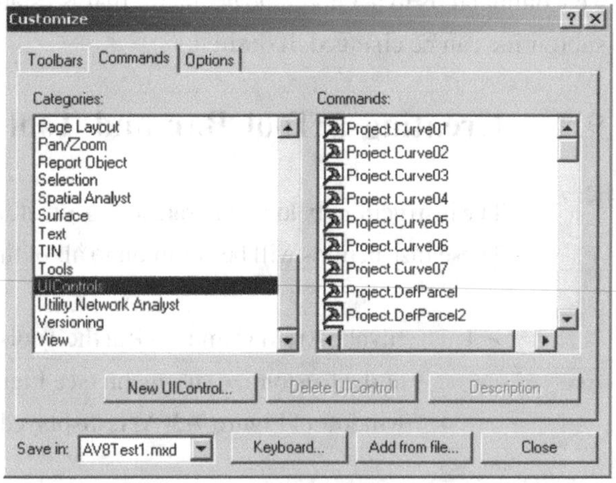

Figure 9-3 Commands Sub-Windows

➢ 6 Click at the New UIControl button below the left sub-window. The New UIControl window of Figure 9-4 is now displayed.

➢ 7 Click at the UIToolControl radio button, and then click at the Create button below it. The New UIControl window now disappears.

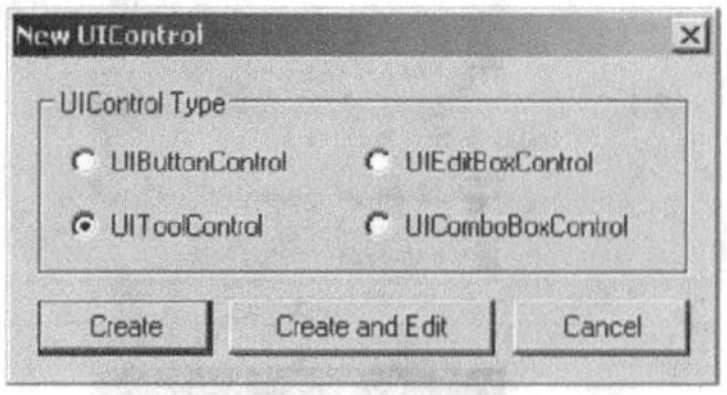

Figure 9-4 New UIControl Window

➢ 8 Scroll down on the Commands (right) sub-window, see Figure 9-5(A). The new tool, Project.UIToolControl2, is shown.

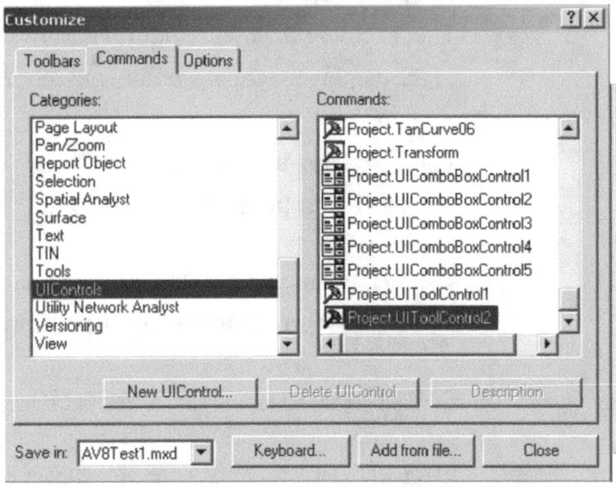

➢ 9 Click at the icon of the new tool (hammer), hold the mouse button down, drag the icon to the tool bar, then release the mouse button. The tool is now added to the tool bar. You may now relocate it by dragging it within the tool bar, see Figure 9-5B. "Right-clicking" on top of a tool enables the user to change the icon that is associated with the tool.

Figure 9-5(A) New Tool Added

9-5 (B)

Create a new tool bar and add a tool

We will now create a new tool bar, and then add a new tool therein.

➢ 10 Click the Toolbars tab to display the Customize window of Figure 9-2, and click the New button in the upper right corner of the window. The New Toolbar window is displayed, see Figure 9-6.

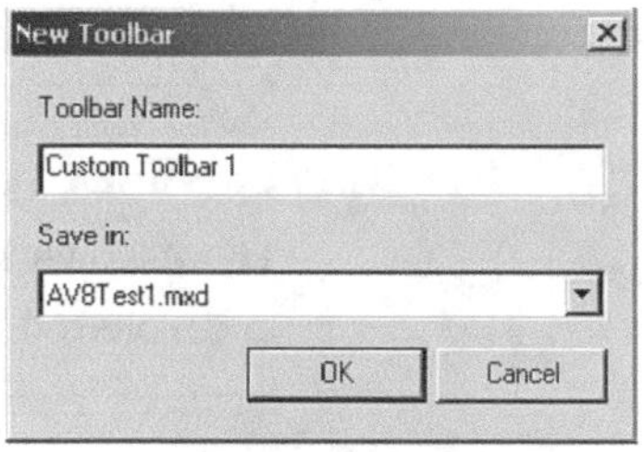

Figure 9-6 New Toolbar Window

➢ 11 Key enter the desired new tool bar name, and scroll down in the "Save in" combo field to save the new tool bar in the desired location.

➢ 12 Click the OK button to create the tool bar, see Figure 9-7.

Figure 9-7 New Tool Bar

➢ 13 Scroll down in the Customize window of Figure 9-2 to see the new tool bar. The new tool bar should be at the bottom of the list of toolbars.

➢ 14 Carry out Steps 4 through 9 above to introduce a new tool in the new tool bar.

DOCUMENT INTERACTION

9.2 Sample Document Interaction

Point Creation

9.2.1 Point Creation

The statements below provide for the user interaction by clicking at a location in the display to return a point, and retrieve its X and Y coordinates.

The corresponding Avenue request is:

```
pt = aDisplay.ReturnUserPoint
```

The VB statement is:

```
Set pt = pmxDoc.CurrentLocation
```

GIVEN: pmxDoc = the IMxDocument interface for the active view

RETURN: pt = the IPoint object

To get the X and Y coordinates, enter the following statements where needed:

```
theX = pt.X
theY = pt.Y
```

The given and returned variables should be declared where first called as:

```
Dim pmxDoc As IMxDocument
Dim pt As IPoint
```

Polyline Creation

9.2.2 Polyline Creation

The statements below provide for the user interaction by clicking at a location to denote the beginning of a polyline, and then successively picking at locations to denote the various vertices of the polyline. Double clicking at a location terminates the polyline.

The corresponding Avenue request is:

```
thePolyline = aDisplay.ReturnUserPolyline
```

The ArcObjects Equivalent is accomplished by creating a UIToolControl, as indicated in Section 9.1, and writing code for the tool's MouseDown event.

The statements below for this event assume that a UIToolControl called MyTool. If the name of the tool is not MyTool, substitute the appropriate name. These statements allow the user to define a polyline as stated in the introduction to this section above.

```
Private Sub MyTool_MouseDown(ByVal Button As Long, _
            ByVal Shift As Long, ByVal X As Long, ByVal Y As Long)
'
   Dim pmxDoc As IMxDocument
   Dim pRubberLine As IRubberBand
   Dim pLine As IPolyline
   Dim pElem As IElement
'
'  ---QI for the MXDocument interface
   Set pmxDoc = ThisDocument
'
'  ---Create a new RubberLine
   Set pRubberLine = New RubberLine
'
'  ---Get a Line from the tracker object using TrackNew
   Set pLine = pRubberLine.TrackNew(pmxDoc.ActiveView. _
                                       ScreenDisplay, Nothing)
'
'  ---Make sure we have a valid line
   If Not pLine Is Nothing Then
'
'     ---Create a new LineElement and set its Geometry
      Set pElem = New LineElement
      pElem.Geometry = pLine
'
'     ---Process the polyline now that it has been defined
      Call MyCommand(pElem)
   End If
End Sub
```

9.2.3 Circle Creation

This subroutine enables the programmer to create a circle by moving the cursor to the location of the circle's center point, and depressing the mouse button. Then, holding down the mouse button, moving the mouse until the desired radius is reached. Releasing the mouse button at that time will complete the operation and create a new circle (CircleElement).

The corresponding Avenue request is:

theCircle = aDisplay.ReturnUserCircle

The ArcObjects Equivalent is accomplished by creating a UIToolControl, as indicated in Section 9.1, and writing code for the tool's MouseDown event.

DOCUMENT INTERACTION

The statements below for this event assume that a UIToolControl called MyTool. If the name of the tool is not MyTool, substitute the appropriate name. These statements allow the user to define a circle as stated in the introduction to this section above.

Circle Creation

```
Private Sub MyTool_MouseDown(ByVal Button As Long, _
            ByVal Shift As Long, ByVal X As Long, ByVal Y As Long)
'
   Dim pmxDoc As IMxDocument
   Dim pRubberCirc As IRubberBand
   Dim pCircArc As ICircularArc
   Dim pElem As IElement
'
'  ---QI for the MXDocument interface
   Set pmxDoc = ThisDocument
'
'  ---Create a new RubberCircle
   Set pRubberCirc = New RubberCircle
'
'  ---Get a Circle from the tracker object using TrackNew
   Set pCircArc = pRubberCirc.TrackNew(pMXDoc.ActiveView. _
                              ScreenDisplay,Nothing)
'
'  ---Make sure we have a valid circle
   If Not pCircArc Is Nothing Then
'
'     ---Create a new CircleElement and set its Geometry
      Set pElem = New CircleElement
      pElem.Geometry = pCircArc
'
'     ---Process the circle now that it has been defined
      Call MyCommand(pElem)
   End If
End Sub
```

9.2.4 Polygon Creation

This subroutine enables the programmer to create a polygon by making a series of picks in the display, terminating the polygon definition by double-clicking the last point in the polygon.

The corresponding Avenue request is:

thePolygon = aDisplay.ReturnUserPolygon

DOCUMENT INTERACTION

The ArcObjects Equivalent is accomplished by creating a UIToolControl, as indicated in Section 9.1, and writing code for the tool's MouseDown event.

The statements below for this event assume that a UIToolControl called MyTool. If the name of the tool is not MyTool, substitute the appropriate name. These statements allow the user to define a polygon as stated in the introduction to this section above.

Polygon Creation

```
Private Sub MyTool_MouseDown(ByVal Button As Long, _
            ByVal Shift As Long, ByVal X As Long, ByVal Y As Long)
'
   Dim pmxDoc As IMxDocument
   Dim pRubberPoly As IRubberBand
   Dim pPoly As IPolygon
   Dim pElem As IElement
'
'  ---QI for the MXDocument interface
   Set pmxDoc = ThisDocument
'
'  ---Create a new RubberPolygon
   Set pRubberPoly = New RubberPolygon
'
'  ---Get a Polygon from the tracker object using TrackNew
   Set pPoly = pRubberPoly.TrackNew(pmxDoc.ActiveView. _
                                 ScreenDisplay, Nothing)
'
'  ---Make sure we have a valid polygon
   If Not pPoly Is Nothing Then
'
'     ---Create a new PolygonElement and set its Geometry
      Set pElem = New PolygonElement
      pElem.Geometry = pPoly
'
'     ---Process the polygon now that it has been defined
      Call MyCommand(pElem)
   End If
End Sub
```

9.2.5 Point or Rectangle Creation

This subroutine enables the programmer to create either (a) a point by making a single pick, or (b) a rectangle by picking at a corner of the rectangle, holding the mouse button down (MouseDown event will execute), and then dragging

DOCUMENT INTERACTION

the cursor to the location of the diagonally opposite corner (MouseMove event executes) of the rectangle, at which point the mouse button is released (MouseUp event executes).

The corresponding Avenue request is:

```
theRect = aDisplay.ReturnUserRect
```

Point or Rectangle Creation

The ArcObjects Equivalent is accomplished by creating a UIToolControl and writing code for the tool's MouseDown, MouseMove and MouseUp events. The global variables shown below will be utilized in the implementation.

```
Public ugpPoint As IPoint
Public ugpRect As IEnvelope
Public ugbIsMouseDown As Boolean
Public ugpFeedbackEnv As INewEnvelopeFeedback
```

The code below, for these events, assumes that a UIToolControl called MyTool has been created by the user as indicated in Section 9.1 and that the subroutine MyCommand will process the point or rectangle that the user defines.

The global variables ugpPoint and ugpRect store the position of a single pick or the rectangle, respectively. When the mouse button is first depressed, ugpPoint is defined, storing the location where the cursor was when the button was depressed. The second variable ugpRect is initialized to be NOTHING thereby signifying that a rectangle has not been defined. As the mouse is moved, with the button depressed, the position is updated and stored in ugpPoint. Once the mouse button is released, the MouseUp event checks if a rectangle was indeed defined, and if so stores the rectangle as an IEnvelope object in ugpRect. In this case, ugpPoint is re-initialized to NOTHING thereby indicating that a rectangle, and not a point, was defined by the user.

It is in the MouseUp event where the user can call a subroutine or function (in this example, MyCommand) to act upon the point or rectangle that was defined using ugpPoint or ugpRect, depending upon which variable is not set to NOTHING.

```
Private Sub MyTool_MouseDown(ByVal Button As Long, _
            ByVal Shift As Long, ByVal X As Long, ByVal Y As Long)
'
   Dim pmxDoc As IMxDocument
   Dim pActiveView As IActiveView
'
'  ---Get the ActiveView for the map
   Set pmxDoc = Application.Document
   Set pActiveView = pmxDoc.FocusMap
'
'  ---Store current point, set mousedown flag
   Set ugpPoint = pActiveView.ScreenDisplay. _
                  DisplayTransformation.ToMapPoint(X, Y)
   ugbIsMouseDown = True
'
'  ---Initialize the rectangle variable
   Set ugpRect = Nothing
End Sub


Private Sub MyTool_MouseMove(ByVal Button As Long, _
            ByVal Shift As Long, ByVal X As Long, ByVal Y As Long)
'
   Dim pmxDoc As IMxDocument
   Dim pActiveView As IActiveView
'
'  ---Handle any errors that may occur
   On Error GoTo Errorhandler
'
   If Not ugbIsMouseDown Then Exit Sub
'
'  ---Get the ActiveView for the map
   Set pmxDoc = Application.Document
   Set pActiveView = pmxDoc.FocusMap
'
'  ---Create a rubber banding box, if it has not been
'  ---created already
   If (ugpFeedbackEnv Is Nothing) Then
      Set ugpFeedbackEnv = New NewEnvelopeFeedback
      Set ugpFeedbackEnv.Display = pActiveView.ScreenDisplay
      ugpFeedbackEnv.Start ugpPoint
   End If
```

DOCUMENT INTERACTION

```
'
'  ---Store current point, and use to move rubberband
   Set ugpPoint = pActiveView.ScreenDisplay. _
                  DisplayTransformation.ToMapPoint(X, Y)
   ugpFeedbackEnv.MoveTo ugpPoint
'
   Exit Sub
'
'  ---Handle any errors that were detected
Errorhandler:
'
'  ---Display the detected error
   MsgBox "An error has occured within MyTool." & vbCr & _
         vbCr & "Error Details : " & Err.Description, _
         vbExclamation + vbOKOnly, "Error"
End Sub


Private Sub MyTool_MouseUp(ByVal Button As Long, _
      ByVal Shift As Long, ByVal X As Long, ByVal Y As Long)
'
   Dim pActiveView As IActiveView
   Dim pmxDoc As IMxDocument
'
'  ---Get the ActiveView for the map
   Set pmxDoc = Application.Document
   Set pActiveView = pmxDoc.FocusMap
'
'  ---Check If user dragged an envelope (created a rectangle)
   If (Not ugpFeedbackEnv Is Nothing) Then
'     ---Store the rectangle that the user has defined
      Set ugpRect = ugpFeedbackEnv.Stop
'     ---Initialize the point pick variable
      Set ugpPoint = Nothing
   End If
'
'  ---Reset rubberband and mousedown state
   Set ugpFeedbackEnv = Nothing
   ugbIsMouseDown = False
'
'  ---Execute the appropriate command, whatever it may be
   Call MyCommand
End Sub
```

CHAPTER 10

GRAPHICS AND SYMBOLS AVENUE WRAPS

This chapter contains Avenue Wraps that enable the user to (a) create and delete user defined graphics layers (annotation targets), (b) empty a graphics layer, and (c) introduce user defined graphics to a graphics layer. User defined graphics may include points, lines, polygons, symbols, text strings, and the like. In introducing such graphics, the programmer has the ability to set the symbology of the graphic including size, color, angle of inclination (orientation), text font and style, and other properties.

These Avenue Wraps include the following:

The source listing of each of the above Avenue Wraps may be found in Appendix D of this publication.

10.1 General Graphics Avenue Wraps

GENERAL GRAPHICS

10.1.1 Subroutine avGetGraphicList

This subroutine enables the programmer to get a collection of the graphics in a graphics layer.

The corresponding Avenue request is:

There is no corresponding Avenue request.

The call to this Avenue Wrap is:

Call **avGetGraphicList**(pCurGraLyr, graList)

avGetGraphicList

GIVEN: pCurGraLyr = the graphics layer containing the user programmed graphics

RETURN: graList = list of graphic elements in the graphics layer

The given and returned variables should be declared where first called as:

```
Dim pCurGraLyr As IGraphicsLayer
Dim graList As New Collection
```

10.1.2 Subroutine avGetSelected

This subroutine enables the programmer to get a collection of the selected graphics in the view.

The corresponding Avenue request is:

graList = aGraphicList.GetSelected

The call to this Avenue Wrap is:

Call **avGetSelected**(pmxDoc, graList)

avGetSelected

GIVEN: pmxDoc = the active view

RETURN: graList = list of selected graphic elements in the view

The given and returned variables should be declared where first called as:

```
Dim pmxDoc As IMxDocument
Dim graList As New Collection
```

GENERAL GRAPHICS

10.1.3 Subroutine avGraphicInvalidate

This subroutine enables the programmer to redraw or update a graphic element. Use this subroutine to redraw a graphic that has been added to a graphics layer and subsequently modified.

The corresponding Avenue request is:

aGraphic.Invalidate

avGraphic Invalidate

The call to this Avenue Wrap is:

Call **avGraphicInvalidate**(pElement)

GIVEN: pElement = graphic to be redrawn due to a change that has been made to it

RETURN: nothing

The given and returned variables should be declared where first called as:

Dim pElement As IElement

10.1.4 Subroutine avGraphicListDelete

This subroutine enables the programmer to delete a graphics layer. Note that the basic graphics layer can not be deleted. In addition, note that:

- Only user created graphics layers can be deleted with this Avenue Wrap.
- To delete the graphics without deleting the graphics layer, use the avGraphicListEmpty

The corresponding Avenue request is:

There is no corresponding Avenue request.

avGraphicList Delete

The call to this Avenue Wrap is:

Call **avGraphicListDelete**(pCurGraLyr)

GIVEN: pCurGraLyr = the graphics layer containing the user programmed graphics

RETURN: nothing

The given and returned variables should be declared where first called as:

Dim pCurGraLyr As IGraphicsLayer

10.1.5 Subroutine avGraphicListEmpty

GENERAL GRAPHICS

This subroutine enables the programmer to delete all graphics from a graphics layer. Note that the graphics within the graphics layer are deleted, but the graphics layer itself is not deleted, so that graphics can be added to the graphics layer at another time. To delete the graphics layer itself, use theavGraphicListDelete Avenue Wrap.

The corresponding Avenue request is:

There is no corresponding Avenue request.

The call to this Avenue Wrap is:

Call **avGraphicListEmpty**(pCurGraLyr)

avGraphicList Empty

GIVEN: pCurGraLyr = the graphics layer containing the user programmed graphics

RETURN: nothing

The given and returned variables should be declared where first called as:

Dim pCurGraLyr As IGraphicsLayer

10.1.6 Subroutine avRemoveGraphic

This subroutine enables the programmer to delete a graphic element from the display. Note that in Avenue, it is required to specify the graphic list to which the element to be removed belongs. In VB code you only specify the element to be removed; you need not specify the graphics layer.

The corresponding Avenue request is:

aGraphicList.RemoveGraphic (pElement)

The call to this Avenue Wrap is:

Call **avRemoveGraphic**(pElement)

avRemove Graphic

GIVEN: pElement = graphic to be deleted

RETURN: nothing pElement is removed from the display

The given and returned variables should be declared where first called as:

pElement As IElement

GENERAL GRAPHICS

10.1.7 Subroutine avSetGraphicsLayer

This subroutine enables the programmer to set the current graphics layer, or to create a new user defined graphics layer (annotation target). In using this subroutine note that:

1. If the specified graphics layer theGLayer name exists, it will not be deleted, but rather, will become the current graphics layer, and a new graphics layer will not be created. Thus any graphics generated will be added to the existing layer.
2. If NULL is specified for the given argument theGLayer, it will indicate that the subsequently created graphic will be placed in the basic graphics layer.

The corresponding Avenue request is:

There is no corresponding Avenue request.

avSetGraphics Layer

The call to this Avenue Wrap is:

Call **avSetGraphicsLayer**(theGLayer, pCurGraLyr)

GIVEN: theGLayer = graphics layer to contain the subsequently created graphics (see notes 1 and 2 above)

RETURN: pCurGraLyr = graphics layer that will contain the user programmed graphics

The given and returned variables should be declared where first called as:

```
Dim theGLayer As Variant
Dim pCurGraLyr As IGraphicsLayer
```

GENERAL GRAPHICS

10.1.8 Subroutine avViewAddGraphic

This subroutine enables the programmer to add a graphic into the current active graphics layer. In using this subroutine, note the following:

1. Use the subroutine avSetGraphicsLayer Avenue Wrap to set the active graphics layer, be it the basic graphics layer or a user defined layer (annotation target layer).
2. Remember that upon refreshing the view display, all graphics in the basic graphics layer will be lost. Hence, if any graphics have to be preserved, they should be assigned to a user defined layer, not the basic graphics layer.

The corresponding Avenue request is:

There is no corresponding Avenue request.

The call to this Avenue Wrap is:

Call **avViewAddGraphic**(pElement)

avView AddGraphic

GIVEN: pElement = graphic to be added

RETURN: nothing

The given and returned variables should be declared where first called as:

Dim pElement As IElement

10.1.9 Subroutine avViewGetGraphics

This subroutine enables the programmer to get a list of all graphics in the map.

The corresponding Avenue request is:

graList = aView.GetGraphics

The call to this Avenue Wrap is:

Call **avViewGetGraphics**(graList)

avView GetGraphics

GIVEN: nothing

RETURN: graList = list of all graphic elements in the map

The given and returned variables should be declared where first called as:

Dim graList As New Collection

SYMBOL PROPERTIES

10.2 Graphic Attribute Assignment Avenue Wraps

10.2.1 Function avGraphicGetShape

This function enables the programmer to get the geometry that is associated with a graphic. This subroutine will process PEN, MARKER, FILL and TEXT symbols since they all share the IElement interface

The corresponding Avenue request is:

```
theShape = pElement.GetShape
```

avGraphic GetShape

The call to this Avenue Wrap is:

Set theShape = **avGraphicGetShape**(pElement)

GIVEN: pElement = the graphic to be processed

RETURN: theShape = geometry describing the graphic

The given and returned variables should be declared where first called as:

```
Dim pElement As IElement
Dim theShape As IGeometry
```

10.2.2 Function avGraphicGetSymbol

This function enables the programmer to get the symbol which is associated with a graphic.

The corresponding Avenue request is:

```
theSymbol = pElement.GetSymbol
```

avGraphic GetSymbol

The call to this Avenue Wrap is:

Set theSymbol = **avGraphicGetSymbol**(aSymTyp, pElement)

GIVEN: aSymTyp = type of graphic to be processed
- PEN: line symbol
- MARKER: point symbol
- FILL: polygon symbol

pElement = graphic element of which the symbol is desired

RETURN: theSymbol = symbol describing the graphic element's properties such as its color

The given and returned variables should be declared where first called as:

```
Dim aSymTyp As String
Dim pElement As IElement
Dim theSymbol As ISymbol
```

SYMBOL PROPERTIES

10.2.3 Subroutine avGraphicSetShape

This subroutine enables the programmer to set the geometry of a new graphic, or to modify that of an existing graphic.

The corresponding Avenue request is:

```
pElement.SetShape (theGeom)
```

The call to this Avenue Wrap is:

Call **avGraphicSetShape**(pElement, theGeom)

avGraphic SetShape

GIVEN: pElement = the graphic to be modified
theGeom = the new geometry that describes the graphic

RETURN: nothing The geometry of pElement is set, if it is a new, or modified, if it is an existing graphic.

The given and returned variables should be declared where first called as:

```
Dim pElement As IElement
Dim theGeom As IGeometry
```

10.2.4 Subroutine avGraphicSetSymbol

This subroutine enables the programmer to set a symbol to a graphic.

The corresponding Avenue request is:

```
pElement.SetSymbol (pSymbol)
```

The call to this Avenue Wrap is:

Call **avGraphicSetSymbol**(aSymTyp, pElement, pSymbol)

avGraphic SetSymbol

GIVEN: aSymTyp = type of graphic to be assigned
PEN: line symbol
MARKER: point symbol
FILL: polygon symbol
pElement = graphic for which the given symbol is to be assigned to
pSymbol = symbol to be assigned to the graphic

RETURN: nothing

SYMBOL PROPERTIES

The given and returned variables should be declared where first called as:
Dim aSymTyp As String
Dim pElement As IElement
Dim pSymbol As ISymbol

10.2.5 Function avGraphicShapeMake

This function enables the programmer to create a graphic shape that can be added to the graphics collection.

The corresponding Avenue request is:

theShape = GraphicShape.Make (aShape)

avGraphic ShapeMake

The call to this Avenue Wrap is:

Set theShape = **avGraphicShapeMake**(aSymTyp, theGeom)

GIVEN: aSymTyp = type of graphic to be assigned
PEN : line symbol
MARKER : point symbol
FILL : polygon symbol
theGeom = geometry describing the graphic

RETURN: theShape = graphic that can be added to the graphics layer

The given and returned variables should be declared where first called as:
Dim aSymTyp As String
Dim theGeom As IGeometry
Dim theShape As IElement

10.2.6 Function avSymbolGetAngle

This function enables the programmer to get the angle of orientation of a graphic symbol. Note that this function processes only MARKER symbols; the PEN and FILL symbols result in a value of zero for theAngle.

The corresponding Avenue request is:

theAngle = aBasicMarker.GetAngle

avSymbol GetAngle

The call to this Avenue Wrap is:

theAngle = **avSymbolGetAngle**(aSymTyp, pSymbol)

GIVEN: aSymTyp = type of symbol to be processed
pSymbol = symbol to be processed

RETURN: theAngle = angle assigned to symbol (degrees)

SYMBOL PROPERTIES

The given and returned variables should be declared where first called as:
Dim aSymTyp As String
Dim pSymbol As ISymbol
Dim theAngle As Double

10.2.7 Function avSymbolGetColor

This function enables the programmer to get the color assigned to a symbol. Since it is possible for avSymbolGetColor to produce a result of NOTHING, because it is possible for a polygon fill to have no color, it is important for the programmer to check for this condition before using the result.

The corresponding Avenue request is:
theColor = pSymbol.GetColor

The call to this Avenue Wrap is:
Set theColor = **avSymbolGetColor**(aSymTyp, pSymbol)

avSymbol GetColor

GIVEN: aSymTyp = type of symbol to be processed
PEN: line symbol
MARKER: point symbol
FILL: polygon symbol
pSymbol = symbol to be processed

RETURN: theColor = color assigned to the symbol (pSymbol)

The given and returned variables should be declared where first called as:
Dim aSymTyp As String
Dim pSymbol As ISymbol
Dim theColor As iColor

10.2.8 Function avSymbolGetOLColor

This function enables the programmer to get the outline color assigned to a graphic symbol. Since it is possible for avSymbolGetOLColor to produce a result of NOTHING, because it is possible for a graphic outline to have no color, it is important for the programmer to check for this condition before using the result.

The corresponding Avenue request is:
theOLColor = pSymbol.GetOLColor

The call to this Avenue Wrap is:
Set theOLColor = **avSymbolGetOLColor**(aSymTyp, pSymbol)

avSymbolGet OLColor

SYMBOL PROPERTIES

GIVEN: aSymTyp = type of symbol to be processed
PEN: line symbol
MARKER: point symbol
FILL: polygon symbol
pSymbol = symbol to be processed

RETURN: theOLColor = outline color assigned to the symbol (pSymbol)

The given and returned variables should be declared where first called as:

```
Dim aSymTyp As String
Dim pSymbol As ISymbol
Dim theOLColor As iColor
```

10.2.9 Function avSymbolGetOLWidth

This function enables the programmer to get the outline width assigned to a graphic symbol. In using this function, note that this function returns for:

1. PEN symbols, the width of the symbol itself, not that of the outline.
2. MARKER symbols, if the outline is to be drawn, the outline size.
3. MARKER symbols, if the outline is not to be drawn, the size of the marker.
4. FILL symbols, the outline width of the symbol.

The corresponding Avenue request is:

```
theOLWidth = pSymbol.GetOLWidth
```

The call to this Avenue Wrap is:

avSymbolGet OLWidth

theOLWidth **= avSymbolGetOLWidth**(aSymTyp, pSymbol)

GIVEN: aSymTyp = type of symbol to be processed
PEN: line symbol
MARKER: point symbol
FILL: polygon symbol
pSymbol = symbol to be processed

RETURN: theOLWidth = outline color assigned to the symbol (pSymbol)

The given and returned variables should be declared where first called as:

```
Dim aSymTyp As String
Dim pSymbol As ISymbol
Dim theOLWidth As Double
```

SYMBOL PROPERTIES

10.2.10 Function avSymbolGetSize

This function enables the programmer to get the size assigned to a graphic symbol.

The corresponding Avenue request is:

```
theSize = pSymbol.GetSize
```

The call to this Avenue Wrap is:

avSymbolGetSize

```
theSize = avSymbolGetSize(aSymTyp, pSymbol)
```

GIVEN:	aSymTyp	= type of symbol to be processed
		PEN: line symbol
		MARKER: point symbol
		FILL: polygon symbol
	pSymbol	= symbol to be processed
RETURN:	theSize	= size assigned to the symbol (pSymbol)

The given and returned variables should be declared where first called as:

```
Dim aSymTyp As String
Dim pSymbol As ISymbol
Dim theSize As Double
```

10.2.11 Function avSymbolGetStipple

This function enables the programmer to get the stipple assigned to a graphic symbol. In using this function, note the following:

1. Since there is no direct correlation between the Avenue request Stipple and an ArcObject method or property, the avSymbolGetStipple Avenue Wrap returns an object of type IMultiLayerFillSymbol, provided the symbol is of that type; otherwise, NOTHING is passed back
2. This function pertains only to FILL type symbols, and not to PEN or MARKER

The corresponding Avenue request is:

```
theStipple = pSymbol.GetStipple
```

(See Note 1 above)

The call to this Avenue Wrap is:

avSymbolGetStipple

```
Set theStipple = avSymbolGetStipple(aSymTyp, pSymbol)
```

GIVEN:	aSymTyp	= symbol to be processed (see Note 2 above)
	pSymbol	= symbol to be processed

SYMBOL PROPERTIES

RETURN: theStipple = see the notes above

The given and returned variables should be declared where first called as:

```
Dim aSymTyp As String
Dim pSymbol As ISymbol
Dim theStipple As IMultiLayerFillSymbol
```

10.2.12 Function avSymbolGetStyle

This function enables the programmer to get the style assigned to a graphic symbol. In using this function, note that the numeric values to be returned correspond to the symbol types as indicated in Table 10-1.

The corresponding Avenue request is:

```
theStyle = aBasicMarker.GetStyle
```

avSymbol GetStyle

The call to this Avenue Wrap is:

theStyle **= avSymbolGetStyle**(aSymTyp, pSymbol)

GIVEN: aSymTyp = symbol word to be processed (see Table 10-1)
pSymbol = symbol to be processed

RETURN: theStyle = numeric symbol code (see Table 10-1)

The given and returned variables should be declared where first called as:

```
Dim aSymTyp As String
Dim pSymbol As ISymbol
Dim theStyle As Variant
```

10.2.13 Function avSymbolMake

This function enables the programmer to create a new graphic symbol.

The corresponding Avenue request is:

```
theSymbol = Symbol.Make (aSymTyp)
```

avSymbolMake

The call to this Avenue Wrap is:

Set theSymbol **= avSymbolMake**(aSymTyp)

GIVEN: aSymTyp = type of graphic to be created

PEN:	line symbol
MARKER:	point symbol
FILL:	polygon symbol

SYMBOL PROPERTIES

TABLE 10-1 SYMBOL TYPE CODE NUMBERS FOR THE INDICATED SYMBOL TYPE KEY WORDS

PEN symbol key word	**MARKER** symbol key word	**FILL** symbol key word
0 : Solid	0 : Circle	0 : Solid
1 : Dashed	1 : Square	1 : Empty
2 : Dotted	2 : Cross	2 : Horizontal hatch
3 : dashes & dots	3 : X	3 : Vertical hatch
4 : dashes & double dots	4 : Diamond	4 : 45° left-to-right hatch
5 : Is invisible		5 : 45° left-to-right hatch
6 : Fit into bounding rectangle		6 : Horz. and vert. crosshatch
		7 : 45° crosshatch

RETURN: theSymbol = symbol describing a graphic that can be added to the graphics layer

The given and returned variables should be declared where first called as:

Dim aSymTyp As String
Dim theSymbol As ISymbol

10.2.14 Subroutine avSymbolSetAngle

This subroutine enables the programmer to set the angle of a symbol.

The corresponding Avenue request is:

aBasicMarker.SetAngle (aAngle)

The call to this Avenue Wrap is:

Call **avSymbolSetAngle**(aSymTyp, pSymbol, aAngle)

avSymbolSet Angle

GIVEN: aSymTyp = type of symbol to be processed
PEN: line symbol
MARKER: point symbol
FILL: polygon symbol
pSymbol = symbol to be processed
aAngle = angle to be assigned (degrees)

RETURN: nothing

The given and returned variables should be declared where first called as:

Dim aSymTyp As String
Dim theSymbol As ISymbol
Dim aAngle As Variant

SYMBOL PROPERTIES

TABLE 10-2 AVENUE WRAP PROGRAMMED COLOR KEYWORDS AND RGB NUMBERS

COLOR WORD	RED #	GREEN #	BLUE #	COLOR WORD	RED #	GREEN #	BLUE #
BLACK	0	0	0	MAGENTA	255	0	255
BLUE	0	0	255	ORANGE	255	115	0
BROWN	128	50	0	RED	255	0	0
CYAN	128	255	255	WHITE	255	255	255
GRAY	128	128	128	YELLOW	255	255	0
GREEN	0	255	0				

10.2.15 Subroutine avSymbolSetColor

This subroutine enables the programmer to set the color for a symbol. In using this subroutine, note the following:

1. If the value of the given argument designating the desired color to be assigned to a symbol is numeric, it will refer to a RGB color index value (see Arc Objects Developer Help for information regarding this number).

2. If said numeric value is an invalid number, it will result to white color.

3. If the value of the given argument designating the desired color to be assigned to a symbol is alphabetic, it should be one of the key words listed Table 10-2.

4. If said alphabetic value is any word other than those shown in Table 10-2, it will result to black color.

The corresponding Avenue request is:

aSymTyp.SetColor (aColor)

avSymbol SetColor

The call to this Avenue Wrap is:

Call **avSymbolSetColor**(aSymTyp, pSymbol, aColor)

GIVEN:	aSymTyp	= type of symbol to be processed	
		PEN:	line symbol
		MARKER:	point symbol
		FILL:	polygon symbol
	pSymbol	= symbol to be processed	
	aColor	= color to be assigned (see notes above)	

SYMBOL PROPERTIES

RETURN: nothing

The given and returned variables should be declared where first called as:

Dim aSymTyp As String
Dim theSymbol As ISymbol
Dim aColor As Variant

10.2.16 Subroutine avSymbolSetOLColor

This subroutine enables the programmer to set the outline color for a symbol. In using this subroutine, note the following:

1. Refer to the notes of the avSymbolSetColor Avenue Wrap regarding the color assignment.
2. If the specified symbol is PEN or MARKER, the specified color will assigned to the PEN or MARKER, while if the symbol is a FILL, the specified color will be assigned to the outline.

The corresponding Avenue request is:

aBasicMarker.SetOlColor (aColor)

The call to this Avenue Wrap is:

Call **avSymbolSetOLColor**(aSymTyp, pSymbol, aColor)

avSymbol SetOLColor

GIVEN: aSymTyp = type of symbol to be processed
- PEN: line symbol
- MARKER: point symbol
- FILL: polygon symbol

pSymbol = symbol to be processed
aColor = color to be assigned (see notes above)

RETURN: nothing

The given and returned variables should be declared where first called as:

Dim aSymTyp As String
Dim theSymbol As ISymbol
Dim aColor As Variant

SYMBOL PROPERTIES

10.2.17 Subroutine avSymbolSetOLWidth

This subroutine enables the programmer to set the outline width for a symbol. In using this function, note that the width to be set for:

1. PEN symbols, is the width of the symbol itself, not that of the outline.
2. MARKER symbols, is the width of the outline that is to be drawn.
3. FILL symbols, is the outline width of the symbol.

The corresponding Avenue request is:

aBasicMarker.SetOlWidth (aAngle)

The call to this Avenue Wrap is:

avSymbol SetOLWidth

Call **avSymbolSetOLWidth**(aSymTyp, pSymbol, aWidth)

GIVEN: aSymTyp = type of symbol to be processed

PEN:	line symbol
MARKER:	point symbol
FILL:	polygon symbol

pSymbol = symbol to be processed

aWidth = outline width to be assigned, a value of zero denotes no outline is to be drawn

RETURN: nothing

The given and returned variables should be declared where first called as:

```
Dim aSymTyp As String
Dim theSymbol As ISymbol
Dim aWidth As Variant
```

10.2.18 Subroutine avSymbolSetSize

This subroutine enables the programmer to set the size of a symbol. Note that for PEN and FILL symbols this subroutine works the same as the subroutine avSymbolSetOLWidth

The corresponding Avenue request is:

pSymbol.SetSize (aSize)

The call to this Avenue Wrap is:

avSymbolSetSize

Call **avSymbolSetSize**(aSymTyp, pSymbol, aSize)

GIVEN: aSymTyp = type of symbol to be processed
PEN: line symbol
MARKER: point symbol
FILL: polygon symbol
pSymbol = symbol to be processed
aSize = size to be assigned (greater than zero)

RETURN: nothing

The given and returned variables should be declared where first called as:

```
Dim aSymTyp As String
Dim pSymbol As ISymbol
Dim aSize As Variant
```

10.2.19 Subroutine avSymbolSetStipple

This subroutine enables the programmer to set the stipple of a symbol. In using this subroutine, note the following:

1. Since there is no direct correlation between the Avenue request Stipple and an ArcObject method or property, the Avenue Wrap avSymbolSetStipple allows the user to change an ISymbol object into a IMultiLayerFillSymbol, provided a valid IMultiLayerFillSymbol object is given
2. This function pertains only to FILL type symbols, and not to PEN or MARKER

The corresponding Avenue request is:

```
aBasicMarker.SetStipple (aStipple)
```

The call to this Avenue Wrap is:

Call **avSymbolSetStipple**(aSymTyp, pSymbol, aStipple)

avSymbolSet Stipple

GIVEN: aSymTyp = type of symbol to be processed (see notes above)
pSymbol = symbol to be processed
aStipple = stipple to be assigned

RETURN: nothing

The given and returned variables should be declared where first called as:

```
Dim aSymTyp As String
Dim pSymbol As ISymbol
Dim aStipple As IMultiLayerFillSymbol
```

avSymbol SetStyle

10.2.20 Subroutine avSymbolSetStyle

This subroutine enables the programmer to set the style of a symbol. In using this subroutine, refer to the function avSymbolGetStyle.

The corresponding Avenue request is:

```
aBasicMarker.SetStyle (aStyle)
```

The call to this Avenue Wrap is:

Call **avSymbolSetStyle**(aSymTyp, pSymbol, aStyle)

GIVEN:

aSymTyp	=	type of symbol to be processed
pSymbol	=	symbol to be processed
aStyle	=	style to be assigned (avSymbolGetStyle)

RETURN: nothing

The given and returned variables should be declared where first called as:

```
Dim aSymTyp As String
Dim pSymbol As ISymbol
Dim aStyle As Variant
```

10.3 Graphic Text Avenue Wraps

GRAPHIC TEXT

10.3.1 Function avGraphicTextGetAngle

This function enables the programmer to get the Cartesian angle of inclination of a graphic text string.

The corresponding Avenue request is:

aAngle = aGraphicText.GetAngle

The call to this Avenue Wrap is:

aAngle = **avGraphicTextGetAngle**(aGraphicText)

avGraphicText GetAngle

GIVEN: aGraphicText = graphic text to be processed

RETURN: aAngle = graphic text angle (degrees)

The given and returned variables should be declared where first called as:

Dim aGraphicText As IElement
Dim aAngle As Double

10.3.2 Subroutine avGraphicTextGetStyle

This subroutine enables the programmer to get the text attributes of a graphic text. The attributes include fort, size, bold, italic and color.

The corresponding Avenue request is:

aFont = aTextSymbol.GetFont and
aSize = aTextSymbol.GetSize and
aStyle = aFont.GetStyle and
aColor = aTextSymbol.GetColor

The call to this Avenue Wrap is:

Call **avGraphicTextGetStyle**(aGraphic, aFont, aSize, aBold, _
aItalic, aColor)

avGraphicText GetStyle

GIVEN: aGraphicText = graphic text to be processed

RETURN: aFont = font name
aSize = font size
aBold = font style (1 = normal, 2 = bold)
aItalic = font style (1 = normal, 2 = italic)
aColor = font color (color object)

GRAPHIC TEXT

The given and returned variables should be declared where first called as:

```
Dim aGraphicText As IElemen
Dim aFont As String
Dim aSize As Double
Dim aBold, aItalic As Integer
Dim aColor As IColor
```

10.3.3 Function avGraphicTextGetSymbol

This function enables the programmer to get the symbol (font, size, style, color) associated with a graphic text string.

The corresponding Avenue request is:

```
textSymbol = aGraphic.GetSymbol
```

avGraphicText GetSymbol

The call to this Avenue Wrap is:

Set textSymbol = **avGraphicTextGetSymbol**(aGraphicText)

GIVEN: aGraphicText = graphic text to be processed

RETURN: textSymbol = symbol describing the graphic text

The given and returned variables should be declared where first called as:

```
Dim aGraphicText As IElement
Dim textSymbol As ISymbol
```

10.3.4 Function avGraphicTextGetText

This function enables the programmer to get the text string associated with a graphic text.

The corresponding Avenue request is:

```
theTextString = aGraphicText.GetText
```

avGraphicText GetText

The call to this Avenue Wrap is:

theTextString = **avGraphicTextGetText**(aGraphicText)

GIVEN: aGraphicText = graphic text to be processed

RETURN: theTextString = the text string

The given and returned variables should be declared where first called as:

```
Dim aGraphicText As IElement
Dim theTextString As String
```

GRAPHIC TEXT

10.3.5 Function avGraphicTextMake

This function enables the programmer to create a graphic text string that can be added to the graphics layer.

The corresponding Avenue request is:

```
theGraphText = GraphicText.Make (aString, theGeom)
```

The call to this Avenue Wrap is:

avGraphicTextMake

Set theGraphText = **avGraphicTextMake**(aString, theGeom)

GIVEN:	aString	= text string comprising the graphic text
	theGeom	= geometry describing the graphic, which is a point defining the low left corner of an inclined rectangle enclosing the graphic text
RETURN:	theGraphText	= the graphic that can be added to the graphics layer

The given and returned variables should be declared where first called as:

```
Dim aString As String
Dim theGeom As IGeometry
Dim theGraphText As IElement
```

10.3.6 Subroutine avGraphicTextSetAngle

This subroutine enables the programmer to set the Cartesian angle of the orientation of a text graphic string.

The corresponding Avenue request is:

```
aGraphicText.SetAngle (aAngle)
```

The call to this Avenue Wrap is:

avGraphicTextSetAngle

Call **avGraphicTextSetAngle**(aGraphicText, aAngle)

GIVEN:	aGraphicText	= graphic text to be processed
	aAngle	= angle to be assigned (degrees)
RETURN:	nothing	

The given and returned variables should be declared where first called as:

```
Dim aGraphicText As IElement
Dim aAngle As Double
```

GRAPHIC TEXT

10.3.7 Subroutine avGraphicTextSetStyle

This Subroutine enables the programmer to set the style attributes associated with a graphic text. The attributes include fort, size, bold, italic and color.

The corresponding Avenue request is:

aTextSymbol.SetFont(aFont) and
aTextSymbol.SetSize(aSize) and
aFont.SetStyle(aFont) and
aTextSymbol.SetColor(aColor)

avGraphicText SetStyle

The call to this Avenue Wrap is:

```
Call avGraphicTextSetStyle(aGraphicTextt, aFont, aSize, _
                           aBold, aItalic, aColor)
```

GIVEN: aGraphicText = graphic text to be processed
aFont = font name
aSize = font size
aBold = font style (1 = normal, 2 = bold)
aItalic = font style (1 = normal, 2 = italic)
aColor = font color (color object)

RETURN: nothing

The given and returned variables should be declared where first called as:

```
Dim aGraphicText As IElement
Dim aFont As String
Dim aSize As Double
Dim aBold, aItalic As Integer
Dim aColor As iColor
```

10.3.8 Subroutine avGraphicTextSetSymbol

This Subroutine enables the programmer to set the symbol associated with a graphic text.

The corresponding Avenue request is:

aGraphicText.SetSymbol(pTextSymbol)

avGraphicText SetSymbol

The call to this Avenue Wrap is:

```
Call avGraphicTextSetSymbol(aGraphicText, pTextSymbol)
```

GRAPHIC TEXT

GIVEN: aGraphicText = graphic text to be processed
pTextSymbol = symbol describing the graphic text, its properties such as font, size, bold, italic and color

RETURN: nothing

The given and returned variables should be declared where first called as:
Dim aGraphicText As IElement
Dim pTextSymbol As ISymbol

10.3.9 Subroutine avGraphicSetText

This subroutine enables the programmer to set the text string associated with a graphic text.

The corresponding Avenue request is:
aGraphicText.SetText (aString)

The call to this Avenue Wrap is:
Call **avGraphicTextSetText**(aGraphicText, aString)

avGraphicText SetText

GIVEN: aGraphicText = graphic text to be processed
aString = the text string to be assigned

RETURN: nothing

The given and returned variables should be declared where first called as:
Dim aGraphicText As IElement
Dim aString As String

10.3.10 Subroutine GetTextFont

This subroutine enables the programmer to determine the current active text font in the map.

The corresponding Avenue request is:
There is no corresponding Avenue request.

The call to this Avenue Wrap is:
Call **GetTextFont**(pmxDoc, fontStrg, currSize, defTFINC, _
defPMODE, defCOLOR)

avGetTextFont

GRAPHIC TEXT

GIVEN: pMxDoc = current active document

RETURN: fontStrg = name of current active font
currSize = font size
defTFINC = font style (1 = normal, 2 = italic)
defPMODE = font style (1 = normal, 3 = bold)
defCOLOR = font color (RGB color index value)

The given and returned variables should be declared where first called as:

```
Dim pMxDoc As IMxDocument
Dim fontStrg As String
Dim currSize As Double
Dim defTFINC, defPMODE As Integer
Dim defCOLOR As Long
```

10.3.11 Function MakeTextElement

This function enables the programmer to create a text element.

The corresponding Avenue request is:
There is no corresponding Avenue request.

MakeTextElement

The call to this Avenue Wrap is:

```
Set theText = MakeTextElement(sText, dX, dY, dAngle, _
                              pTextSymbol)
```

GIVEN: sText = text string to appear
dX, dY = low left corner coordinates
dAngle = text angle of inclination (degrees)
pTextSymbol = text symbol reflecting font, size and color

RETURN: theText = text element representing the text

The given and returned variables should be declared where first called as:

```
Dim sText As String
Dim dX, dY, dAngle As Double
Dim pTextSymbol As ITextSymbol
Dim theText As ITextElement
```

GRAPHIC TEXT

10.3.12 Function MakeTextSymbol

This function enables the programmer to create a text symbol, that is to set the font and its size, style and color.

The corresponding Avenue request is:

There is no corresponding Avenue request

The call to this Avenue Wrap is:

MakeTextSymbol

```
Set theFont = MakeTextSymbol(strFont, dFontSize, _
                             iItalic, iBold, iColor)
```

GIVEN:	strFont	= text font
	dFontSize	= font size
	iItalic	= font style (1 = normal, 2 = italic)
	iBold	= font style (1 = normal, 3 = bold)
	iColor	= font color (RGB color index value)
RETURN:	theFont	= text symbol representing the text

The given and returned variables should be declared where first called as:

```
Dim strFont As String
Dim dFontSize As Double
Dim iItalic, iBold As Integer
Dim iColor As Long
Dim theFont As ITextSymbol
```

SAMPLE GRAPHICS CODE

10.4 Sample Graphics Code

Presented below is a sample code example on how to process graphics and symbols. Note that the graphics in this example are in an arbitrary coordinate system.

```
'
' ----
' ---Sample illustrating how to process graphics and symbols,
' ---the graphics that are created are based upon an arbitrary
' ---coordinate system
' ----
'
Dim pMxApp As IMxApplication
Dim pmxDoc As IMxDocument
Dim pActiveView As IActiveView
Dim pMap As IMap
Dim usrRect As IPolygon
Dim iok As Integer
Dim newRect As IEnvelope
Dim pCurGraLyr1 As IGraphicsLayer
Dim pCurGraLyr2 As IGraphicsLayer
Dim pCurGraLyr3 As IGraphicsLayer
Dim pCurGraLyr4 As IGraphicsLayer
Dim pLine As IPolyline
Dim pPoint As IPoint
Dim pPolygon As IPolygon
Dim pGraphic1 As IElement
Dim pGraphic2 As IElement
Dim pGraphic3 As IElement
Dim pGraphic4 As IElement
Dim pSymbol1 As ISymbol
Dim pSymbol2 As ISymbol
Dim pSymbol3 As ISymbol
Dim pSymbol4 As ISymbol
Dim i, j As Long
Dim X, Y, X1, Y1, X2, Y2, ang, angX As Double
Dim aColor As iColor
Dim aOLcolor As iColor
Dim aOLwidth, aAngle, aSize As Double
Dim aStyle, aText As Variant
Dim aList As New Collection
Dim thePalette As IStyleGallery
Dim theFills As New Collection
Dim aSym As ISymbol
Dim aFont As String
Dim aBold, aItalic As Integer
'
' ---Get the active view
Call avGetActiveDoc(pMxApp, pmxDoc, pActiveView, pMap)
```

SAMPLE GRAPHICS CODE

```
'   ---Change the view so that the graphics can be seen by defining
'   ---the extent of the view explicitly
    Set usrRect = avRectMakeXY(10000#, 12000#, 40000#, 32000#)
    Call ChangeView(pmxDoc, 3, 1#, 0#, 0#, usrRect, iok, newRect)
'
'   ----
'   ---Part 1: Process Graphic Lines
'   ---        results shown in middle column of Figure 10-2
'   ----
'
'   ---Set the current active graphics layer (annotation target) to
'   ---be the basic graphics layer
    Call avSetGraphicsLayer(Null, pCurGraLyr1)
'
'   ---Generate seven horizontal lines in different line styles
'   ---illustrating the various style patterns for ISimpleLineSymbol
    X1 = 20000
    Y1 = 16000
    X2 = 25000
    Y2 = 16000
    For i = 0 To 6
        Y1 = Y1 + 2000
        Y2 = Y2 + 2000
'       ---Create a line
        Set pLine = avPolyline2Pt(X1, Y1, X2, Y2)
'       ---Create an element using the line as its geometry
        Set pGraphic1 = avGraphicShapeMake("PEN", pLine)
'       ---Create a line symbol
        Set pSymbol1 = avSymbolMake("PEN")
'       ---Assign the line symbol: color, width and style attributes
        Call avSymbolSetColor("PEN", pSymbol1, "blue")
        Call avSymbolSetOLWidth("PEN", pSymbol1, 1)
        Call avSymbolSetStyle("PEN", pSymbol1, i)
'       ---Associate the line symbol with the element
        Call avGraphicSetSymbol("PEN", pGraphic1, pSymbol1)
'       ---Add the element to the current active graphics layer
        Call avViewAddGraphic(pGraphic1)
'       ---Get the attributes associated with the line symbol
        Set aColor = avSymbolGetColor("PEN", pSymbol1)
        aOLwidth = avSymbolGetOLWidth("PEN", pSymbol1)
        aSize = avSymbolGetSize("PEN", pSymbol1)
        aStyle = avSymbolGetStyle("PEN", pSymbol1)
'       ---Echo the symbol attributes
        MsgBox "RGB Color = " + CStr(aColor.RGB) + Chr(13) + _
               "Outline width = " + CStr(aOLwidth) + Chr(13) + _
               "Size = " + CStr(aSize) + Chr(13) + _
               "Style = " + CStr(aStyle)
    Next
'
```

SAMPLE GRAPHICS CODE

```
'   ----
'   ---Part 2: Process Graphic Markers
'   ---         results shown in left most column of Figure 10-2
'   ----
'
'   ---Set the current active graphics layer to be PointMarkers
'   ---since it does not exist, it will be created
    Call avSetGraphicsLayer("PointMarkers", pCurGraLyr2)
'
'   ---Generate five point markers using different markers
'   ---illustrating the various style patterns for ISimpleMarkerSymbol
    X = 15000
    Y = 16000
    ang = -45
    For i = 0 To 4
        Y = Y + 2500
        ang = ang + 35
'       ---Create a point
        Set pPoint = avPointMake(X, Y)
'       ---Create an element using the point as its geometry
        Set pGraphic2 = avGraphicShapeMake("MARKER", pPoint)
'       ---Create a point symbol
        Set pSymbol2 = avSymbolMake("MARKER")
'       ---Assign the point symbol: color, outline color, outline width,
'       ---style, size and angle attributes
        Call avSymbolSetColor("MARKER", pSymbol2, "green")
        Call avSymbolSetOLColor("MARKER", pSymbol2, "blue")
        Call avSymbolSetOLWidth("MARKER", pSymbol2, 2)
        Call avSymbolSetStyle("MARKER", pSymbol2, i)
        Call avSymbolSetSize("MARKER", pSymbol2, (i + 1) * 3)
        Call avSymbolSetAngle("MARKER", pSymbol2, ang)
'       ---Associate the point symbol with the element
        Call avGraphicSetSymbol("MARKER", pGraphic2, pSymbol2)
'       ---Add the element to the current active graphics layer
        Call avViewAddGraphic(pGraphic2)
'       ---Get the attributes associated with the point symbol
        Set aColor = avSymbolGetColor("MARKER", pSymbol2)
        Set aOLcolor = avSymbolGetOLColor("MARKER", pSymbol2)
        aOLwidth = avSymbolGetOLWidth("MARKER", pSymbol2)
        aAngle = avSymbolGetAngle("MARKER", pSymbol2)
        aSize = avSymbolGetSize("MARKER", pSymbol2)
        aStyle = avSymbolGetStyle("MARKER", pSymbol2)
'       ---Echo the symbol attributes
        MsgBox "RGB Color = " + CStr(aColor.RGB) + Chr(13) + _
               "Outline RGB color = " + CStr(aOLcolor.RGB) + Chr(13) + _
               "Outline width = " + CStr(aOLwidth) + Chr(13) + _
               "Angle = " + CStr(aAngle) + Chr(13) + _
               "Size = " + CStr(aSize) + Chr(13) + _
               "Style = " + CStr(aStyle)
    Next
'
```

SAMPLE GRAPHICS CODE

```
'  ----
'  ---Part 3: Process Graphic Polygons
'  ---        results shown in right most column of Figure 10-2
'  ----
'
'  ---Set the current active graphics layer to be PolygonMarkers
'  ---since it does not exist, it will be created
   Call avSetGraphicsLayer("PolygonMarkers", pCurGraLyr3)
'
'  ---Generate eight polygons using different types of fills
'  ---illustrating the various style patterns for ISimpleFillSymbol
   X1 = 33000
   Y1 = 16000
   X2 = 38000
   Y2 = 17000
   For i = 0 To 7
       Y1 = Y1 + 1000
       Y2 = Y1 + 1250
'      ---Create a rectangle
       Set pPolygon = avRectMakeXY(X1, Y1, X2, Y2)
'      ---Create an element using the polygon as its geometry
       Set pGraphic3 = avGraphicShapeMake("fill", pPolygon)
'      ---Create a polygon symbol
       Set pSymbol3 = avSymbolMake("fill")
'      ---Assign the polygon symbol: color, outline color, outline width
'      ---and style attributes
       Call avSymbolSetColor("fill", pSymbol3, "magenta")
       Call avSymbolSetOLColor("fill", pSymbol3, "black")
       Call avSymbolSetOLWidth("fill", pSymbol3, 2)
       Call avSymbolSetStyle("fill", pSymbol3, i)
'      ---Associate the polygon symbol with the element
       Call avGraphicSetSymbol("fill", pGraphic3, pSymbol3)
'      ---Add the element to the current active graphics layer
       Call avViewAddGraphic(pGraphic3)
'      ---Get the attributes associated with the polygon symbol
       Set aColor = avSymbolGetColor("fill", pSymbol3)
       Set aOLcolor = avSymbolGetOLColor("fill", pSymbol3)
       aOLwidth = avSymbolGetOLWidth("fill", pSymbol3)
       aStyle = avSymbolGetStyle("fill", pSymbol3)
'      ---Echo the symbol attributes
       MsgBox "RGB Color = " + CStr(aColor.RGB) + Chr(13) + _
              "Outline RGB color = " + CStr(aOLcolor.RGB) + Chr(13) + _
              "Outline width = " + CStr(aOLwidth) + Chr(13) + _
              "Style = " + CStr(aStyle)
       Y1 = Y1 + 750
   Next
'
```

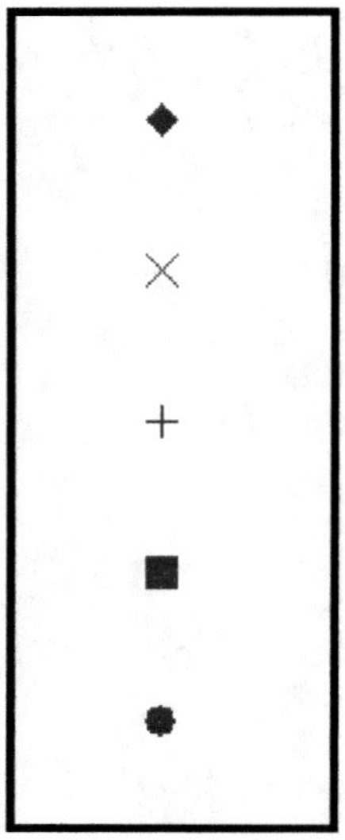

Figure 10-1 Basic markers with no rotation and no outline

SAMPLE GRAPHICS CODE

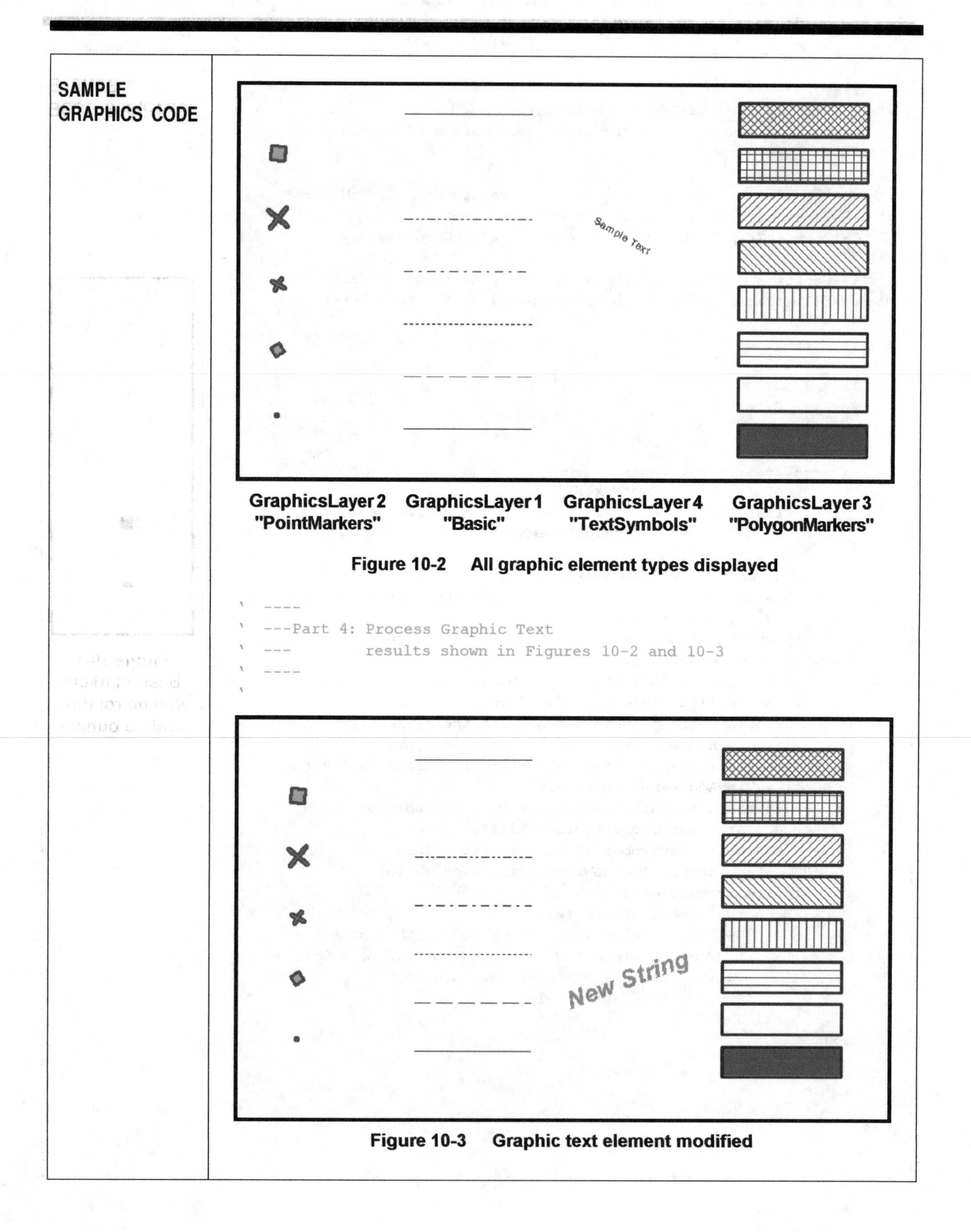

Figure 10-2 All graphic element types displayed

```
' ----
' ---Part 4: Process Graphic Text
' ---        results shown in Figures 10-2 and 10-3
' ----
'
```

Figure 10-3 Graphic text element modified

SAMPLE GRAPHICS CODE

```
' ---Set the current active graphics layer to be TextSymbols
' ---since it does not exist, it will be created
  Call avSetGraphicsLayer("TextSymbols", pCurGraLyr4)
'
' ---Generate a sample graphic text
' ---Create a point denoting the location of the text
  Set pPoint = avPointMake(27500, 26000)
' ---Create an element using the point as its geometry
  Set pGraphic4 = avGraphicTextMake("Sample Text", pPoint)
' ---Get the text symbol associated with the element
  Set pSymbol4 = avGraphicTextGetSymbol(pGraphic4)
' ---Assign the text an angle
  Call avGraphicTextSetAngle(pGraphic4, 330)
' ---Add the element to the current active graphics layer
  Call avViewAddGraphic(pGraphic4)
' ---Get the attributes associated with the point symbol
  Set aColor = avSymbolGetColor("TEXT", pSymbol4)
  angX = avGraphicTextGetAngle(pGraphic4)
  aSize = avSymbolGetSize("TEXT", pSymbol4)
  aText = avGraphicTextGetText(pGraphic4)
  Call avGraphicTextGetStyle(pGraphic4, aFont, aSize, _
                             aBold, aItalic, aColor)
' ---Echo the symbol attributes
  MsgBox "RGB Color = " + CStr(aColor.RGB) + Chr(13) + _
         "Angle = " + CStr(angX) + Chr(13) + _
         "Size = " + CStr(aSize) + Chr(13) + _
         "Text = " + CStr(aText) + Chr(13) + _
         "Font = " + CStr(aFont) + Chr(13) + _
         "Bold = " + CStr(aBold) + Chr(13) + _
         "Italic = " + CStr(aItalic)
' ---Assign new attributes to the text
  Call avGraphicTextSetAngle(pGraphic4, 20)
  Call avSymbolSetSize("TEXT", pSymbol4, aSize * 2.2)
  Call avSymbolSetColor("TEXT", pSymbol4, "ORANGE")
  Call avGraphicTextSetText(pGraphic4, "New String")
' ---Set the text symbol that is associated with the element
  Call avGraphicTextSetSymbol(pGraphic4, pSymbol4)
' ---Relocate the text
  Set pPoint = avPointMake(26500, 20000)
  Call avGraphicSetShape(pGraphic4, pPoint)
' ---Make the text bold and italic and keep the other attributes
' ---the same
  Call avGraphicTextGetStyle(pGraphic4, aFont, aSize, _
                             aBold, aItalic, aColor)
  Call avGraphicTextSetStyle(pGraphic4, aFont, aSize, 2, 2, aColor)
  Call avGraphicInvalidate(pGraphic4)
' ---Get the attributes associated with the point symbol
  angX = avGraphicTextGetAngle(pGraphic4)
  aText = avGraphicTextGetText(pGraphic4)
  Call avGraphicTextGetStyle(pGraphic4, aFont, aSize, _
                             aBold, aItalic, aColor)
```

SAMPLE GRAPHICS CODE

```
'    ---Echo the symbol attributes
     MsgBox "RGB Color = " + CStr(aColor.RGB) + Chr(13) + _
            "Angle = " + CStr(angX) + Chr(13) + _
            "Size = " + CStr(aSize) + Chr(13) + _
            "Text = " + CStr(aText) + Chr(13) + _
            "Font = " + CStr(aFont) + Chr(13) + _
            "Bold = " + CStr(aBold) + Chr(13) + _
            "Italic = " + CStr(aItalic)
'
'    ---
'    ---Part 5: Deleting Graphics
'    ---
'
'    ---Get the total number of graphics drawn
     Call avViewGetGraphics(aList)
     MsgBox "Total graphics drawn = " + CStr(aList.Count)
'
'    ---Delete the last polygon that was created above
     Call avRemoveGraphic(pGraphic3)
'
'    ---Get the current list of graphics in the map
     Call avViewGetGraphics(aList)
     MsgBox "A graphic has been deleted, now there" + Chr(13) + _
            "are " + CStr(aList.Count) + " graphics drawn"
'
'    ----
'    ---Part 6: Manipulating Graphics Layers
'    ----
'
'    ---Determine the number of graphics in each graphics layer
'    ---echo the value and then empty the graphics layer which
'    ---will remove the graphics from the display
     Call avGetGraphicList(pCurGraLyr1, aList)
     MsgBox CStr(aList.Count) + " graphics drawn in layer 1"
     Call avGraphicListEmpty(pCurGraLyr1)
'
     Call avGetGraphicList(pCurGraLyr2, aList)
     MsgBox CStr(aList.Count) + " graphics drawn in layer 2"
     Call avGraphicListEmpty(pCurGraLyr2)
'
     Call avGetGraphicList(pCurGraLyr3, aList)
     MsgBox CStr(aList.Count) + " graphics drawn in layer 3"
     Call avGraphicListEmpty(pCurGraLyr3)
'
     Call avGetGraphicList(pCurGraLyr4, aList)
     MsgBox CStr(aList.Count) + " graphics drawn in layer 4"
     Call avGraphicListEmpty(pCurGraLyr4)
'
```

```
'   ---Delete the graphics layers created above except the
'   ---basic graphics layer, which can not be deleted
    Call avGraphicListDelete(pCurGraLyr2)
    Call avGraphicListDelete(pCurGraLyr3)
    Call avGraphicListDelete(pCurGraLyr4)
'
'   ----
'   ---Part 7: Processing IMultiLayerFillSymbol objects
'   ---        results shown in Figure 10-4
'   ----
'
'   ---Get the application palette
    Set thePalette = avGetPalette(pmxDoc)
'
'   ---From the application palette extract the available fills
    Call avPaletteGetList("FILL", thePalette, theFills)
'
'   ---Set the current active graphics layer to PolygonFills
    Call avSetGraphicsLayer("PolygonFills", pCurGraLyr3)
'
'   ---Draw various multilayerfill symbols
    X1 = 33000
    Y1 = 16000
    X2 = 38000
    Y2 = 17000
    j = 42
    For i = 0 To 7
        Y1 = Y1 + 1000
        Y2 = Y1 + 1250
'       ---Create a rectangle
        Set pPolygon = avRectMakeXY(X1, Y1, X2, Y2)
'       ---Create an element using the polygon as its geometry
        Set pGraphic3 = avGraphicShapeMake("fill", pPolygon)
'       ---Create a polygon symbol
        Set pSymbol3 = avSymbolMake("fill")
'       ---Get a fill from the list of available fills
        Set aSym = theFills.Item(j + i)
'       ---Assign the fill to the symbol
        Call avSymbolSetStipple("fill", pSymbol3, aSym)
'       ---Assign the polygon symbol: color, outline width and
'       ---outline color attributes
        Call avSymbolSetColor("fill", pSymbol3, "blue")
        Call avSymbolSetOLWidth("fill", pSymbol3, i + 1)
        Call avSymbolSetOLColor("fill", pSymbol3, "green")
'       ---Associate the polygon symbol with the element
        Call avGraphicSetSymbol("fill", pGraphic3, pSymbol3)
'       ---Add the element to the current active graphics layer
        Call avViewAddGraphic(pGraphic3)
'       ---Get the attributes associated with the polygon symbol
        Set aColor = avSymbolGetColor("fill", pSymbol3)
        Set aOLcolor = avSymbolGetOLColor("fill", pSymbol3)
```

SAMPLE GRAPHICS CODE

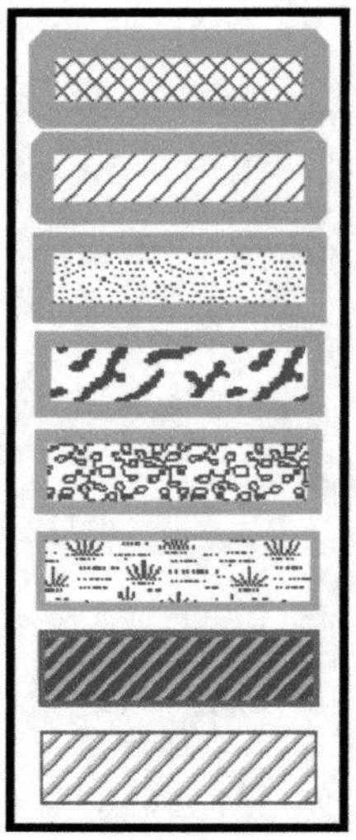

Figure 10-4
Multi-layer fill symbols

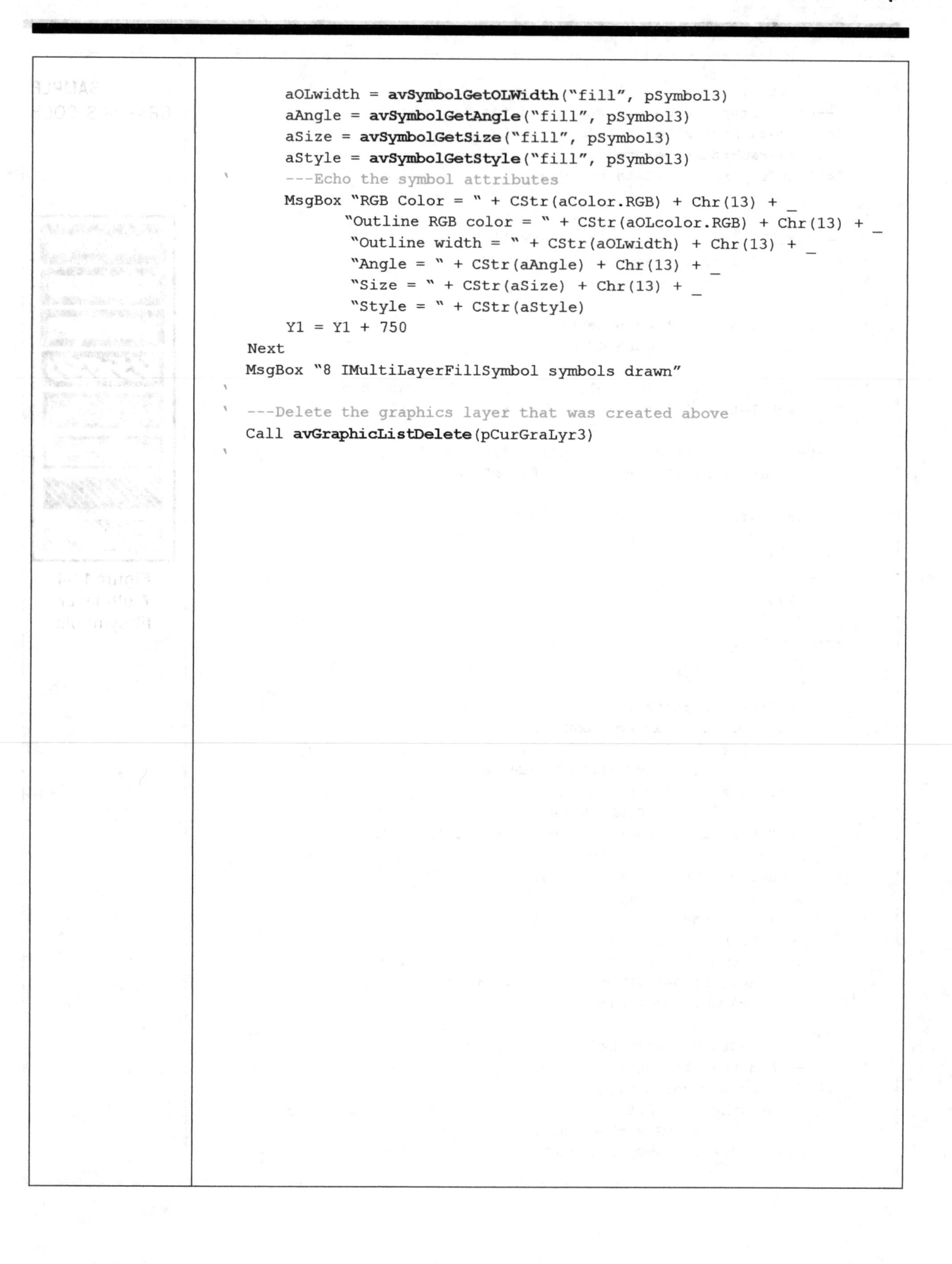

```
        aOLwidth = avSymbolGetOLWidth("fill", pSymbol3)
        aAngle = avSymbolGetAngle("fill", pSymbol3)
        aSize = avSymbolGetSize("fill", pSymbol3)
        aStyle = avSymbolGetStyle("fill", pSymbol3)
'       ---Echo the symbol attributes
        MsgBox "RGB Color = " + CStr(aColor.RGB) + Chr(13) + _
               "Outline RGB color = " + CStr(aOLcolor.RGB) + Chr(13) + _
               "Outline width = " + CStr(aOLwidth) + Chr(13) + _
               "Angle = " + CStr(aAngle) + Chr(13) + _
               "Size = " + CStr(aSize) + Chr(13) + _
               "Style = " + CStr(aStyle)
        Y1 = Y1 + 750
    Next
    MsgBox "8 IMultiLayerFillSymbol symbols drawn"
'
'   ---Delete the graphics layer that was created above
    Call avGraphicListDelete(pCurGraLyr3)
'
```

CHAPTER 11

CLASSIFICATION AND LEGEND AVENUE WRAPS

This chapter contains Avenue Wraps that enable the user to (a) classify features based upon various attributes, (b) create symbol legends, (c) extract symbols and attributes thereof from a legend, (d) update a legend, (e) interact with the ArcMap palettes. These Avenue Wraps include the following:

The source listing of each of the above Avenue Wraps may be found in Appendix D of this publication.

CLASSIFICATION

11.1 Classification Avenue Wraps

Note that the classification Avenue Wraps of this section operate on a map layer (theTheme), while the corresponding Avenue requests operate on a theme's legend (aLegend), which implies that the classification request must have been preceded somewhere in the Avenue code with the following statement:

```
aLegend = aTheme.GetLegend
```

11.1.1 Subroutine avInterval

This subroutine enables the programmer to set the classification of the legend associated with a layer (theme) to be of the equal interval type. In using this subroutine, note the following:

1. The subroutine divides the features in the layer (theme) into numClass classifications of equal intervals calculated from the given argument, aField.

2. The subroutine is applicable only to numeric fields.

3. This type of classification could also be accomplished with the Avenue Wrap avQuantileX

The corresponding Avenue request is:

```
aLegend.Interval (theTheme, aField, numClass)
```

The call to this Avenue Wrap is:

Call **avInterval**(pmxDoc, theTheme, aField, numClass)

avInterval

GIVEN:	pmxDoc	= the active view
	theTheme	= theme to be processed
	aField	= field name that theme is to be classified upon
	numClass	= number of classes to be generated

RETURN: nothing

The given and returned variables should be declared where first called as:

```
Dim pmxDoc As IMxDocument
Dim theTheme As Variant
Dim aField As String
Dim numClass As Long
```

CLASSIFICATION

11.1.2 Subroutine avNatural

This subroutine enables the programmer to set the classification of the legend associated with a layer (theme) to be of the natural interval type. In using this subroutine, note the following:

1. The subroutine divides the features in the layer (theme) into numClass classifications using the ArcObject's Natural Break algorithm applied to the given argument afield.
2. The subroutine is applicable only to numeric fields.
3. This type of classification could also be accomplished with the Avenue Wrap avQuantileX

The corresponding Avenue request is:

aLegend.Natural (theTheme, aField, numClass)

The call to this Avenue Wrap is:

avNatural

Call **avNatural**(pmxDoc, theTheme, aField, numClass)

GIVEN: pmxDoc = the active view
theTheme = theme to be processed
aField = field name that theme is to be classified upon
numClass = number of classes to be generated

RETURN: nothing

The given and returned variables should be declared where first called as:

Dim pmxDoc As IMxDocument
Dim theTheme As Variant
Dim aField As String
Dim numClass As Long

11.1.3 Subroutine avQuantile

This subroutine enables the programmer to set the classification of the legend associated with a theme to be of the quantile interval type. In using this subroutine, note the following:

1. The subroutine divides the features in the layer (theme) into numClass classifications of equal size based upon the given argument aField.
2. The subroutine is applicable only to numeric fields.
3. This type of classification could also be accomplished with the Avenue Wrap avQuantileX

CLASSIFICATION

The corresponding Avenue request is:

aLegend.Quantile (aTheme, aField, numClass)

The call to this Avenue Wrap is:

Call **avQuantile**(pmxDoc, theTheme, aField, numClass)

avQuantile

GIVEN: pmxDoc = the active view
theTheme = theme to be processed
aField = field name that theme is to be classified upon
numClass = number of classes to be generated

RETURN: nothing

The given and returned variables should be declared where first called as:

Dim pmxDoc As IMxDocument
Dim theTheme As Variant
Dim aField As String
Dim numClass As Long

11.1.4 Subroutine avQuantileX

This subroutine enables the programmer to set the classification of the legend associated with a theme to be of the interval, natural or quantile interval type. In using this subroutine, note the following:

1. The subroutine combines the function of avInterval, avNatural and avQuantile into one by providing the classification type choice with the given argument method.
2. The subroutine is applicable only to numeric fields.

The corresponding Avenue request is:

There is no corresponding Avenue request.

The call to this Avenue Wrap is:

Call **avQuantileX**(pmxDoc, theTheme, aField, numClass, method)

avQuantileX

GIVEN: pmxDoc = the active view
theTheme = theme to be processed
aField = field name that theme is to be classified upon
numClass = number of classes to be generated
method = classification method

CLASSIFICATION

1 : Defined Interval (not implemented yet)
2 : Equal Interval
3 : Natural Breaks
4 : Quantile
5 : Standard Deviation (not implemented yet)

RETURN: nothing

The given and returned variables should be declared where first called as:

```
Dim pmxDoc As IMxDocument
Dim theTheme As Variant
Dim aField As String
Dim numClass, method As Long
```

11.1.5 Subroutine avSingleSymbol

This subroutine enables the programmer to set the classification of the legend associated with a layer (theme) to be of a single symbol type. In using this subroutine, note the following:

1. All features in the theme will be classified such that every feature is drawn using the same symbology, color, etc.
2. If default values are to be used for the given arguments pDesc, pLabel and pSym, then NULL can be specified for pDesc and pLabel, and NOTHING can be specified for pSym.

avSingleSymbol

The corresponding Avenue request is:

aLegend.SingleSymbol

The call to this Avenue Wrap is:

Call **avSingleSymbol**(pmxDoc, theTheme, pDesc, pLabel, pSym)

GIVEN:	pmxDoc	= the active view
	theTheme	= theme to be processed
	pDesc	= renderer description (see Note 2 above)
	pLabel	= label (this text will appear in the Table of Contents) (see Note 2 above)
	pSym	= symbol used to draw every feature in theme (see Note 2 above)

RETURN: nothing

The given and returned variables should be declared where first called as:

```
Dim pmxDoc As IMxDocument
Dim theTheme As Variant
Dim pDesc As String
Dim pLabel As String
Dim pSym As ISymbol
```

11.1.6 Subroutine avUnique

This subroutine enables the programmer to set the classification of the legend associated with a layer (theme) to be of the unique interval type. All features in the theme will be classified such that each feature in the layer (theme), with a unique value of an attribute, will be assigned a unique symbol, or a different symbol from the other features in the layer (theme) with different values of the attribute.

The corresponding Avenue request is:

```
aLegend.Unique (theTheme, aField)
```

The call to this Avenue Wrap is:

Call **avUnique**(pmxDoc, theTheme, aField, showNulls)

GIVEN:

pmxDoc	=	the active view
theTheme	=	theme to be processed
aField	=	field name that theme is to be classified upon
showNulls	=	flag denoting whether features that have not been assigned a value for aField should be drawn or not (true, false)

RETURN: nothing

The given and returned variables should be declared where first called as:

```
Dim pmxDoc As IMxDocument
Dim theTheme As Variant
Dim aField As String
Dim showNulls As Boolean
```

GENERAL
GEOMETRY

11.2 Legend Avenue Wraps

LEGEND

11.2.1 Subroutine avGetClassifications

This subroutine enables the programmer to get the list of classes used in a classification.

The corresponding Avenue request is:

classList = theLegend.GetClassifications

The call to this Avenue Wrap is:

Call **avGetClassifications**(theLegend, classList)

avGet Classifications

GIVEN: theLegend = legend to be processed

RETURN: classList = list of classifications

The given and returned variables should be declared where first called as:

Dim theLegend As IFeatureRenderer
Dim classList As New Collection

11.2.2 Function avGetClassType

This function enables the programmer to determine the type of classification that has been used to create a legend. Note that the type of classification could be one of the following

DefinedInterval	EqualInterval
Manual (either SingleSymbol or Unique)	NaturalBreaks
Quantile	StandardDeviation

The corresponding Avenue request is:

theClassType = theLegend.GetClassType

The call to this Avenue Wrap is:

theClassType = **avGetClassType**(theLegend)

avGetClassType

GIVEN: theLegend = legend to be processed

RETURN: theClassType = type of classification used in the generation of a legend (renderer). See Note above.

LEGEND

The given and returned variables should be declared where first called as:
Dim theLegend As IFeatureRenderer
Dim avGetClassType As String

11.2.3 Subroutine avGetFieldNames

This subroutine enables the programmer to get a collection of the field names used to classify a layer (theme).

The corresponding Avenue request is:

nameList = theLegend.GetFieldNames

avGetFieldNames

The call to this Avenue Wrap is:

Call **avGetFieldNames**(theLegend, nameList)

GIVEN: theLegend = legend to be processed

RETURN: nameList = list of field names used in a classification (will be empty for SingleSymbol legends)

The given and returned variables should be declared where first called as:
Dim theLegend As IFeatureRenderer
Dim nameList As New Collection

11.2.4 Subroutine avGetLegend

This subroutine enables the programmer to get the legend associated with a layer (theme).

The corresponding Avenue request is:

aLegend = aTheme.GetLegend.

avGetLegend

The call to this Avenue Wrap is:

Call **avGetLegend**(pmxDoc, theTheme, aLegend)

GIVEN: pmxDoc = the active view
theTheme = theme to be processed

RETURN: aLegend = theme legend

The given and returned variables should be declared where first called as:
Dim pmxDoc As IMxDocument
Dim theTheme As Variant
Dim aLegend As IFeatureRenderer

LEGEND

11.2.5 Function avGetLegendType

This function enables the programmer to determine the type of legend in use. Note that the type of legend could be one of the following

BIVARIATE:	Bivariate	CHART:	Chart
CLASS:	Class Breaks	DOT:	Dot Density
PROPORTIONAL:	Proportional Symbol	SIMPLE:	Simple
SCALE:	ScaleDependent	UNIQUE:	Unique Value

The corresponding Avenue request is:

aLegendType = theLegend.GetLegendType

The call to this Avenue Wrap is:

aLegendType = **avGetLegendType**(theLegend)

avGetLegendType

GIVEN: theLegend = legend to be processed

RETURN: aLegendType = type of legend (renderer) in use. See Note above.

The given and returned variables should be declared where first called as:

Dim theLegend As IFeatureRenderer
Dim aLegendType As String

11.2.6 Function avGetNumClasses

This function enables the programmer to determine the number of classes in the legend.

The corresponding Avenue request is:

NumClasses = aLegend.GetNumClasses

The call to this Avenue Wrap is:

NumClasses = **avGetNumClasses**(theLegend)

avGetNumClasses

GIVEN: theLegend = legend to be processed

RETURN: NumClasses = number of classifications in legend

The given and returned variables should be declared where first called as:

Dim theLegend As IFeatureRenderer
Dim NumClasses As Long

LEGEND

11.2.7 Subroutine avLegendGetSymbols

This subroutine enables the programmer to get a collection of symbols appearing in a legend.

The corresponding Avenue request is:

```
symbList = theLegend.GetSymbols
```

avLegend GetSymbols

The call to this Avenue Wrap is:

Call **avLegendGetSymbols**(theLegend, symbList)

GIVEN: theLegend = legend to be processed

RETURN: symbList = list of ISymbol objects used in the legend

The given and returned variables should be declared where first called as:

```
Dim theLegend As IFeatureRenderer
Dim symbList As New Collection
```

11.2.8 Subroutine avLegendSetSymbols

This subroutine enables the programmer to set the symbols to be used in the classifications in a legend.

The corresponding Avenue request is:

```
symbList = theLegend.SetSymbols
```

avLegend SetSymbols

The call to this Avenue Wrap is:

Call **avLegendSetSymbols**(theLegend, symbList)

GIVEN: theLegend = legend to be processed
symbList = list of ISymbol objects that are to be assigned to the classifications in a legend

RETURN: nothing

The given and returned variables should be declared where first called as:

```
Dim theLegend As IFeatureRenderer
Dim symbList As New Collection
```

11.2.9 Subroutine avUpdateLegend

LEGEND

This subroutine enables the programmer to update a layer (theme) to reflect any changes made to its legend. Both, the layer (theme) and the table of contents will be updated (redrawn). Note that it may be necessary to follow the call to the subroutine avUpdateLegend with a call to avDisplayInvalidate in order to refresh the display so that the changes made to the theme are properly displayed. These calls are not made in avUpdateLegend in order to eliminate multiple screen redraws.

The corresponding Avenue request is:

```
theTheme.UpdateLegend
```

The call to this Avenue Wrap is:

Call **avUpdateLegend**(pmxDoc, theTheme)

avUpdateLegend

GIVEN: pmxDoc = the active view
theTheme = layer (theme) to be processed

RETURN: nothing

The given and returned variables should be declared where first called as:

```
Dim pmxDoc As IMxDocument
Dim theTheme As Variant
```

SYMBOL PALETTE

11.3 Symbol Palette Avenue Wraps

11.3.1 Function avGetPalette

This function enables the programmer to get the various style galleries (palettes) that are available in the application. In using this function, note the following:

1. Depending upon the installation, the total number of galleries and items (symbols) within each gallery may vary.
2. The returned argument, thePalette, contains the galleries (palettes) in the application, and their default number of items (symbols) shown in Table 11-1. Reference is also made to Appendix A, in which these galleries and their individual default item names are listed.
3. While the Avenue request operates on a symbol window, avGetPalette operates on the document,

TABLE 11-1 GALLERY AND DEFAULT NUMBER OF ITEMS THEREIN

Gallery Name	Gallery Keyword	Num. Items	Gallery Name	Gallery Keyword	Numb. Items
Area Patches	AREAPATCHES	8	Line Symbols	PEN	86
Backgrounds	BACKGROUNDS	18	Marker Symbols	MARKER	114
Borders	BORDERS	16	North Arrows	NORTHARROWS	97
Colors	COLOR	120	Reference Systems	REFERENCE	12
Color Ramps	COLORAMPS	78	Scale Bars	SCALEBARS	11
Fill Symbols	FILL	53	Scale Texts	SCALETEXT	7
Labels	LABELS	20	Shadows	SHADOWS	12
Legend Items	LEGENDITEMS	18	Text Symbols	TEXT	35
Line Patches	LINEPATCHES	9			

The corresponding Avenue request is:

```
thePalette = aSymbolWin.GetPalette
```

The call to this Avenue Wrap is:

thePalette = **avGetPalette**(pmxDoc)

avGetPalette

GIVEN: pmxDoc = the active view

RETURN: thePalette = the available style galleries (see notes above)

SYMBOL PALETTE

The given and returned variables should be declared where first called as:

```
Dim pmxDoc As IMxDocument
Dim thePalette As IStyleGallery
```

11.3.2 Subroutine avPaletteGetList

This subroutine enables the programmer to get a collection of the items within a specific gallery (palette). In using this function, note the following:

1. Depending upon the installation, the total number of galleries and items within each gallery may vary.
2. The default galleries (palettes) in an application, and the number of items therein could be one of one of those shown in Table 11-1.
3. The given argument aGalleryType should be one of the keywords shown in Table 11-1.
4. While the Avenue request operates on a symbol window, avGetPalette operates on the document

The corresponding Avenue request is:

```
aGalleryList = thePalette.GetList (aGalleryType)
```

avGetPaletteList

The call to this Avenue Wrap is:

Call **avPaletteGetList**(aGalleryType, thePalette, aGalleryList)

GIVEN: aGalleryType = type of gallery to be processed. Specify the appropriate keyword shown in Table 11-1.

thePalette = style gallery, contains all galleries

(a) Solid Fill

(b) Pattern Fill

Figure 11-1 Sample Polygon Fill Palette

SYMBOL PALETTE

RETURN: aGalleryList = list of items in the desired gallery

The given and returned variables should be declared where first called as:

```
Dim aGalleryType As String
Dim thePalette As IStyleGallery
Dim aGalleryList As New Collection
```

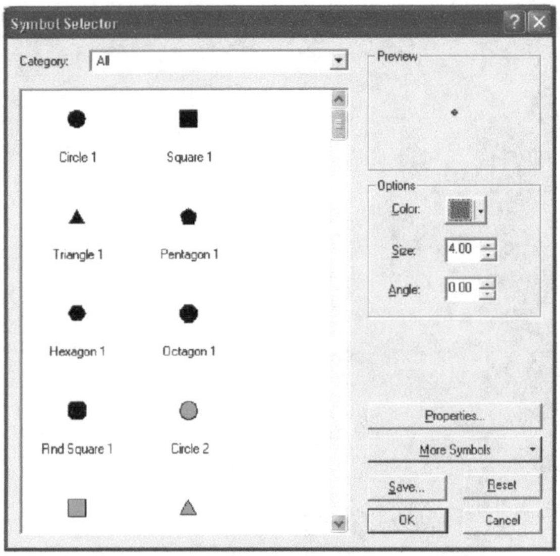

Symbol Selector
Category: All
Preview
Circle 1
Square 1
Triangle 1
Pentagon 1
Hexagon 1
Octagon 1
Rnd Square 1
Circle 2
Options
Color:
Size: 4.00
Angle: 0.00
Properties...
More Symbols
Save...
Reset
OK
Cancel

Figure 11-2
Sample Marker Symbol Palette

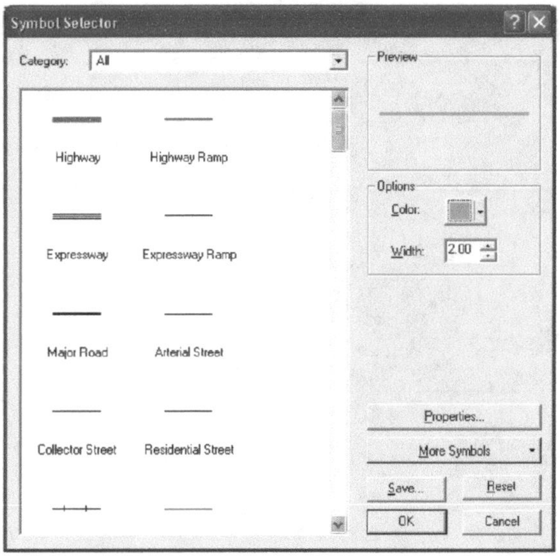

Symbol Selector
Category: All
Preview
Highway
Highway Ramp
Expressway
Expressway Ramp
Major Road
Arterial Street
Collector Street
Residential Street
Options
Color:
Width: 2.00
Properties...
More Symbols
Save...
Reset
OK
Cancel

Figure 11-3
Sample Pen Symbol Palette

CHAPTER 12

UTILITY MACROS

This chapter contains certain programs of general or special utility function such as enabling the programmer to (a) mass import VBA subroutines and functions and related forms from a specified directory into an ArcMap™ project, (b) to mass export the same from an ArcMap™ project to a specified directory, and (c) provide support for certain Avenue Wraps presented in previous chapters. The support, or background programs as referred to herein, may also be used to perform certain other geometric and other types of functions. These utility programs include the following:

The source listing of each of the above Avenue Wraps may be found in Appendix D of this publication.

12.1 Import and Export VBA Code

Import/Export VBA Code

12.1.1 Subroutine ExportVBAcode

This subroutine enables the programmer to export all VBA program components from the current project into a hard coded directory. The ExportVBAcode subroutine has no given and no returned arguments. It is executed from within the ArcMap™ Visual Basic Editor. To execute it:

ExportVBAcode

➢ 1 Display the ExportVBA code subroutine in the Module window of the ArcMap™ Visual Basic Editor.

➢ 2 Locate the first executable statement below the declaration Dim statements which should read as follows:

```
aDIR = "c:\Directory\SubDirectory\vbaCode"
```

➢ 3 Change Directory, SubDirectory and vbaCode to the appropriate directory path where the VBA subroutines, functions and forms are to be stored.

➢ 4 Click at the Run Sub/UserForm tool (▶).

➢ 5 Using Windows Explorer, navigate to the directory to which the VBA program components are to be stored.

All subroutines, functions and forms are exported to the specified directory.

12.1.2 Subroutine LoadVBAcode

This subroutine enables the programmer to import all VBA program components from a hard coded specified directory into the current project. This subroutine has no given and no returned arguments. It is executed from within the ArcMap™ Visual Basic Editor. To execute it:

LoadVBAcode

➢ 1 Display the ExportVBA code subroutine in the Module window of the ArcMap™ Visual Basic Editor.

➢ 2 Locate the first executable statement below the declaration Dim statements which should read as follows:

```
aDIR = "c:\Directory\SubDirectory\vbaCode"
```

Import/Export VBA Code

➢ 3 Change Directory, SubDirectory and vbaCode to the appropriate directory path where the VBA subroutines, functions and forms are to be stored.

➢ 4 Click at the Run Sub/UserForm tool (▶).

All subroutines, functions and forms are imported into the current project.

12.2 Background Scripts

BACKGROUND SCRIPTS

The subroutines and functions of this section are used transparently to the programmer (unless the programmer decides to peruse the code) by certain of the Avenue Wraps presented in the other chapters of this book. This, however, does not mean that these subroutines and functions can not be used in other programs as a programmer may deem appropriate.

12.2.1 Subroutine avConvertArea

This subroutine enables the programmer to convert a value representing an area from one unit of measure into another. For example, an area value in square feet can be converted into square meters using this procedure. Note that the input argument, fromMeasure, is modified by this subroutine.

The corresponding Avenue request is:

```
anArea = Units.ConvertArea (fromMeasure, fromUnits, toUnits)
```

The call to this Avenue Wrap is:

```
Call avConvertArea(fromMeasure, fromUnits, toUnits)
```

avConvertArea

GIVEN: fromMeasure = area value to be converted
fromUnits = units of measure the area value is presently in
toUnits = units of measure the area value is to be converted to

RETURN: nothing

The given and returned variables should be declared where first called as:

```
Dim fromMeasure As Variant
Dim fromUnits As esriUnits
Dim toUnits As esriUnits
```

12.2.2 Subroutine avConvertDistance

This subroutine enables the programmer to convert a value representing a length or distance from one unit of measure into another. For example, a distance value in feet can be converted into meters using this procedure. Note that the input argument, fromMeasure, is modified by this subroutine.

BACKGROUND SCRIPTS

The corresponding Avenue request is:

aDist = Units.Convert (fromMeasure, fromUnits, toUnits)

avConvertDistance

The call to this Avenue Wrap is:

Call **avConvertDistance**(fromMeasure, fromUnits, toUnits)

GIVEN: fromMeasure = distance value to be converted
fromUnits = units of measure the distance value is presently in
toUnits = units of measure the distance value is to be converted to

RETURN: nothing

The given and returned variables should be declared where first called as:

Dim fromMeasure As Variant
Dim fromUnits As esriUnits
Dim toUnits As esriUnits

12.2.3 Subroutine GetTextRect

This subroutine enables the programmer to get the angle of inclination, height and width of a rectangle that encloses a text element. In using this subroutine, note the following:

1. The Avenue request GetBounds is not a direct counterpart, even though it is identified as such below.
2. The returned arguments reflect those of an inclined, and not an orthogonal, enclosing rectangle that circumscribes the text element.

The corresponding Avenue request is:

theRect = aGraphic.GetBounds

GetTextRect

The call to this Avenue Wrap is:

Call **GetTextRect**(pTextElement, pScreenDisplay, _
x1, y1, aAngle, aWidth, aHeight)

GIVEN: pTextElement = text element representing the text
pScreenDisplay = display which text element appears in

RETURN:	x1,y1	= low left corner coordinates of text
	aAngle	= text angle (degrees)
	aWidth	= text width (along the text angle)
	aHeight	= text height (perpendicular to angle)

The given and returned variables should be declared where first called as:
Dim pTextElement As ITextElement
Dim pScreenDisplay As IScreenDisplay
Dim x1, y1, aAngle, aWidth, aHeight As Double

12.2.4 Function icasinan

This function enables the programmer to compute the arcsine of a numeric double precision value.

The corresponding Avenue request is:
theAngle = angleX.ASin

The call to this Avenue Wrap is:
theAngle = **icasinan**(angleX)

icasinan

GIVEN: angleX = sine value of an angle

RETURN: theAngle = angle in radians whose sine is angleX

The given and returned variables should be declared where first called as:
Dim angleX, theAngle As Double

12.2.5 Subroutine icatan

This subroutine enables the programmer to compute the arctangent of a numeric double precision value.

The corresponding Avenue request is:
theAngle = D.ATan

The call to this Avenue Wrap is:
theAngle = **icatan**(D)

icatan

GIVEN: D = tangent value of an angle

RETURN: theAngle = angle in radians whose tangent is D

BACKGROUND SCRIPTS

The given and returned variables should be declared where first called as:
Dim D, theAngle As Double

12.2.6 Function iccomdis

This function enables the programmer to compute the distance between two points, the coordinates of which are known.

The corresponding Avenue request is:
There is no corresponding Avenue request.

iccomdis

The call to this Avenue Wrap is:
theDist = **iccomdis**(X1, Y1, X2, Y2)

GIVEN:	X1,Y1	= coordinates of the first point
	X2,Y2	= coordinates of the second point

RETURN:	theDist	= distance between the two points

The given and returned variables should be declared where first called as:
Dim X1, Y1, X2, Y2, theDist As Double

12.2.7 Subroutine iccomppt

This subroutine enables the programmer to check if a point's coordinates match those of another point within the tolerance of the display. In using this subroutine, note that the coordinates of the first given point are matched against the coordinates of the second given point (base point). If the two given point coordinates are the same within the point snapping tolerance, then the coordinates of the first given point are set to be exactly the same as those of the second given point; otherwise the given points are passed back unchanged.

The corresponding Avenue request is:
There is no corresponding Avenue request.

iccomppt

The call to this Avenue Wrap is:
Call **iccomppt**(XCORD, YCORD, X2, Y2, XCRD2, YCRD2, NOFND)

GIVEN:	XCORD,YCORD	= coordinates of point to be matched against
	X2,Y2	= coordinates of base point

BACKGROUND SCRIPTS

RETURN: XCRD2, YCRD2 = input values, if NOFND = 0
= X2, Y2 values, if NOFND = 1

NOFND = flag whether a match was found (= 1), or not found (= 0) within the point snapping tolerance

The given and returned variables should be declared where first called as:
Dim XCORD, YCORD, X2, Y2, XCRD2, YCRD2 As Double
Dim NOFND As Integer

12.2.8 Function icdegrad

This function enables the programmer to convert an angle from degrees to radians.

The corresponding Avenue request is:
theRadians = theAngle.AsRadians

The call to this Avenue Wrap is:
theRadians = **icdegrad**(theAngle)

icdegrad

GIVEN: theAngle = angle in degrees (decimal)

RETURN: theRadians = angle in radians

The given and returned variables should be declared where first called as:
Dim theAngle As Double
Dim theRadians As Double

12.2.9 Subroutine icforce

This subroutine enables the programmer to inverse from one point of known coordinates to another point of known coordinates. That is to find the azimuth and distance from one point to another.

The corresponding Avenue request is:
There is no corresponding Avenue request.

The call to this Avenue Wrap is:
Call **icforce**(PTN1, PTE1, PTN2, PTE2, D, AZ)

icforce

BACKGROUND SCRIPTS

GIVEN:	PTN1,PTE1	= north-east coordinates of point 1
	PTN2,PTE2	= north-east coordinates of point 2
RETURN:	D	= distance from point 1 to point 2
	AZ	= azimuth (radians) from point 1 to 2

The given and returned variables should be declared where first called as:
Dim PTN1, PTE1, PTN2, PTE2, D, AZ As Double

12.2.10 Function icmakdir

This function enables the programmer to compute the Cartesian direction between two points.

The corresponding Avenue request is:
There is no corresponding Avenue request.

icmakdir

The call to this Avenue Wrap is:
theCartDir = **icmakdir**(X1, Y1, X2, Y2)

GIVEN:	X1,Y1	= coordinates of the first point
	X2,Y2	= coordinates of the second point
RETURN:	theCartDir	= Cartesian direction from point 1 to point 2 in radians

The given and returned variables should be declared where first called as:
Dim X1, Y1, X2, Y2, theCartDir As Double

12.2.11 Function icraddeg

This function enables the programmer to convert an angle from radians to degrees in decimal form.

The corresponding Avenue request is:
AngleDegs = AngleRads.AsDegrees

icraddeg

The call to this Avenue Wrap is:
AngleDegs = **icraddeg**(AngleRads)

BACKGROUND SCRIPTS

GIVEN: AngleRads = angle in radians

RETURN: AngleDegs = angle in degrees (decimal form)

The given and returned variables should be declared where first called as:
Dim AngleRads, AngleDegs As Double

12.2.12 Subroutine SetViewSnapTol

This subroutine enables the programmer to set the snap tolerance for the current view.

The corresponding Avenue request is:
There is no corresponding Avenue request.

The call to this Avenue Wrap is:

SetViewSnapTol

```
Call SetViewSnapTol(theView, xP, yP, _
                    viewRect, thePoint, difxxx, difzzz, difwww)
```

GIVEN: theView = the current active view
xP = x coordinate of given point
yP = y coordinate of given point

RETURN: viewRect = the width of the current display
thePoint = point in the projected coordinate system
difxxx = tolerance as a percentage of the view width based upon user-defined value for ugsnapTol
difzzz = smaller tolerance (difxxx * 0.1)
difwww = tolerance which will be: (a) the same as difxxx if the tolerance is defined as a percentage (ugsnapTolMode = "P") or (b) equal to the absolute tolerance value (ugsnapTol), which is converted into the projected environment, if the tolerance is defined to be absolute (ugsnapTolMode = "A").

The given and returned variables should be declared where first called as:
Dim theView As IMxDocument
Dim xP, yP As Double
Dim viewRect, difxxx, difzzz, difwww As Double
Dim thePoint As IPoint

BACKGROUND SCRIPTS	

CHAPTER 13

AVENUE WRAPS FORMS

This chapter discusses the creation and operation of the forms, which are used by the Avenue Wraps. The forms described herein are **VBA**, Visual Basic for Applications, forms, but can be created in the **VB**, Visual Basic, environment just as easily. There are four forms employed by the Avenue Wraps. The *first* is the **EnterFiles** form that controls the selection of a single, as well as, multiple files. This form is used by the **avFileDialogPut** and **avFileDialogReturnFiles** Avenue Wraps. The *second* is the **VDialogBox** form that is a general purpose form used by the **avMsgBoxChoice**, **avMsgBoxInput**, **avMsgBoxList** and **avMsgBoxMultiInput** Avenue Wraps, while the *third* is the **VDialogBox2** form, which is used by the **avMsgBoxMultiInput2** Avenue Wrap. The fourth form is the **HDialogBox** form that is a general purpose form used in the creation of a horizontal dialog box.

The second, third and fourth forms are generic forms that are customized programmatically by the **VDBbuild**, **VDBbuild2** and **HDBbuild** Avenue Wraps. That is to say, these subroutines add certain controls to the forms, the developer does not need to add all of the controls when the form is initially created. The developer will only add the basic controls when initially creating the form. The first form, however, is not customized programmatically, but rather, the developer will fully design the form with all of the necessary controls. Note that the approach described below can be applied to creating other types of forms.

13.1 Creating a VBA Form

The steps presented below describe how the developer, in the **VBA** environment, can create a form. To add a form, the developer should:

➢ **1** Open the ArcMap™ project file, which is to have the forms discussed herein added to it. This can be either an existing project file or a new project file.

➢ **2** Click at the **Tools** menu and then at the **Macros** and **Visual Basic Editor** sub-menus, see Figure 13-1. Once these selections have been made the VBE, Visual Basic Editor, interface will appear.

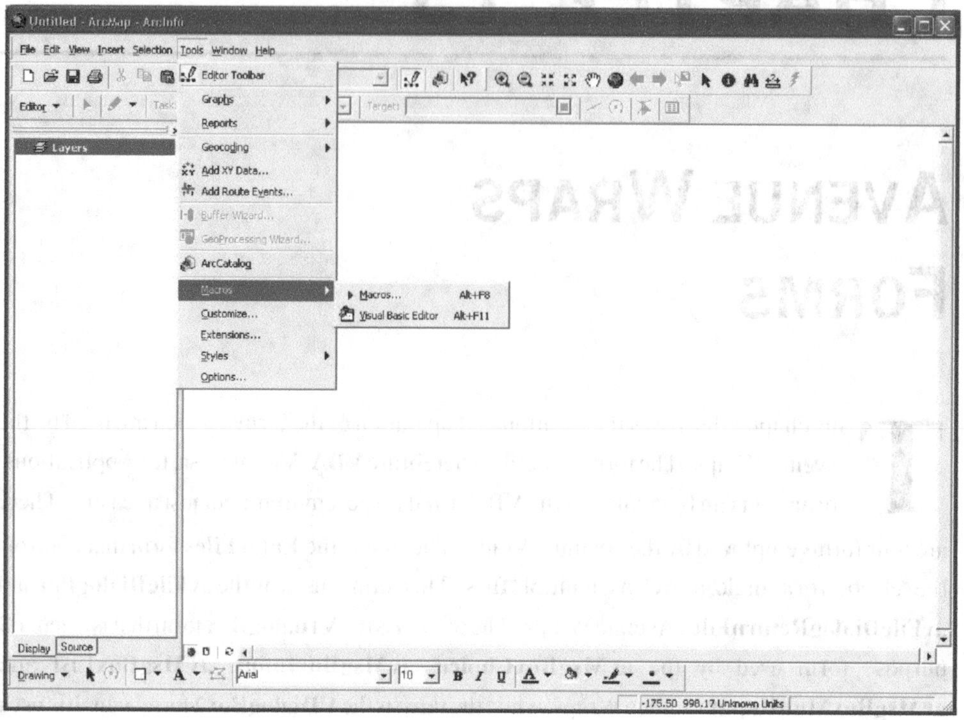

Figure 13-1 Menu Item Selection to Enter the Visual Basic Editor

➤ 3 Click at the **Insert** menu and then at the **UserForm** sub-menu, see Figure 13-2. Once these selections have been made a blank form as shown in Figure 13-3 will appear. At this point we have created a blank form upon which various controls can be added, either explicitly or programmatically. **Reposition** the blank form and Toolbox panel to be as shown in Figure 13-3.

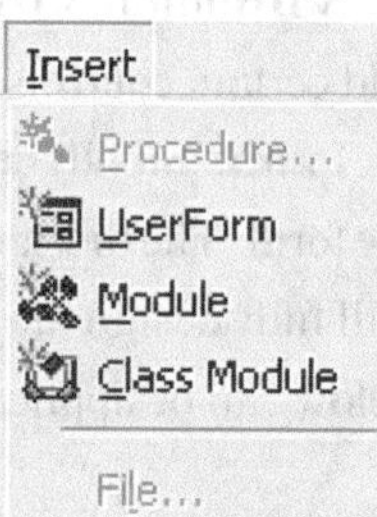

Figure 13-2 Menu Selection to Create a Form

Shown in Figure 13-3, to the left of the blank form named UserForm1, is the default Toolbox panel. The user specifies which controls appear on the form by selecting the appropriate control from the Toolbox panel followed by either (a) making a pick, or (b) dragging a rectangle, on the form denoting the initial size and position of the control. The user can then modify the control (a) graphically or (b) by explicitly entering the appropriate property value in the Properties panel, which appears on the left side of Figure 13-3.

Additional types of controls can be added to the Toolbox panel by using the Tools menu and Additional Controls... sub-menu, as shown in Figure 13-4. So that:

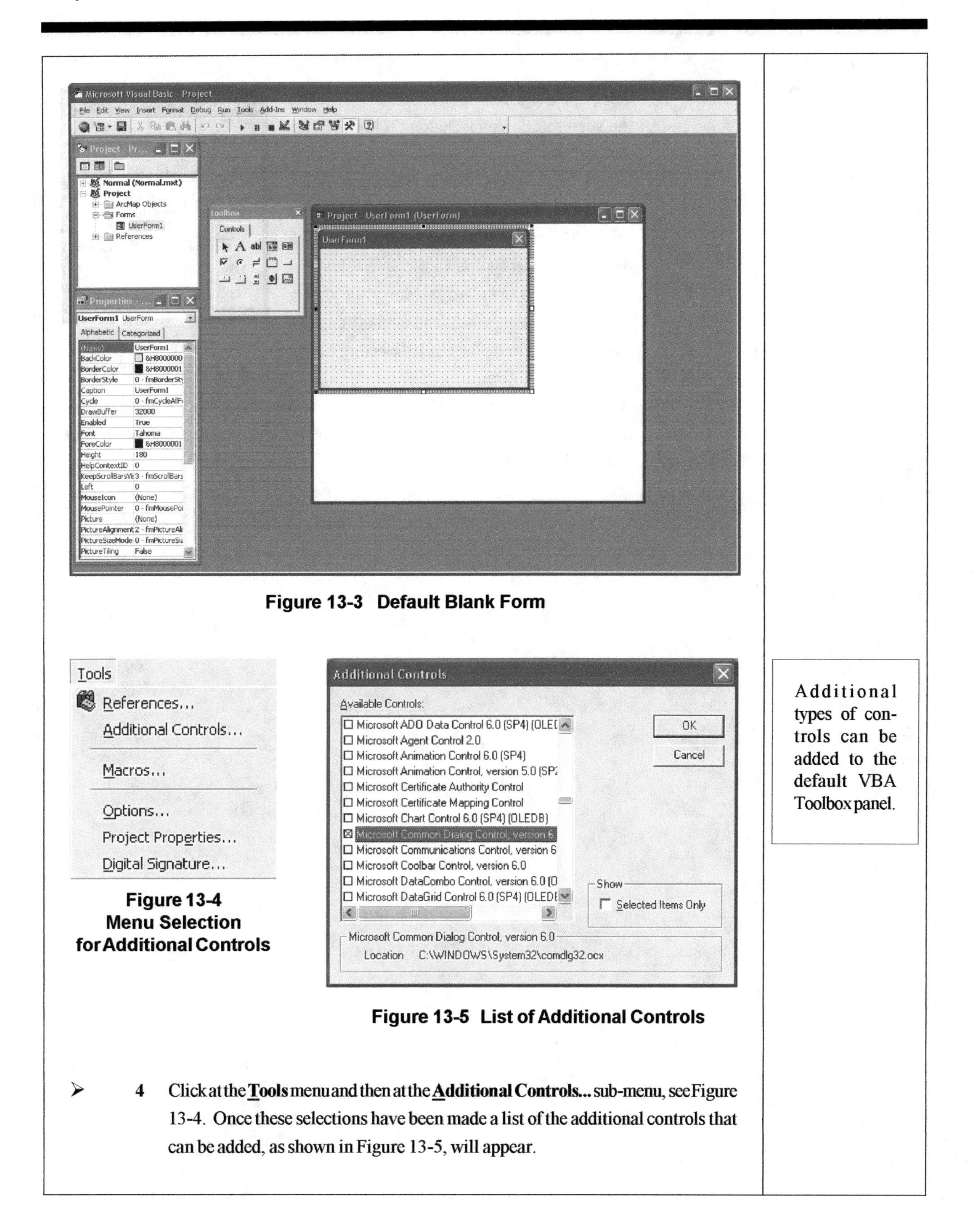

Figure 13-3 Default Blank Form

Figure 13-4 Menu Selection for Additional Controls

Figure 13-5 List of Additional Controls

➢ 4 Click at the **Tools** menu and then at the **Additional Controls...** sub-menu, see Figure 13-4. Once these selections have been made a list of the additional controls that can be added, as shown in Figure 13-5, will appear.

Additional types of controls can be added to the default VBA Toolbox panel.

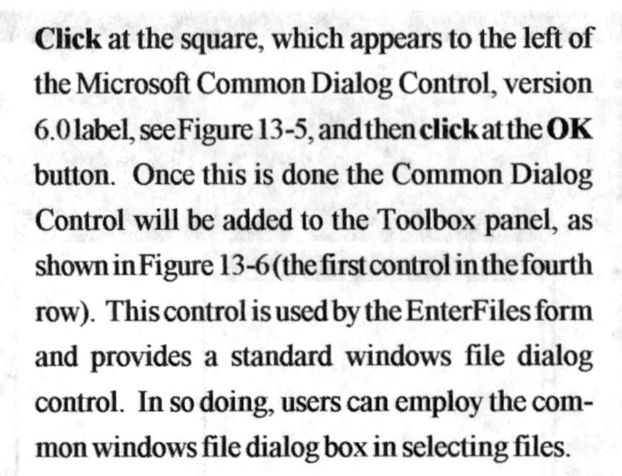

➤ 5 **Click** at the square, which appears to the left of the Microsoft Common Dialog Control, version 6.0 label, see Figure 13-5, and then **click** at the **OK** button. Once this is done the Common Dialog Control will be added to the Toolbox panel, as shown in Figure 13-6 (the first control in the fourth row). This control is used by the EnterFiles form and provides a standard windows file dialog control. In so doing, users can employ the common windows file dialog box in selecting files.

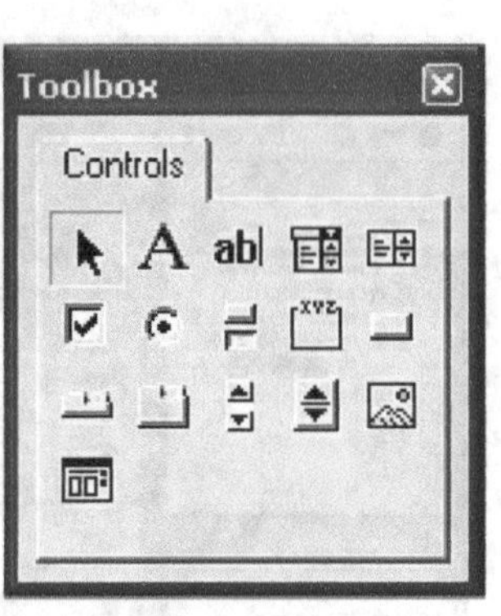

Figure 13-6
Additional Controls

Note that once the Common Dialog Control has been added, the developer will not have to add the control again. It will appear automatically the next time the Toolbox panel is displayed.

At this point we are now ready to begin creating the forms used by the Avenue Wraps.

13.2 Creating the EnterFiles Form

As shown in Figure 13-7, the EnterFiles form consists of a TextBox control, a CommandButton control and the CommonDialog Control. The Common Dialog Control will appear on the form during "design" mode, but during "run" mode, it will be invisible to the user. The form that is created, by following the steps below, should look like Figure 13-7. Shown in Figure 13-8, are the properties for the EnterFiles form. The properties that are displayed are categorized, based upon type, because the Categorized tab is active. If the user wishes to have the properties displayed alphabetically, then the Alphabetic tab should be selected.

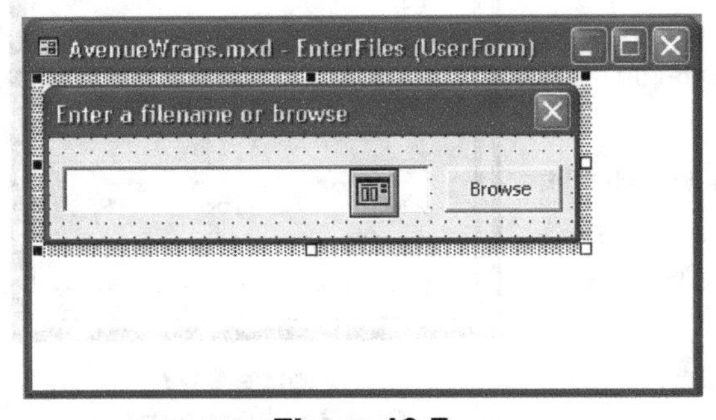

Figure 13-7
EnterFiles Form - Full Design

Properties - EnterFiles

EnterFiles UserForm

Alphabetic | Categorized

Appearance	
BackColor	&H800(
BorderColor	&H8000001
BorderStyle	0 - fmBorderSt
Caption	Enter a filenam
ForeColor	&H8000001
SpecialEffect	0 - fmSpecialEf
Behavior	
Cycle	0 - fmCycleAllF
Enabled	True
RightToLeft	False
ShowModal	True
Font	
Font	Tahoma
Misc	
(Name)	EnterFiles
DrawBuffer	32000
HelpContextID	0
MouseIcon	(None)
MousePointer	0 - fmMousePc
Tag	
WhatsThisButto	False
WhatsThisHelp	False
Zoom	100
Picture	
Picture	(None)
PictureAlignmen	2 - fmPictureAl
PictureSizeMode	0 - fmPictureSi
PictureTiling	False
Position	
Height	66.75
Left	0
StartUpPosition	1 - CenterOwr
Top	0
Width	228
Scrolling	
KeepScrollBarsVi	3 - fmScrollBar
ScrollBars	0 - fmScrollBar
ScrollHeight	0
ScrollLeft	0
ScrollTop	0
ScrollWidth	0

Figure 13-8
EnterFiles Properties

➢ 6 **Click** at the **TextBox** control in the Toolbox panel (third control from the left in the first row), then **make a pick** in the blank form. A textbox control should appear.

➢ 7 **Click** at the **CommandButton** control in the Toolbox panel (fifth control from the left in the second row), then **make a pick** in the blank form. The control should now appear.

➢ 8 **Click** at the **Pointer** tool in the Toolbox panel (first control from the left in the first row), then **click** at the textbox control in the blank form. The control should now become highlighted.

➢ **9** **Click** at the **Categorized** tab in the Properties panel (left side of the VBE), then **enter** the following values for Height, Left, Top and Width, respectively, **20**, **6**, **12** and **156**, see Figure 13-9.

➢ **10** **Click** at the title bar of the window, where the text "Project - UserForm1" appears, to make the window active. The Toolbox panel should appear.

➢ **11** **Click** at the **Pointer** tool in the Toolbox panel (first control from the left in the first row), then **click** at the command button control in the blank form. The control should now become highlighted.

➢ **12** **Enter** the following values for the Height, Left, Top and Width properties, respectively, **20**, **168**, **12** and **50**, in the Properties panel.

➢ **13** **Enter** for the Caption property, **Browse**, in the Properties panel.

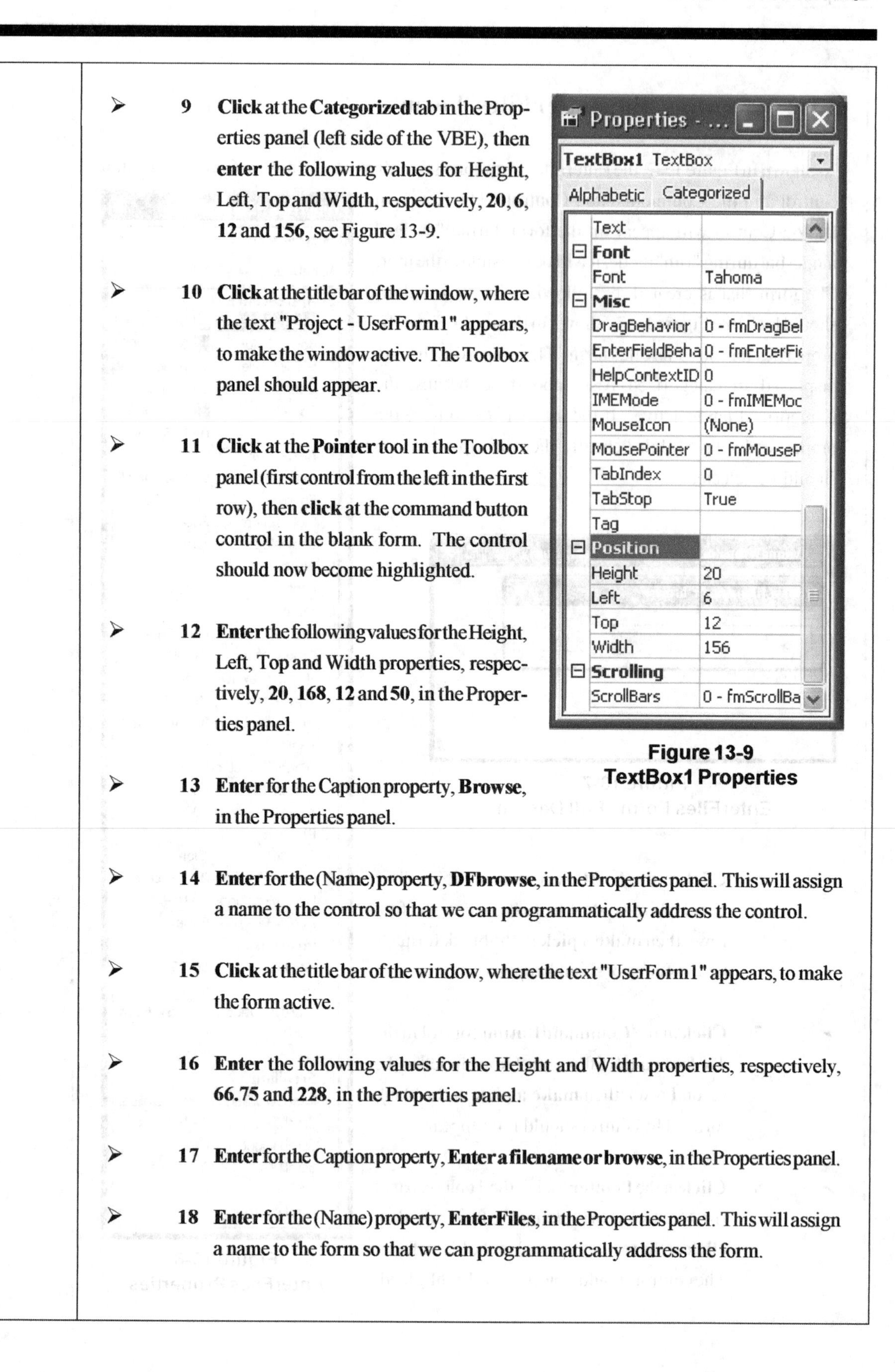

Figure 13-9
TextBox1 Properties

➢ **14** **Enter** for the (Name) property, **DFbrowse**, in the Properties panel. This will assign a name to the control so that we can programmatically address the control.

➢ **15** **Click** at the title bar of the window, where the text "UserForm1" appears, to make the form active.

➢ **16** **Enter** the following values for the Height and Width properties, respectively, **66.75** and **228**, in the Properties panel.

➢ **17** **Enter** for the Caption property, **Enter a filename or browse**, in the Properties panel.

➢ **18** **Enter** for the (Name) property, **EnterFiles**, in the Properties panel. This will assign a name to the form so that we can programmatically address the form.

➢ **19** **Click** at the title bar of the window, where the text "Enter a filename or browse" appears, to make the form active.

➢ **20** **Click** at the **CommonDialog** control in the Toolbox panel (first control in the fourth row), then **make a pick** in the textbox control, see Figure 13-7.

➢ **21** **Click** at the title bar of the window, where the text "Enter a filename or browse" appears, to make the form active.

At this point the form has been designed with all of the necessary controls. It is now time to write the code that is associated with the various controls. Each type of control will support different types of "events". What we will do now is to insert the code for the "events" that the controls on the EnterFiles form will support.

➢ **22** **Double-click** in the gray area just below the command button. Do not double-click on the command button, but rather, just below it. This will invoke the code window for the EnterFiles form, see Figure 13-10. Insert the code shown below into the code window.

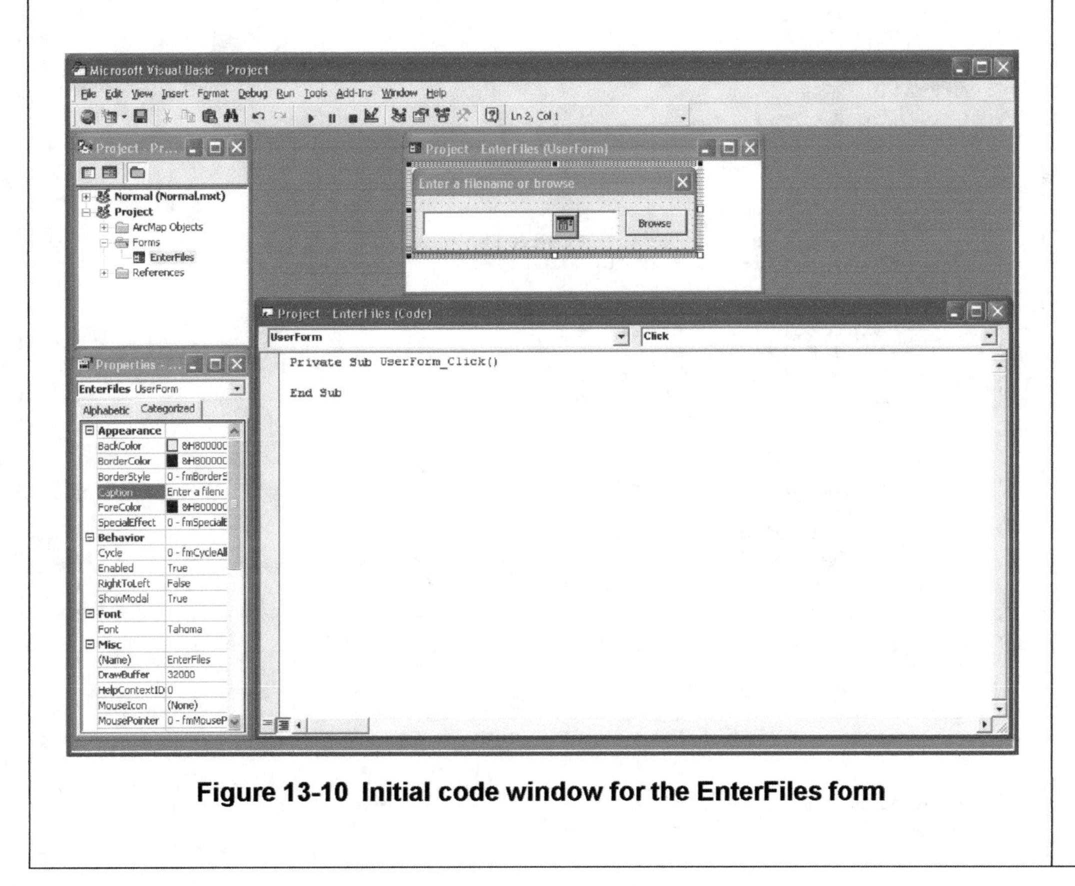

Figure 13-10 Initial code window for the EnterFiles form

```
Option Explicit
'
'  * * * * * * * * * * * * * * * * * * * * * * * * * * * * * * * * *
'  *                                                               *
'  *  Name: EnterFiles                    File Name: enterfls.frm  *
'  *                                                               *
'  * * * * * * * * * * * * * * * * * * * * * * * * * * * * * * * * *
'  *                                                               *
'  *  PURPOSE:  MULTI-FILE SELECTION CONTROLS AND EVENTS CODE      *
'  *                                                               *
'  *  GIVEN:    Handled by VB                                      *
'  *                                                               *
'  *  RETURN:   Handled by VB                                      *
'  *                                                               *
'  *  NOTE:     Global variables are used to transfer information  *
'  *            from the form to the modules, global variables     *
'  *            begin with the characters ug                       *
'  *                                                               *
'  * * * * * * * * * * * * * * * * * * * * * * * * * * * * * * * * *
'

Private Sub DFbrowse_Click()
'
   Dim opmode As Integer
'
'  ---Handle any errors that may occur
   On Error GoTo Errorhandler
'
'  ---Get rid of the dialog box
   Unload EnterFiles
'
'  ---Set property that selecting Cancel is an error
   CommonDialog2.CancelError = True
'
'  ---Define mode of operation to be to open a file or group of files
   opmode = 1
'
'  ---Check if a file is to be created
   If (ugfilterIndex = -1) Then
      opmode = 2
   End If
'
'  ---Set the properties associated with the file dialog for
'  ---opening a file or a group of files
   If (opmode = 1) Then
'     ---Enable multiselect
'     ---Explorer-like Interface
'     ---Long Filenames
'     ---Allow invalid characters
'     ---Hide read-only check box
      CommonDialog2.Flags = &H200 Or &H80000 Or &H200000 Or _
                            &H100 Or &H4
   Else
'     ---Explorer-like Interface
'     ---Long Filenames
'     ---Hide read-only check box
'     ---Prompt to confirm overwrite of existing file
      CommonDialog2.Flags = &H80000 Or &H200000 Or &H4 Or &H2
   End If
'
'  ---Initialize the name of the selected file
   If (opmode = 1) Then
      ugsearchstring = " "
   Else
      CommonDialog2.fileName = ugsearchstring
   End If
'
'  ---Assign the file filter
   CommonDialog2.Filter = ugfilterStr
```

```
'
'  ---Specify default file type to be displayed
   If (opmode = 1) Then
      CommonDialog2.FilterIndex = ugfilterIndex + 1
   Else
      CommonDialog2.FilterIndex = 1
   End If
'
'  ---Define the message box title
   CommonDialog2.DialogTitle = ugFDtitle
'
'  ---Define the initial directory
   CommonDialog2.InitDir = ugcwDirName
'
'  ---Initialize the name of the selected file to be passed back
   ugsearchstring = " "
'
'  ---Handle opening a file or group of files
   If (opmode = 1) Then
'
'     ---Increase allocated memory for multiple file selection
      CommonDialog2.MaxFileSize = 32000
'
'     ---Allow the user the ability to pick the file
      CommonDialog2.ShowOpen
   Else
'
'     ---Allow the user the ability to pick the file
      CommonDialog2.ShowSave
   End If
'
'  ---Get the name of the file
   ugsearchstring = CommonDialog2.fileName
'
'  ---Our work is done
   Exit Sub
'
'  ---Handle any errors that were detected
Errorhandler:
   Unload EnterFiles
   Exit Sub
'
End Sub

Private Sub TextBox1_Exit(ByVal Cancel As MSForms.ReturnBoolean)
'
   Dim aName As String
'
'  ---Get the string presently in the text box
   aName = Trim(TextBox1.Value)
'
'  ---Make sure something is present
   If (Len(aName) > 0) Then
'
'     ---Preserve the name of the file
      ugsearchstring = aName
'
'     ---Get rid of the dialog box
      Unload EnterFiles
   End If
'
End Sub

Private Sub UserForm_Initialize()
'
'  ---Position the form at the previous location
   EnterFiles.left = ugvFileFrmLeft
   EnterFiles.top = ugvFileFrmRight
```

```
'
'   ---Check if the text box is to contain a default value
    If (ugfilterIndex = -1) Then
       TextBox1.Value = ugsearchstring
    End If
'
End Sub

Private Sub UserForm_MouseMove(ByVal button As Integer, _
                               ByVal shift As Integer, _
                               ByVal x As Single, ByVal y As Single)
'
'   ---Get the current position of the window
    ugvFileFrmLeft = EnterFiles.left
    ugvFileFrmRight = EnterFiles.top
'
End Sub
```

In reviewing the code window for the EnterFiles form, the developer will notice two drop-downs at the top of the window. The drop-down on the left contains the controls, which appear in the form, as well as the form itself (UserForm), while the drop-down of the right contains the events that are associated with a control.

In addition, the developer will notice that the global variable, ugsearchstring, will contain the response which was entered by the user in the Common Dialog. This variable will contain a blank character or an empty string ("") if the user cancels out of the Common Dialog.

➤ **23** **Close** the code window and the form.

At this point we are ready to create the HDialogBox form.

13.3 Creating the HDialogBox Form

As shown in Figure 13-11, the HDialogBox form consists of two CommandButton controls. Shown in Figure 13-12, are the properties for the HDialogBox form. The properties that are displayed are listed alphabetically, because the Alphabetic tab is active. If the user wishes to have the properties displayed according to category, then the Categorized tab should be selected. For this form all the developer needs to do is to create the two controls. As discussed at the beginning of this Chapter, all other controls will be added programmatically.

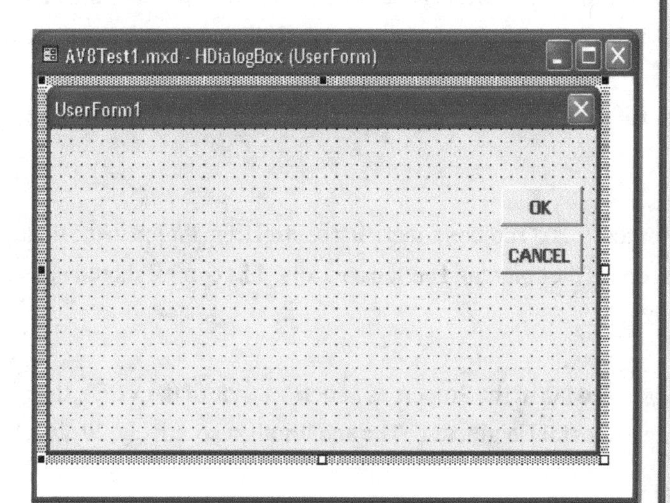

Figure 13-11
HDialogBox Form - Full Design

➤ **24** Click at the **Insert** menu and then at the **UserForm** sub-menu. **Reposition** the blank form and Toolbox panel as desired. Make sure the form window is large enough to accommodate the form, as shown in Figure 13-11.

➤ **25** **Click** at the **CommandButton** control in the Toolbox panel (fifth control from the left in the second row), then **make a pick** in the blank form. The control should now appear.

Properties - ...

HDialogBox UserForm

Alphabetic | Categorized

(Name)	HDialogBox
BackColor	&H80000C
BorderColor	&H800000128
BorderStyle	0 - fmBorderStyle
Caption	UserForm1
Cycle	0 - fmCycleAllForr
DrawBuffer	32000
Enabled	True
Font	Tahoma
ForeColor	&H800000128
Height	185.25
HelpContextID	0
KeepScrollBarsVisib	3 - fmScrollBarsBc
Left	0
MouseIcon	(None)
MousePointer	0 - fmMousePoint
Picture	(None)
PictureAlignment	2 - fmPictureAlign
PictureSizeMode	0 - fmPictureSizeN
PictureTiling	False
RightToLeft	False
ScrollBars	0 - fmScrollBarsNc
ScrollHeight	0
ScrollLeft	0
ScrollTop	0
ScrollWidth	0
ShowModal	True
SpecialEffect	0 - fmSpecialEffec
StartUpPosition	1 - CenterOwner
Tag	
Top	0
WhatsThisButton	False
WhatsThisHelp	False
Width	324.75
Zoom	100

Figure 13-12
HDialogBox Properties

➢ **26** **Click** at the **CommandButton** control in the Toolbox panel, then **make a pick** in the blank form.

➢ **27** **Click** at the title bar of the window, where the text "UserForm1" appears, to make the form active.

➢ **28** **Enter** the following values for the Height and Width properties, respectively, **185.25** and **324.75**, in the Properties panel.

➢ **29** **Set** the StartUpPosition property to be **0 - Manual**.

➢ **30** **Enter** for the (Name) property, **HDialogBox**, in the Properties panel. This will assign a name to the control so that we can programmatically address the control.

➢ **31** **Click** at the title bar of the window, where the text "UserForm1" appears, to make the form active.

➢ **32** **Click** at the **Pointer** tool in the Toolbox panel (first control from the left in the first row), then **click** at one of the command button controls. The control should now become highlighted.

➢ **33** **Click** at the **Categorized** tab in the Properties panel (left side of the VBE), then **enter** the following values for Height, Left, Top and Width, respectively, **20**, **264**, **28.5** and **48**.

➢ **34** **Enter** for the Caption property, **OK**, in the Properties panel.

➢ **35** **Enter** for the (Name) property, **HDBOK**, in the Properties panel. This will assign a name to the control so that we can programmatically address the control.

➢ **36** **Click** at the title bar of the window, where the text "UserForm1" appears, to make the form active.

➢ **37** **Click** at the **Pointer** tool in the Toolbox panel (first control from the left in the first row), then **click** at the other command button control. The control should now become highlighted.

➢ **38** **Enter** the following values for the Height, Left, Top and Width properties, respectively, **20**, **264**, **52.5** and **48**, in the Properties panel.

➢ 39 **Enter** for the Caption property, **CANCEL**, in the Properties panel.

➢ 40 **Enter** for the (Name) property, **HDBCancel**, in the Properties panel. This will assign a name to the control so that we can programmatically address the control.

At this point the form has been designed with all of the necessary controls. It is now time to write the code that is associated with the various controls. Each type of control will support different types of "events". What we will do now is to insert the code for the "events" that the controls on the HDialogBox form will support.

➢ 41 **Double-click** anywhere in the gray area. Do not double-click on any of the command buttons. This will invoke the code window for the HDialogBox form. Insert the code shown below into the code window.

```
Option Explicit
'
'  * * * * * * * * * * * * * * * * * * * * * * * * * * * * * * * * *
'  *                                                               *
'  *  Name: HDialogBox                    File Name: hdbbuild.frm  *
'  *                                                               *
'  * * * * * * * * * * * * * * * * * * * * * * * * * * * * * * * * *
'  *                                                               *
'  *  PURPOSE:  CUSTOMIZABLE HORIZONTAL DIALOG BOX CONTROLS AND    *
'  *            EVENTS CODE                                        *
'  *                                                               *
'  *  GIVEN:    Handled by VB                                      *
'  *                                                               *
'  *  RETURN:   Handled by VB                                      *
'  *                                                               *
'  *  NOTE:     Global variables are used to transfer information  *
'  *            from the form to the modules, global variables     *
'  *            begin with the characters ug                       *
'  *                                                               *
'  * * * * * * * * * * * * * * * * * * * * * * * * * * * * * * * * *
'

Private Sub UserForm_Initialize()
'
'
'  ---Position the form at the previous location
   HDialogBox.left = ughDialogBoxLeft
   HDialogBox.top = ughDialogBoxRight
'
End Sub

Private Sub HDBCancel_Click()
'
   Dim ii, jj As Long
'
'  ---Remove any data from the list to be passed back
   If (ughDialogBoxData.Count > 0) Then
      jj = ughDialogBoxData.Count + 1
      For ii = 1 To ughDialogBoxData.Count
          jj = jj - 1
          ughDialogBoxData.Remove (jj)
      Next
   End If
'
```

```
'  ---Get the current position of the window
   ughDialogBoxLeft = HDialogBox.left
   ughDialogBoxRight = HDialogBox.top
'
'  ---Remove the form from the display
   Unload HDialogBox
'
End Sub

Private Sub HDBOK_Click()
'
   Dim aCntrl As Control
   Dim aTag As Variant
   Dim aString As Variant
   Dim aValue As Variant
   Dim i As Long
'
'  ---Cycle thru the controls in the dialog box
   For Each aCntrl In HDialogBox.Controls
'
'      ---Get the tag that is associated with the control
       aTag = aCntrl.Tag
'
'      ---Check if a TextBox control has been found
       If (Len(aTag) > 6) Then
          aString = Mid(aTag, 1, 7)
          If (aString = "TextBox") Then
             aValue = aCntrl.Value
             ughDialogBoxData.Add (aValue)
          End If
       End If
'
'      ---Check if a ComboBox control has been found
       If (Len(aTag) > 7) Then
          aString = Mid(aTag, 1, 8)
          If (aString = "ComboBox") Then
             aValue = aCntrl.Value
             ughDialogBoxData.Add (aValue)
          End If
       End If
   Next
'
'  ---Get the current position of the window
   ughDialogBoxLeft = HDialogBox.left
   ughDialogBoxRight = HDialogBox.top
'
'  ---Remove the form from the display
   Unload HDialogBox
'
End Sub
```

➤ **42** **Close** the code window and the form.

The developer will notice that the global collection, ughDialogBoxData, will contain the responses entered by the user in the dialog. This collection will be empty if the user selects the cancel button. Note that the form can contain textbox and combo-box controls.

At this point we are ready to create the VDialogBox form.

13.4 Creating the VDialogBox Form

As shown in Figure 13-13, the VDialogBox form consists of two CommandButton controls Shown in Figure 13-14, are the properties for the VDialogBox form. The properties that are displayed are categorized, based upon type, because the Categorized tab is active. If the user wishes to have the properties displayed alphabetically, then the Alphabetic tab should be selected. For this form all the developer needs to do is to create the two controls. As discussed at the beginning of this Chapter, all other controls will be added programmatically.

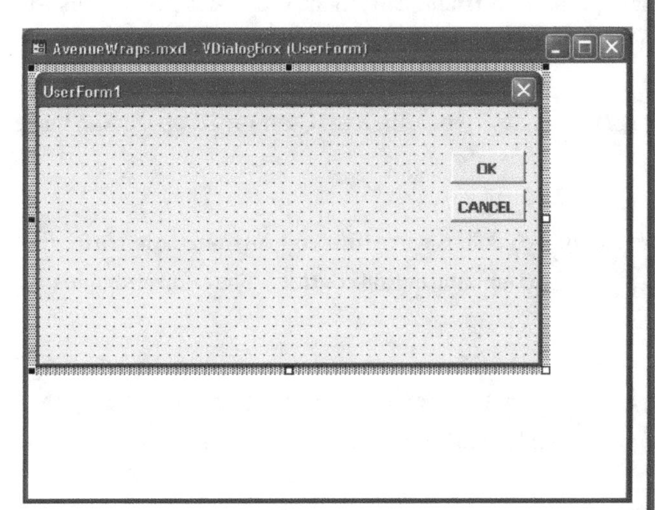

Figure 13-13
VDialogBox Form - Full Design

➤ **43** Click at the **Insert** menu and then at the **UserForm** sub-menu. **Reposition** the blank form and Toolbox panel as desired. Make sure the form window is large enough to accommodate the form, as shown in Figure 13-13.

➤ **44** **Click** at the **CommandButton** control in the Toolbox panel (fifth control from the left in the second row), then **make a pick** in the blank form. The control should now appear.

Properties - VDialogBox

VDialogBox UserForm

Alphabetic | Categorized

Appearance	
BackColor	&H8000000F&
BorderColor	&H8000001
BorderStyle	0 - fmBorderSt
Caption	UserForm1
ForeColor	&H8000001
SpecialEffect	0 - fmSpecialEf
Behavior	
Cycle	0 - fmCycleAllF
Enabled	True
RightToLeft	False
ShowModal	True
Font	
Font	Tahoma
Misc	
(Name)	VDialogBox
DrawBuffer	32000
HelpContextID	0
MouseIcon	(None)
MousePointer	0 - fmMousePc
Tag	
WhatsThisButto	False
WhatsThisHelp	False
Zoom	100
Picture	
Picture	(None)
PictureAlignmen	2 - fmPictureAl
PictureSizeMode	0 - fmPictureSi
PictureTiling	False
Position	
Height	185.25
Left	0
StartUpPosition	0 - Manual
Top	0
Width	324.75
Scrolling	
KeepScrollBarsVi	3 - fmScrollBar
ScrollBars	0 - fmScrollBar
ScrollHeight	0
ScrollLeft	0
ScrollTop	0
ScrollWidth	0

Figure 13-14
VDialogBox Properties

- ➢ **45** **Click** at the **CommandButton** control in the Toolbox panel, then **make a pick** in the blank form.
- ➢ **46** **Click** at the title bar of the window, where the text "UserForm1" appears, to make the form active.
- ➢ **47** **Enter** the following values for the Height and Width properties, respectively, **185.25** and **324.75**, in the Properties panel.
- ➢ **48** **Set** the StartUpPosition property to be **0 - Manual**.
- ➢ **49** **Enter** for the (Name) property, **VDialogBox**, in the Properties panel. This will assign a name to the control so that we can programmatically address the control.
- ➢ **50** **Click** at the title bar of the window, where the text "UserForm1" appears, to make the form active.
- ➢ **51** **Click** at the **Pointer** tool in the Toolbox panel (first control from the left in the first row), then **click** at one of the command button controls. The control should now become highlighted.
- ➢ **52** **Click** at the **Categorized** tab in the Properties panel (left side of the VBE), then **enter** the following values for Height, Left, Top and Width, respectively, **20**, **264**, **28.5** and **48**.
- ➢ **53** **Enter** for the Caption property, **OK**, in the Properties panel.
- ➢ **54** **Enter** for the (Name) property, **VDBOK**, in the Properties panel. This will assign a name to the control so that we can programmatically address the control.
- ➢ **55** **Click** at the title bar of the window, where the text "UserForm1" appears, to make the form active.
- ➢ **56** **Click** at the **Pointer** tool in the Toolbox panel (first control from the left in the first row), then **click** at the other command button control. The control should now become highlighted.
- ➢ **57** **Enter** the following values for the Height, Left, Top and Width properties, respectively, **20**, **264**, **52.5** and **48**, in the Properties panel.

➢ **58** **Enter** for the Caption property, **CANCEL**, in the Properties panel.

➢ **59** **Enter** for the (Name) property, **VDBCancel**, in the Properties panel. This will assign a name to the control so that we can programmatically address the control.

At this point the form has been designed with all of the necessary controls. It is now time to write the code that is associated with the various controls. Each type of control will support different types of "events". What we will do now is to insert the code for the "events" that the controls on the VDialogBox form will support.

➢ **60** **Double-click** anywhere in the gray area. Do not double-click on any of the command buttons. This will invoke the code window for the VDialogBox form. Insert the code shown below into the code window.

```
Option Explicit
'
'  * * * * * * * * * * * * * * * * * * * * * * * * * * * * * * * * *
'  *                                                                 *
'  *  Name: VDialogBox                      File Name: vdbbuild.frm  *
'  *                                                                 *
'  * * * * * * * * * * * * * * * * * * * * * * * * * * * * * * * * *
'  *                                                                 *
'  *  PURPOSE:  CUSTOMIZABLE VERTICAL DIALOG BOX CONTROLS AND        *
'  *            EVENTS CODE                                          *
'  *                                                                 *
'  *  GIVEN:    Handled by VB                                        *
'  *                                                                 *
'  *  RETURN:   Handled by VB                                        *
'  *                                                                 *
'  *  NOTE:     Global variables are used to transfer information    *
'  *            from the form to the modules, global variables       *
'  *            begin with the characters ug                         *
'  *                                                                 *
'  * * * * * * * * * * * * * * * * * * * * * * * * * * * * * * * * *
'

Private Sub UserForm_Initialize()
'
'
'  ---Position the form at the previous location
   VDialogBox.left = ugvDialogBoxLeft
   VDialogBox.top = ugvDialogBoxRight
'
End Sub

Private Sub VDBCancel_Click()
'
   Dim ii, jj As Long
'
'  ---Remove any data from the list to be passed back
   If (ugvDialogBoxData.Count > 0) Then
      jj = ugvDialogBoxData.Count + 1
      For ii = 1 To ugvDialogBoxData.Count
          jj = jj - 1
          ugvDialogBoxData.Remove (jj)
      Next
   End If
```

```
    '
    '   ---Get the current position of the window
        ugvDialogBoxLeft = VDialogBox.left
        ugvDialogBoxRight = VDialogBox.top
    '
    '   ---Remove the form from the display
        Unload VDialogBox
    '
    End Sub

    Private Sub VDBOK_Click()
    '
        Dim aCntrl As Control
        Dim aTag As Variant
        Dim aString As Variant
        Dim aValue As Variant
    '
    '   ---Cycle thru the controls in the dialog box
        For Each aCntrl In VDialogBox.Controls
    '
    '       ---Get the tag that is associated with the control
            aTag = aCntrl.Tag
    '
    '       ---Check if a TextBox control has been found
            If (Len(aTag) > 6) Then
               aString = Mid(aTag, 1, 7)
               If (aString = "TextBox") Then
                  aValue = aCntrl.Value
                  ugvDialogBoxData.Add (aValue)
               End If
            End If
    '
    '       ---Check if a ComboBox control has been found
            If (Len(aTag) > 7) Then
               aString = Mid(aTag, 1, 8)
               If (aString = "ComboBox") Then
                  aValue = aCntrl.Value
                  ugvDialogBoxData.Add (aValue)
               End If
            End If
        Next
    '
    '   ---Get the current position of the window
        ugvDialogBoxLeft = VDialogBox.left
        ugvDialogBoxRight = VDialogBox.top
    '
    '   ---Remove the form from the display
        Unload VDialogBox
    '
    End Sub
```

➢ **61** **Close** the code window and the form.

The developer will notice that the global collection, ugvDialogBoxData, will contain the responses entered by the user in the dialog. This collection will be empty if the user selects the cancel button. Note that the form can contain textbox and combo-box controls.

At this point we are ready to create the VDialogBox2 form.

13.5 Creating the VDialogBox2 Form

As shown in Figure 13-15, the VDialogBox2 form consists of three CommandButton controls Shown in Figure 13-16, are the properties for the VDialogBox form. The properties that are displayed are categorized, based upon type, because the Categorized tab is active. If the user wishes to have the properties displayed alphabetically, then the Alphabetic tab should be selected. For this form all the developer needs to do is to create the three controls. As discussed at the beginning of this Chapter, all other controls will be added programmatically.

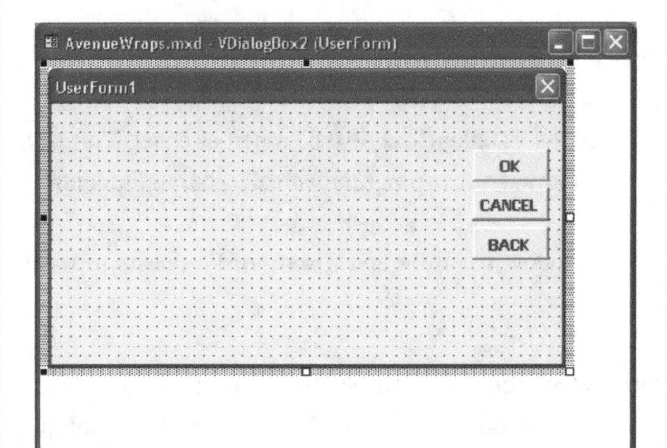

Figure 13-15
VDialogBox2 Form - Full Design

Properties - VDialogBox2

VDialogBox2 UserForm

Alphabetic | Categorized

Appearance	
BackColor	000F8
BorderColor	&H8000001
BorderStyle	0 - fmBorderSt
Caption	UserForm1
ForeColor	&H8000001
SpecialEffect	0 - fmSpecialEf
Behavior	
Cycle	0 - fmCycleAllF
Enabled	True
RightToLeft	False
ShowModal	True
Font	
Font	Tahoma
Misc	
(Name)	VDialogBox2
DrawBuffer	32000
HelpContextID	0
MouseIcon	(None)
MousePointer	0 - fmMousePc
Tag	
WhatsThisButto	False
WhatsThisHelp	False
Zoom	100
Picture	
Picture	(None)
PictureAlignmen	2 - fmPictureAl
PictureSizeMode	0 - fmPictureSi
PictureTiling	False
Position	
Height	185.25
Left	0
StartUpPosition	0 - Manual
Top	0
Width	324.75
Scrolling	
KeepScrollBarsVi	3 - fmScrollBar
ScrollBars	0 - fmScrollBar
ScrollHeight	0
ScrollLeft	0
ScrollTop	0
ScrollWidth	0

Figure 13-16
VDialogBox2 Properties

➤ **62** Click at the **Insert** menu and then at the **UserForm** sub-menu. **Reposition** the blank form and Toolbox panel as desired. Make sure the form window is large enough to accommodate the form, as shown in Figure 13-15.

➤ **63** **Click** at the **CommandButton** control in the Toolbox panel (fifth control from the left in the second row), then **make a pick** in the blank form. The control should now appear.

- ➢ 64 **Click** at the **CommandButton** control in the Toolbox panel, then **make a pick** in the blank form.

- ➢ 65 **Click** at the **CommandButton** control in the Toolbox panel, then **make a pick** in the blank form.

- ➢ 66 **Click** at the title bar of the window, where the text "UserForm1" appears, to make the form active.

- ➢ 67 **Enter** the following values for the Height and Width properties, respectively, **185.25** and **324.75**, in the Properties panel.

- ➢ 68 **Set** the StartUpPosition property to be **0 - Manual**.

- ➢ 69 **Enter** for the (Name) property, **VDialogBox2**, in the Properties panel. This will assign a name to the control so that we can programmatically address the control.

- ➢ 70 **Click** at the title bar of the window, where the text "UserForm1" appears, to make the form active.

- ➢ 71 **Click** at the **Pointer** tool in the Toolbox panel (first control from the left in the first row), then **click** at the first command button control that was created. The control should now become highlighted.

- ➢ 72 **Click** at the **Categorized** tab in the Properties panel (left side of the VBE), then **enter** the following values for Height, Left, Top and Width, respectively, **20**, **264**, **28.5** and **48**.

- ➢ 73 **Enter** for the Caption property, **OK**, in the Properties panel.

- ➢ 74 **Enter** for the (Name) property, **VDB2OK**, in the Properties panel. This will assign a name to the control so that we can programmatically address the control.

- ➢ 75 **Click** at the title bar of the window, where the text "UserForm1" appears, to make the form active.

- ➢ 76 **Click** at the **Pointer** tool in the Toolbox panel (first control from the left in the first row), then **click** at the second command button control that was created. The control should now become highlighted.

➢ 77 **Enter** the following values for the Height, Left, Top and Width properties, respectively, **20**, **264**, **52.5** and **48**, in the Properties panel.

➢ 78 **Enter** for the Caption property, **CANCEL**, in the Properties panel.

➢ 79 **Enter** for the (Name) property, **VDB2Cancel**, in the Properties panel. This will assign a name to the control so that we can programmatically address the control.

➢ 80 **Click** at the title bar of the window, where the text "UserForm1" appears, to make the form active.

➢ 81 **Click** at the **Pointer** tool in the Toolbox panel (first control from the left in the first row), then **click** at the third command button control that was created. The control should now become highlighted.

➢ 82 **Enter** the following values for the Height, Left, Top and Width properties, respectively, **20**, **264**, **76.5** and **48**, in the Properties panel.

➢ 83 **Enter** for the Caption property, **BACK**, in the Properties panel.

➢ 84 **Enter** for the (Name) property, **VDB2Back**, in the Properties panel. This will assign a name to the control so that we can programmatically address the control.

At this point the form has been designed with all of the necessary controls. It is now time to write the code that is associated with the various controls. Each type of control will support different types of "events". What we will do now is to insert the code for the "events" that the controls on the VDialogBox2 form will support.

➢ 85 **Double-click** anywhere in the gray area. Do not double-click on any of the command buttons. This will invoke the code window for the VDialogBox2 form. Insert the code shown below into the code window.

```
Option Explicit
'
' * * * * * * * * * * * * * * * * * * * * * * * * * * * * * * * * *
' *                                                               *
' *  Name: VDialogBox2                    File Name: vdbbuld2.frm *
' *                                                               *
' * * * * * * * * * * * * * * * * * * * * * * * * * * * * * * * *
' *                                                               *
' *  PURPOSE:  CUSTOMIZABLE VERTICAL DIALOG BOX WITH A BACK       *
' *            BUTON CONTROLS AND EVENTS CODE                     *
' *                                                               *
' *  GIVEN:    Handled by VB                                      *
' *                                                               *
' *  RETURN:   Handled by VB                                      *
' *                                                               *
' *  NOTE:     Global variables are used to transfer information *
' *            from the form to the modules, global variables    *
' *            begin with the characters ug                       *
' *                                                               *
' * * * * * * * * * * * * * * * * * * * * * * * * * * * * * * * *
'

Private Sub UserForm_Initialize()
'
'
'  ---Position the form at the previous location
   VDialogBox2.left = ugvDialogBoxLeft
   VDialogBox2.top = ugvDialogBoxRight
'
End Sub

Private Sub VDB2Back_Click()
'
   Dim ii, jj As Long
'
'  ---Remove any data from the list to be passed back
   If (ugvDialogBoxData.Count > 0) Then
      jj = ugvDialogBoxData.Count + 1
      For ii = 1 To ugvDialogBoxData.Count
          jj = jj - 1
          ugvDialogBoxData.Remove (jj)
      Next
   End If
'
'  ---Add one item into the list with a special code denoting
'  ---the BACK button was selected
   ugvDialogBoxData.Add ("BACK_BUTTON")
'
'  ---Get the current position of the window
   ugvDialogBoxLeft = VDialogBox2.left
   ugvDialogBoxRight = VDialogBox2.top
'
'  ---Remove the form from the display
   Unload VDialogBox2
'
End Sub

Private Sub VDB2Cancel_Click()
'
   Dim ii, jj As Long
'
'  ---Remove any data from the list to be passed back
   If (ugvDialogBoxData.Count > 0) Then
      jj = ugvDialogBoxData.Count + 1
      For ii = 1 To ugvDialogBoxData.Count
          jj = jj - 1
          ugvDialogBoxData.Remove (jj)
      Next
   End If
```

```
'
'   ---Get the current position of the window
    ugvDialogBoxLeft = VDialogBox2.left
    ugvDialogBoxRight = VDialogBox2.top
'
'   ---Remove the form from the display
    Unload VDialogBox2
'
End Sub

Private Sub VDB2OK_Click()
'
    Dim aCntrl As Control
    Dim aTag As Variant
    Dim aString As Variant
    Dim aValue As Variant
'
'   ---Cycle thru the controls in the dialog box
    For Each aCntrl In VDialogBox2.Controls
'
'       ---Get the tag that is associated with the control
        aTag = aCntrl.Tag
'
'       ---Check if a TextBox control has been found
        If (Len(aTag) > 6) Then
           aString = Mid(aTag, 1, 7)
           If (aString = "TextBox") Then
              aValue = aCntrl.Value
              ugvDialogBoxData.Add (aValue)
           End If
        End If
'
'       ---Check if a ComboBox control has been found
        If (Len(aTag) > 7) Then
           aString = Mid(aTag, 1, 8)
           If (aString = "ComboBox") Then
              aValue = aCntrl.Value
              ugvDialogBoxData.Add (aValue)
           End If
        End If
    Next
'
'   ---Get the current position of the window
    ugvDialogBoxLeft = VDialogBox2.left
    ugvDialogBoxRight = VDialogBox2.top
'
'   ---Remove the form from the display
    Unload VDialogBox2
'
End Sub
```

➢ **86** **Close** the code window and the form.

➢ **87** **Click** at the **File** menu and then the **Close and Return to ArcMap** sub-menu.

The steps presented below assume that the work to date has been done in a new project file, not an existing one. So that:

➢ **88** **Click** at the **File** menu and then the **Save As...** sub-menu.

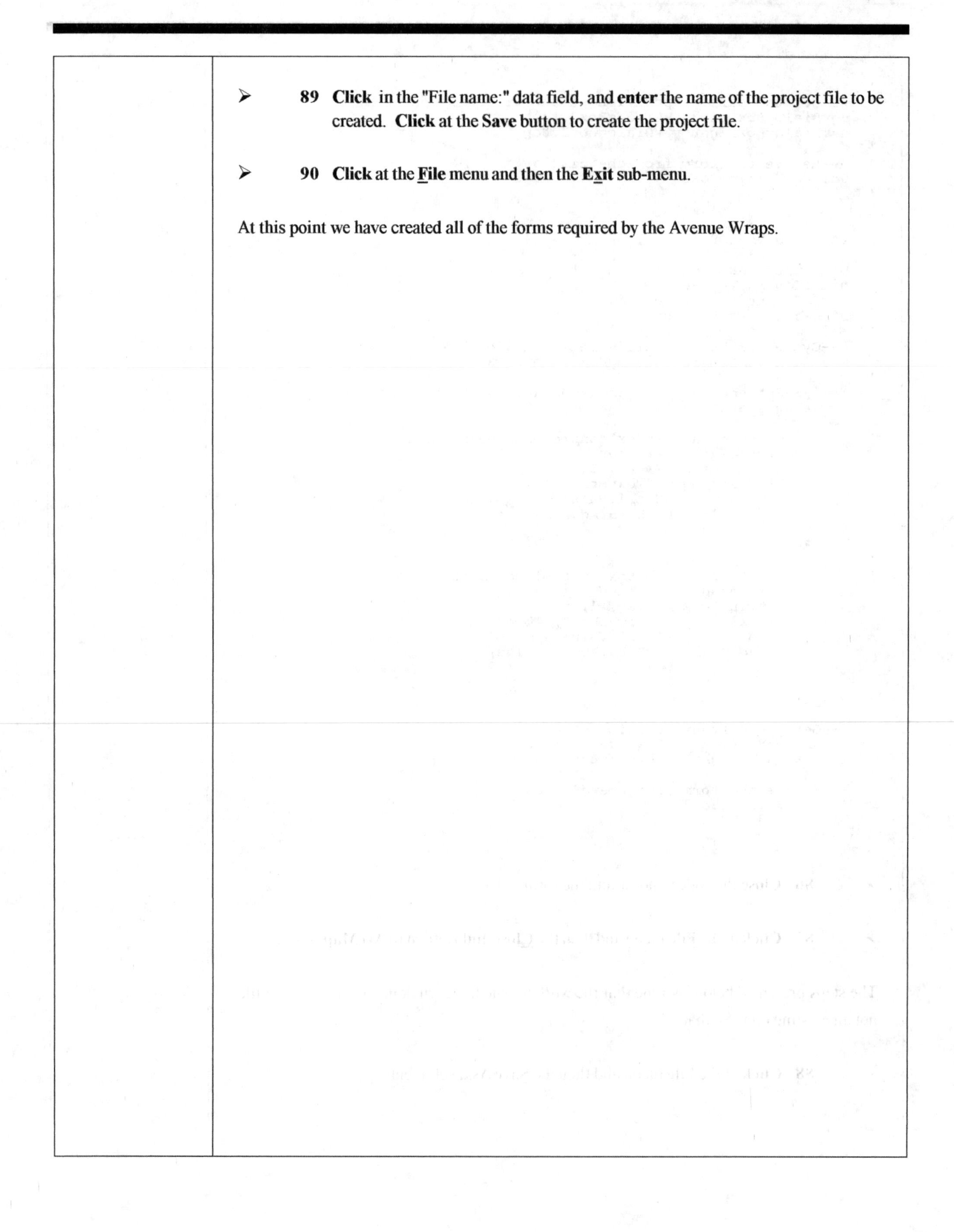

➢ **89** **Click** in the "File name:" data field, and **enter** the name of the project file to be created. **Click** at the **Save** button to create the project file.

➢ **90** **Click** at the **File** menu and then the **Exit** sub-menu.

At this point we have created all of the forms required by the Avenue Wraps.

CHAPTER 14

APPLICATION DEPLOYMENT METHODS

APPLICATION DEPLOYMENT METHODS

This chapter discusses the possible methods in deploying an ArcMap™ based application. When working in the **VBA**, Visual Basic for Applications, environment, the programmer can deploy an application in the form of an ArcMap™ template (*.mxt) or an ArcMap™ project file, which references an ArcMap™ template. In this mode the source for the application is packaged with the ArcMap™ template. To ensure that the source is not modified or viewed, the programmer can password protect the ArcMap™ template. In the **VB**, Visual Basic, environment, the programmer will create DLL(s) for the application.

From a performance point of view, creating DLL(s) for an application will provide slightly better performance over creating an ArcMap™ project file, which references an ArcMap™ template. The downside is that this process is much more involved than creating an ArcMap™ template.

14.1 Creating an ArcMap Template File

The steps presented below describe how the developer can create an ArcMap™ template file. In the **VBA** environment, the developer works within an ArcMap™ project file. This project file will contain the VBA code, forms, combo boxes, tools, etc. that the application uses. Once the application has been tested and is ready for deployment, the developer should:

➢ 1 Open the ArcMap™ project file, which is to be converted into an ArcMap™ template file.

Figure 14-1 Save As Dialog Box

➢ 2 **Click** at the **File** menu and the **Save As...** sub-menu. The conventional file browsing window of Figure 14-1 is displayed.

Create the ArcMap™ template file

Apply password protection to the template file

➢ 3 **Scroll** down in the "Save as type:" combo box, and **click** at the ***.mxt** option.

➢ 4 **Click** in the "File name:" data field, and **enter** the name of the template file to be created. **Click** at the **Save** button to create the template file.

➢ 5 **Click** at the **File** menu and then at the **Exit** sub menu to exit ArcMap™.

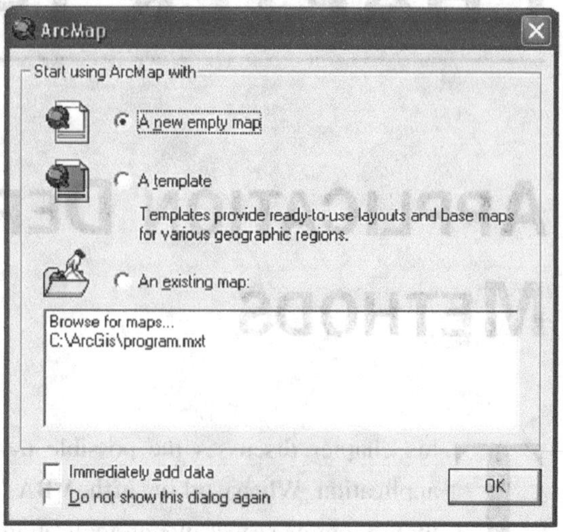

Figure 14-2 ArcMap Initial Dialog Box

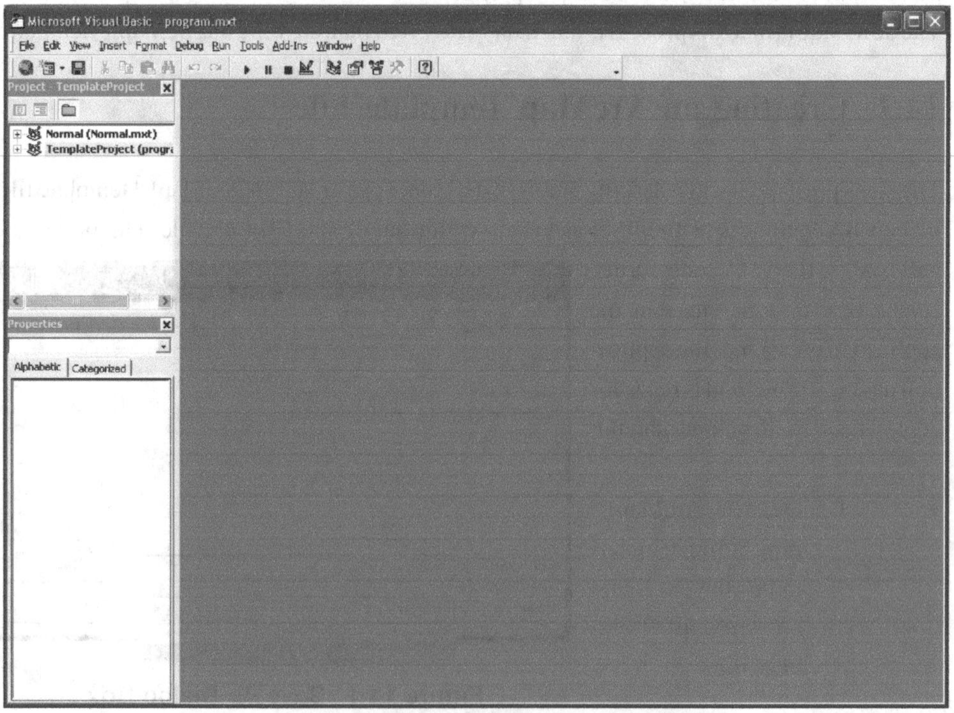

➢ 6 **Invoke** the ArcMap™ program. The "Start using ArcMap with" selection box of Figure 14-2 is displayed.

➢ 7 **Click** at the radio button, which appears, to the left of the label "A new empty map", and then **click** at the **OK** button to confirm the selection. This will close the dialog box.

Figure 14-3 Visual Basic Editor Interface

➤ 8 **Click** at the **File** menu.

➤ 9 **Click** at the first name which appears under the Export Map... sub menu. This should be the name of the template file that was created above.

➤ 10 Click at the **Tools** menu and then at the **Macros** and **Visual Basic Editor** sub-menus, see Figure 14-3.

➤ 11 **Right-click** on the **TemplateProject** name to invoke the properties pop-up window, see Figure 14-4.

➤ 12 **Click** at the **TemplateProject Properties...** command. The Project Properties window of Figure 14-5 should now appear.

➤ 13 **Click** on the **Protection** tab.

View Code
View Object
TemplateProject Properties...
Insert
Import File...
Export File...
Remove
Print...
Dockable
Hide

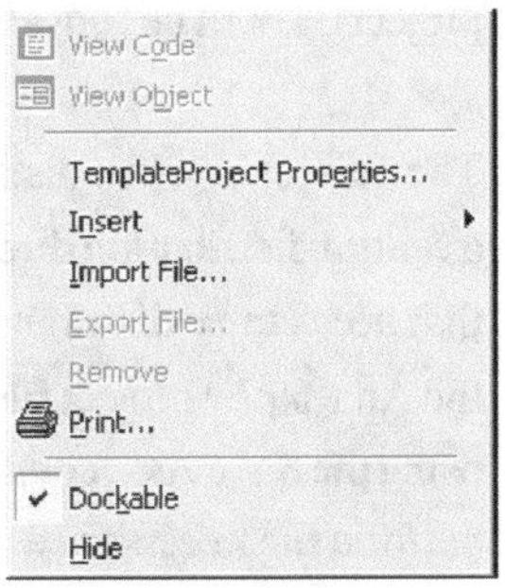

Figure 14-4 Properties Pop-Up

➤ 14 **Click** on the square to the left of the label "Lock project for viewing".

➤ 15 **Click** in the "Password" data field, and **enter** the password that is to be used to restrict access to the customizations provided in the template.

Passwords are case sensitive, that is, there is a difference between upper and lower case characters.

➤ 16 **Click** in the "Confirm password" data field, and **enter** the same password to confirm the password and then **click** at the **OK** button to complete the password protection phase.

TemplateProject - Project Properties
General Protection
Lock project
Lock project for viewing
Password to view project properties
Password
Confirm password
OK Cancel Help

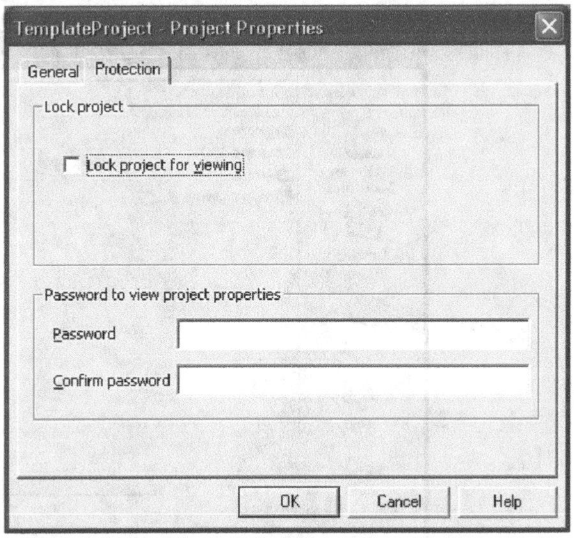

Figure 14-5 Template Protection Dialog Box

➤ 17 **Click** at the **File** menu and then the **Close and Return to ArcMap** sub-menu.

➤ 18 **Click** at the **File** menu and then the **Save** sub-menu.

➤ 19 **Click** at the **File** menu and then the **Exit** sub-menu.

The template file has now been applied password protection and is ready to be referenced by an ArcMap™ project file.

14.2 Creating an ArcMap Project File referencing an ArcMap Template File

Once an ArcMap™ template file has been created, the developer can create an ArcMap™ project file which references the template file. In so doing, the developer can distribute the project file, along with the template file. Note that the template file must be included with the project file. The project file merely contains a reference to the template file. That is why the file size of the project file will be considerably smaller than the file size of the template file.

In distributing an ArcMap™ project file, consideration should be given to establishing a central distribution directory where all files that are associated with the application should reside.

The developer when distributing the application will want to give consideration to setting up a central distribution directory where the project and template files, along with any other data that needs to be distributed, should reside. The developer can then instruct the end user that the ArcMap™ project file can be initially opened from the central distribution directory, then using the Save As... command, a new project file can be created, which the end user can use to perform the necessary work.

➢ 1 **Invoke** the ArcMap™ program. The "Start using ArcMap with" selection box of Figure 14-2 is displayed.

➢ 2 **Click** at the radio button, which appears, to the left of the label "An existing map" of Figure 14-2, and then **click** at the **OK** button to confirm the selection. This will close the dialog box, and will display the conventional file browsing window, similar to that of Figure 14-6

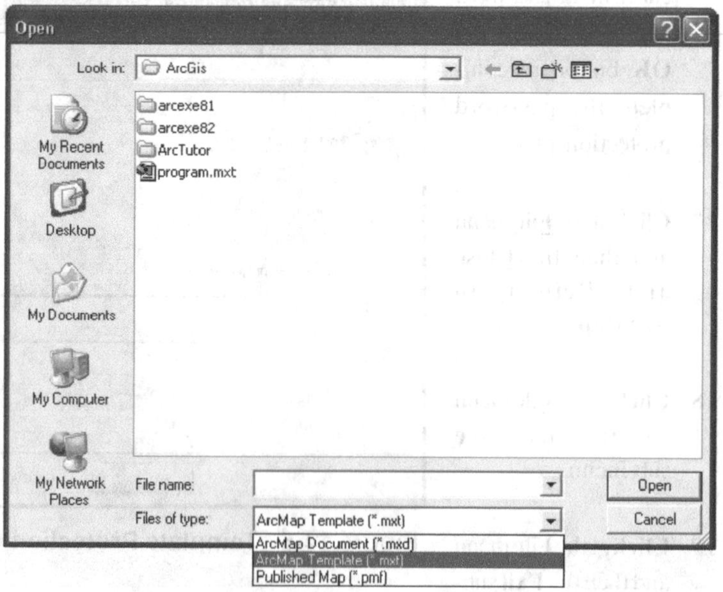

Figure 14-6 Open An Existing Map Dialog Box

➢ 3 **Navigate** to the directory where the appropriate template file resides, **Scroll** down in the "Files of type:" combo box, and **click** at the ***.mxt** option. The template file should now appear.

➢ 4 **Click** at the name of the template file, in the file display area, and then **click** at the **Open** button to open it.

The typical ArcMap™ interface window is displayed. At this point we have created a new project file, which only has a reference to the template file. The project file has not been assigned a name.

➢ 5 **Click** at the **File** menu and then the **Save As...** sub-menu. The conventional file browsing window, similar to that of Figure 14-7 is now displayed.

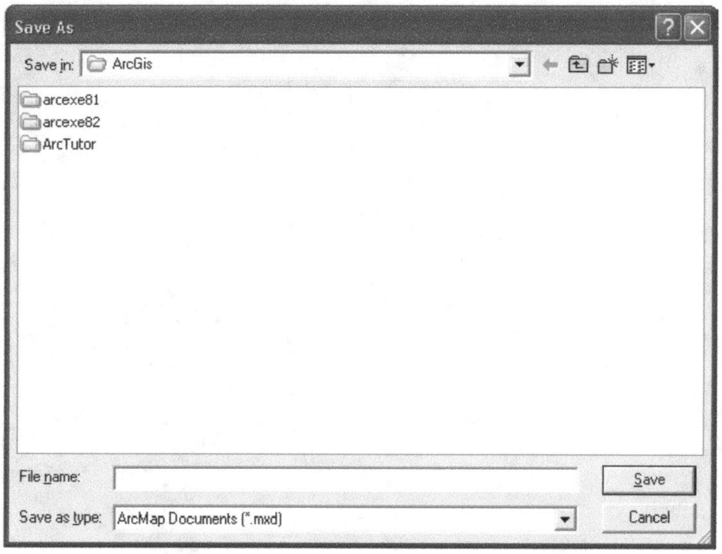

Figure 14-7 Save As Dialog Box

➢ 6 **Navigate** to the directory where the project file is to be stored, **click** in the "File name:" data field, and **enter** the desired project file name omitting any file name extension.

➢ 7 **Click** at the **Save** button to create the project file.

➢ 8 **Click** at the **File** menu and then at the **Exit** sub menu to exit ArcMap™.

The project file which has been created contains only a reference to the template file. The end user can begin working with this project file, adding data if need be, utilizing the customizations available in the template file. Since the template file is password protected, the customizations are safe from tampering.

In distributing an ArcMap™ project file, which references a template file, the template file must be included in the distribution.

APPENDIX A

PALETTE INDEX VALUES

This appendix contains the 17 default ArcMap™ galleries (palettes) and their corresponding default symbol index numbers. For each of these galleries, the keyword code name, as used in the in Avenue Wraps, is shown to the right of the gallery number. Following the listings the reader will find sample images for a few of the galleries. Most of the samples were created using the Style Dump tool, which is included in the **ArcObjects Developer Kit** sub-directory within the ArcGIS™ distribution directory.

Gallery: 1 Reference Systems

Index = 1 Graticule
Index = 2 Graticule with sub ticks
Index = 3 Graticule with inset labels
Index = 4 Graticule with Calibrated Border
Index = 5 Graticule with Border
Index = 6 Measured Grid 1
Index = 7 Measured Grid 2
Index = 8 Button Index Grid
Index = 9 Square Index Grid
Index = 10 Rounded Index Grid
Index = 11 Continuous Index Grid
Index = 12 RoundRect Index Grid

Gallery: 2 Shadows

Index = 1 Grey 20%
Index = 2 Grey 30%
Index = 3 Grey 40%
Index = 4 Grey 50%
Index = 5 Grey 60%
Index = 6 Black
Index = 7 Sienna
Index = 8 Med Sand
Index = 9 Sand
Index = 10 Aqua
Index = 11 Blue
Index = 12 Leaf

Gallery: 3 Area Patches

Index = 1 Rectangle
Index = 2 Rounded Rectangle
Index = 3 Ellipse
Index = 4 Diamond
Index = 5 Park or Preserve
Index = 6 Urbanized Area
Index = 7 Water Body
Index = 8 Natural Area

Gallery: 4 Line Patches

Index = 1 Horizontal
Index = 2 Flowing Water
Index = 3 Incline
Index = 4 Decline
Index = 5 ZigZag
Index = 6 S Curve
Index = 7 Chevron Down
Index = 8 Chevron Up
Index = 9 Arc

Gallery: 5 Labels

Index = 1	U.S. Route
Index = 2	U.S. Interstate HWY
Index = 3	Country 1
Index = 4	Country 2
Index = 5	Country 3
Index = 6	Capital
Index = 7	County
Index = 8	Large City
Index = 9	City
Index = 10	Town
Index = 11	Street
Index = 12	Physical Region
Index = 13	Historic Region
Index = 14	Coastal Region
Index = 15	Ocean
Index = 16	Sea
Index = 17	River
Index = 18	Stream
Index = 19	Banner
Index = 20	Banner, Rounded

Gallery: 6 North Arrows

Index = 1	ESRI North 1
Index = 2	ESRI North 2
Index = 3	ESRI North 3
Index = 4	ESRI North 4
Index = 5	ESRI North 5
Index = 6	ESRI North 6
Index = 7	ESRI North 7
Index = 8	ESRI North 8
Index = 9	ESRI North 9
Index = 10	ESRI North 10
Index = 11	ESRI North 11
Index = 12	ESRI North 12
Index = 13	ESRI North 13
Index = 14	ESRI North 14
Index = 15	ESRI North 15
Index = 16	ESRI North 16
Index = 17	ESRI North 17
Index = 18	ESRI North 18
Index = 19	ESRI North 19
Index = 20	ESRI North 20
Index = 21	ESRI North 21
Index = 22	ESRI North 22
Index = 23	ESRI North 23
Index = 24	ESRI North 24
Index = 25	ESRI North 25
Index = 26	ESRI North 26
Index = 27	ESRI North 27
Index = 28	ESRI North 28
Index = 29	ESRI North 29
Index = 30	ESRI North 30
Index = 31	ESRI North 31
Index = 32	ESRI North 32
Index = 33	ESRI North 33
Index = 34	ESRI North 34
Index = 35	ESRI North 35
Index = 36	ESRI North 36
Index = 37	ESRI North 37
Index = 38	ESRI North 38
Index = 39	ESRI North 39
Index = 40	ESRI North 40
Index = 41	ESRI North 41
Index = 42	ESRI North 42
Index = 43	ESRI North 43
Index = 44	ESRI North 44
Index = 45	ESRI North 45
Index = 46	ESRI North 46
Index = 47	ESRI North 47
Index = 48	ESRI North 48
Index = 49	ESRI North 49
Index = 50	ESRI North 50
Index = 51	ESRI North 51
Index = 52	ESRI North 52
Index = 53	ESRI North 53
Index = 54	ESRI North 54
Index = 55	ESRI North 55

Index=56 ESRI North 56
Index=57 ESRI North 57
Index=58 ESRI North 58
Index=59 ESRI North 59
Index=60 ESRI North 60
Index=61 ESRI North 61
Index=62 ESRI North 62
Index=63 ESRI North 63
Index=64 ESRI North 64
Index=65 ESRI North 65
Index=66 ESRI North 66
Index=67 ESRI North 67
Index=68 ESRI North 68
Index=69 ESRI North 69
Index=70 ESRI North 70
Index=71 ESRI North 71
Index=72 ESRI North 72
Index=73 ESRI North 73
Index=74 ESRI North 74
Index=75 ESRI North 75
Index=76 ESRI North 76
Index=77 ESRI North 77
Index=78 ESRI North 78
Index=79 ESRI North 79
Index=80 ESRI North 80
Index=81 ESRI North 81
Index=82 ESRI North 82
Index=83 ESRI North 83
Index=84 ESRI North 84
Index=85 ESRI North 85
Index=86 ESRI North 86
Index=87 ESRI North 87
Index=88 ESRI North 88
Index=89 ESRI North 89
Index=90 ESRI North 90
Index=91 ESRI North 91
Index=92 ESRI North 92
Index=93 ESRI North 93
Index=94 ESRI North 94
Index=95 ESRI North 95
Index=96 ESRI North 96
Index=97 ESRI North 97

Gallery: 7 Scale Bars

Index=1 Scale Line 1
Index=2 Scale Line 2
Index=3 Scale Line 3
Index=4 Stepped Scale Line
Index=5 Alternating Scale Bar 1
Index=6 Alternating Scale Bar 2
Index=7 Single Division Scale Bar
Index=8 Hollow Scale Bar 1
Index=9 Hollow Scale Bar 2
Index=10 Double Alternating Scale Bar 1
Index=11 Double Alternating Scale Bar 2

Gallery: 8 Legend Items

Index=1 Horizontal Bar with Heading, Labels, and Description
Index=2 Horizontal Single Symbol Description Only
Index=3 Horizontal Single Symbol Label Only
Index=4 Horizontal Single Symbol Layer Name and Description
Index=5 Horizontal Single Symbol Layer Name and Label
Index=6 Horizontal with Heading and Labels
Index=7 Horizontal with Heading, Labels, and Description
Index=8 Horizontal with Layer Name, Heading and Label
Index=9 Horizontal with Layer Name, Heading, Label, and Description
Index=10 Nested with Heading, Labels, and Description

Index=11 Vertical Single Symbol Description Only
Index=12 Vertical Single Symbol Label Only
Index=13 Vertical Single Symbol Layer Name and Description
Index=14 Vertical Single Symbol Layer Name and Label
Index=15 Vertical with Heading and Labels
Index=16 Vertical with Heading, Labels, and Description
Index=17 Vertical with Layer Name, Heading and Label
Index=18 Vertical with Layer Name, Heading, Label, and Description

Gallery: 9 Scale Texts

Index=1 Absolute Scale
Index=2 Centimeters=Kilometers
Index=3 Centimeters = Meters
Index=4 Inches = Feet
Index=5 Inches = Miles
Index=6 Inches = Yards
Index=7 Relative Scale

Gallery: 10 Color Ramps

Index=1 Yellow to Dark Red
Index=2 Blue Light to Dark
Index=3 Purple-Blue Light to Dark
Index=4 Blue-Green Light to Dark
Index=5 Green Light to Dark
Index=6 Purple-Red Light to Dark
Index=7 Red Light to Dark
Index=8 Yellow-Green Light to Dark
Index=9 Gray Light to Dark
Index=10 Brown Light to Dark
Index=11 Orange Light to Dark
Index=12 Blue Bright
Index=13 Purple-Blue Bright
Index=14 Blue-Green Bright
Index=15 Green Bright
Index=16 Purple Bright
Index=17 Purple-Red Bright
Index=18 Red Bright
Index=19 Yellow-Green Bright
Index=20 Orange Bright
Index=21 White to Black
Index=22 Black to White
Index=23 Spectrum-Full Bright
Index=24 Spectrum-Full Light
Index=25 Spectrum-Full Dark
Index=26 Red to Green
Index=27 Cyan to Purple
Index=28 Yellow to Red
Index=29 Partial Spectrum
Index=30 Cyan-Light to Blue-Dark
Index=31 Green to Blue
Index=32 Yellow to Green to Dark Blue
Index=33 Precipitation
Index=34 Temperature
Index=35 Elevation #1
Index=36 Elevation #2
Index=37 Brown to Blue Green Diverging, Bright
Index=38 Brown to Blue Green Diverging, Dark
Index=39 Red to Blue Diverging, Dark
Index=40 Red to Blue Diverging, Bright
Index=41 Purple to Green Diverging, Dark
Index=42 Purple to Green Diverging, Bright
Index=43 Partial Spectrum 1 Diverging
Index=44 Partial Spectrum 2 Diverging
Index=45 Pink to YellowGreen Diverging, Dark
Index=46 Pink to YellowGreen Diverging, Bright

Index=47	Red to Green Diverging, Dark
Index=48	Red to Green Diverging, Bright
Index=49	Distance
Index=50	Surface
Index=51	Slope
Index=52	Aspect
Index=53	Pastels
Index=54	Muted Pastels
Index=55	Enamel
Index=56	Dark Glazes
Index=57	Cool Tones
Index=58	Warm Tones
Index=59	Pastels Blue to Red
Index=60	Verdant Tones
Index=61	Terra Tones
Index=62	Basic Random
Index=63	Pastel Terra Tones
Index=64	Reds
Index=65	Oranges
Index=66	Yellows
Index=67	Yellow Greens
Index=68	Greens
Index=69	Green Blues
Index=70	Cyans
Index=71	Blues
Index=72	Purple Blues
Index=73	Purples
Index=74	Purple Reds
Index=75	Magentas
Index=76	Warm Grey
Index=77	Cool Grey
Index=78	Black and White

Gallery: 11 Borders

Index=1	0.5 Point
Index=2	1.0 Point
Index=3	1.5 Point
Index=4	2.0 Point
Index=5	2.5 Point
Index=6	3.0 Point
Index=7	3.5 Point
Index=8	4.0 Point
Index=9	Double Line
Index=10	Double, Graded
Index=11	Triple Line
Index=12	Triple, Graded
Index=13	Triple, Ctr-W
Index=14	Triple Yellow Black
Index=15	Single, Nautical Dash
Index=16	Double, Nautical Dash

Gallery: 12 Backgrounds

Index=1	Blue
Index=2	Lt Cyan
Index=3	Lt Blue
Index=4	Green
Index=5	Olive
Index=6	Yellow
Index=7	Sand
Index=8	Sepia
Index=9	Sienna
Index=10	Grey 40%
Index=11	Grey 10%
Index=12	Hollow
Index=13	Black
Index=14	Linear Gradient
Index=15	Rectangular Gradient
Index=16	Circular Gradient
Index=17	Cretean Blue
Index=18	White

Gallery: 13 Colors

Index=1	Arctic White
Index=2	Rose Quartz
Index=3	Sahara Sand
Index=4	Topaz Sand
Index=5	Yucca Yellow
Index=6	Olivine Yellow

Index=7	Tzavorite Green	Index=46	Lapis Lazuli
Index=8	Indicolite Green	Index=47	Anemone Violet
Index=9	Sodalite Blue	Index=48	Peony Pink
Index=10	Sugilite Sky	Index=49	Gray 40%
Index=11	Lepidolite Lilac	Index=50	Tuscan Red
Index=12	Rhodolite Rose	Index=51	Cherry Cola
Index=13	Gray 10%	Index=52	Raw Umber
Index=14	Medium Coral Light	Index=53	Olivenite Green
Index=15	Cantaloupe	Index=54	Tarragon Green
Index=16	Mango	Index=55	Leaf Green
Index=17	Autunite Yellow	Index=56	Malachite Green
Index=18	Lemongrass	Index=57	Delft Blue
Index=19	Light Apple	Index=58	Ultra Blue
Index=20	Beryl Green	Index=59	Dark Amethyst
Index=21	Apatite Blue	Index=60	Cattleya Orchid
Index=22	Yogo Blue	Index=61	Gray 50%
Index=23	Heliotrope	Index=62	Dark Umber
Index=24	Fushia Pink	Index=63	Cherrywood Brown
Index=25	Gray 20%	Index=64	Burnt Umber
Index=26	Mars Red	Index=65	Dark Olivenite
Index=27	Fire Red	Index=66	Spruce Green
Index=28	Electron Gold	Index=67	Fir Green
Index=29	Solar Yellow	Index=68	Peacock Green
Index=30	Peridot Green	Index=69	Steel Blue
Index=31	Medium Apple	Index=70	Dark Navy
Index=32	Tourmaline Green	Index=71	Ultramarine
Index=33	Big Sky Blue	Index=72	Purple Heart
Index=34	Cretean Blue	Index=73	Gray 60%
Index=35	Amethyst	Index=74	Rose Dust
Index=36	Ginger Pink	Index=75	Soapstone Dust
Index=37	Gray 30%	Index=76	Tecate Dust
Index=38	Poinsettia Red	Index=77	Lime Dust
Index=39	Flame Red	Index=78	Apple Dust
Index=40	Seville Orange	Index=79	Sage Dust
Index=41	Citroen Yellow	Index=80	Turquoise Dust
Index=42	Macaw Green	Index=81	Blue Gray Dust
Index=43	Quetzel Green	Index=82	Violet Dust
Index=44	Chrysophase	Index=83	Lilac Dust
Index=45	Moorea Blue	Index=84	Tudor Rose Dust

Index=85 Gray 70%
Index=86 Medium Coral
Index=87 Orange Dust
Index=88 Medium Sand
Index=89 Medium Yellow
Index=90 Medium Lime
Index=91 Medium Key Lime
Index=92 Light Vert
Index=93 Oxide Blue
Index=94 Medium Azul
Index=95 Medium Lilac
Index=96 Medium Fushia
Index=97 Gray 80%
Index=98 Tulip Pink
Index=99 Nubuck Tan
Index=100 Light Sienna
Index=101 Light Olivenite
Index=102 Medium Olivenite
Index=103 Fern Green
Index=104 Jadeite
Index=105 Atlantic Blue
Index=106 Pacific Blue
Index=107 Aster Purple
Index=108 Protea Pink
Index=109 Black
Index=110 Cordovan Brown
Index=111 Cocoa Brown
Index=112 Leather Brown
Index=113 Lichen Green
Index=114 Moss Green
Index=115 Lotus Pond Green
Index=116 Deep Forest
Index=117 Larkspur Blue
Index=118 Glacier Blue
Index=119 Blackberry
Index=120 Cabernet

Gallery: 14 Fill Symbols

Index=1 Green
Index=2 Blue
Index=3 Sun
Index=4 Hollow
Index=5 Lake
Index=6 Rose
Index=7 Beige
Index=8 Yellow
Index=9 Olive
Index=10 Green
Index=11 Jade
Index=12 Blue
Index=13 Med Blue
Index=14 Lilac
Index=15 Violet
Index=16 Grey
Index=17 Orange
Index=18 Coral
Index=19 Pink
Index=20 Tan
Index=21 Lt Orange
Index=22 Med Green
Index=23 Med Yellow
Index=24 100 Year Flood Overlay
Index=25 500 Year Flood Overlay
Index=26 Potential Flood Overlay
Index=27 Biohazard Overlay
Index=28 Chemical Overlay
Index=29 Radiation Overlay
Index=30 Poison Overlay
Index=31 Noise Overlay
Index=32 Historic Site
Index=33 Cropland
Index=34 Open Pasture
Index=35 Orchard or Nursery
Index=36 Vineyard
Index=37 Scrub 1
Index=38 Grassland

Index=39 Scattered Trees 1
Index=40 Sand
Index=41 Water Intermittent
Index=42 Reservoir
Index=43 Wetlands
Index=44 Swamp
Index=45 Mangrove
Index=46 Glacier
Index=47 Snowfield/Ice
Index=48 10% Simple hatch
Index=49 10% Crosshatch
Index=50 10% Ordered Stipple
Index=51 Linear Gradient
Index=52 Rectangular Gradient
Index=53 Circular Gradient

Gallery: 15 Line Symbols

Index=1 Highway
Index=2 Highway Ramp
Index=3 Expressway
Index=4 Expressway Ramp
Index=5 Major Road
Index=6 Arterial Street
Index=7 Collector Street
Index=8 Residential Street
Index=9 Railroad
Index=10 River
Index=11 Boundary, National
Index=12 Boundary, State
Index=13 Boundary, County
Index=14 Boundary, City
Index=15 Boundary, Military Installation
Index=16 Boundary, Neighborhood
Index=17 Boundary, Township
Index=18 Freeway
Index=19 Freeway Ramp
Index=20 Freeway, Under Construction
Index=21 Freeway, Proposed
Index=22 Stacked Multi Roadway
Index=23 Stacked Multi Roadway Ramp
Index=24 Toll Road
Index=25 High Occupancy Lane
Index=26 High Occupancy Lane Ramp
Index=27 Bus Route
Index=28 Bicycle Route
Index=29 Mass Transit
Index=30 New Road, Under Construction
Index=31 Existing Road Under Construction
Index=32 Existing Road Needs Repair
Index=33 Road, Unpaved
Index=34 Road, Undefined
Index=35 Road, Proposed
Index=36 Automobile Tunnel
Index=37 Railroad, Multi-Track
Index=38 Railroad, Under Construction
Index=39 Railroad, Abandoned
Index=40 Railroad, In Street
Index=41 Railroad, Narrow Gauge
Index=42 Railroad, Narrow Gauge Multi-Track
Index=43 Railroad, Trunkline
Index=44 Ferry
Index=45 Contour, Topographic, Intermediate
Index=46 Contour, Topographic, Index
Index=47 Contour, Topographic, Supplementary
Index=48 Contour, Topographic, Depression
Index=49 Contour, Topographic, Cut
Index=50 Contour, Bathymetric, Intermediate
Index=51 Contour, Bathymetric, Index
Index=52 Contour, Bathymetric, Primary
Index=53 Contour, Bathymetric, Index Primary

Index=54	Contour, Bathymetric, Supplementary
Index=55	Coastline
Index=56	River, Navigable
Index=57	Stream or Creek
Index=58	Stream, Intermittent
Index=59	Canal
Index=60	Aqueduct
Index=61	Single, Narrow
Index=62	Single, Wide
Index=63	Single, Nautical Dashed
Index=64	Double, Plain
Index=65	Double, Graded
Index=66	Double, Nautical Dashed
Index=67	Triple, Plain
Index=68	Triple, Wide Center
Index=69	Triple, Graded
Index=70	Dashed 6:1
Index=71	Dashed 4:1
Index=72	Dashed 2:1
Index=73	Dashed 6:6
Index=74	Dashed 4:4
Index=75	Dashed 2:2
Index=76	Dashed 1 Long 1 Short
Index=77	Dashed 1 Long 2 Short
Index=78	Dashed 1 Long 3 Short
Index=79	Dashed with 1 Dot
Index=80	Dashed with 2 Dots
Index=81	Dashed with 3 Dots
Index=82	Arrow at End
Index=83	Arrow at Start
Index=84	Arrows at Start and End
Index=85	Dam
Index=86	Single, Nautical Dashed 2

Gallery: 16 Marker Symbols

Index=1	Circle 1
Index=2	Square 1
Index=3	Triangle 1
Index=4	Pentagon 1
Index=5	Hexagon 1
Index=6	Octagon 1
Index=7	Rnd Square 1
Index=8	Circle 2
Index=9	Square 2
Index=10	Triangle 2
Index=11	Pentagon 2
Index=12	Hexagon 2
Index=13	Octagon 2
Index=14	Rnd Square 2
Index=15	Circle 3
Index=16	Square 3
Index=17	Triangle 3
Index=18	Pentagon 3
Index=19	Hexagon 3
Index=20	Octagon 3
Index=21	Rnd Square 3
Index=22	Star 1
Index=23	Star 2
Index=24	Star 3
Index=25	Star 4
Index=26	Star 5
Index=27	Star 6
Index=28	Diamond 1
Index=29	Diamond 2
Index=30	Diamond 3
Index=31	Diamond 4
Index=32	Diamond 5
Index=33	X
Index=34	Cross 1
Index=35	Cross 2
Index=36	Cross 3
Index=37	Cross 4
Index=38	Pushpin 1
Index=39	Pushpin 2
Index=40	Pushpin 3
Index=41	Bolt
Index=42	Identify

Index=43	Question
Index=44	Inquiry
Index=45	Check 1
Index=46	Check 2
Index=47	Asterisk 1
Index=48	Asterisk 2
Index=49	Asterisk 3
Index=50	Asterisk 4
Index=51	School 1
Index=52	School 2
Index=53	Airplane
Index=54	Airfield
Index=55	Airport
Index=56	Handicapped 1
Index=57	Handicapped 2
Index=58	Hospital 1
Index=59	Hospital 2
Index=60	Interstate HWY 1
Index=61	Interstate HWY 2
Index=62	U.S. Route 1
Index=63	U.S. Route 2
Index=64	Trans Canada HWY
Index=65	Circle 4
Index=66	Circle 5
Index=67	Circle 6
Index=68	Circle 7
Index=69	Circle 8
Index=70	Circle 9
Index=71	Circle 10
Index=72	Circle 11
Index=73	Circle 12
Index=74	Circle 13
Index=75	Circle 14
Index=76	Circle 15
Index=77	Circle 16
Index=78	Circle 17
Index=79	Circle 18
Index=80	Circle 19
Index=81	Circle 20
Index=82	Circle 21
Index=83	Circle 22
Index=84	Circle 23
Index=85	Circle 24
Index=86	Square 6
Index=87	Square 7
Index=88	Square 8
Index=89	Square 9
Index=90	Square 10
Index=91	Square 11
Index=92	Square 12
Index=93	Triangle 4
Index=94	Triangle 5
Index=95	Triangle 6
Index=96	Triangle 7
Index=97	Triangle 8
Index=98	Pentagon 4
Index=99	Pentagon 5
Index=100	Pentagon 6
Index=101	Pentagon 7
Index=102	Hexagon 4
Index=103	Hexagon 5
Index=104	Hexagon 6
Index=105	Hexagon 7
Index=106	Octagon 4
Index=107	Octagon 5
Index=108	Octagon 6
Index=109	Octagon 7
Index=110	Rnd Square 4
Index=111	Rnd Square 5
Index=112	Rnd Square 6
Index=113	Rnd Square 7
Index=114	DamLock

Gallery: 17 Text Symbols

Index=1	Country 1
Index=2	Country 2
Index=3	Country 3
Index=4	Capital

Index=5	County
Index=6	Large City
Index=7	City
Index=8	Town
Index=9	Street
Index=10	Physical Region
Index=11	Historic Region
Index=12	Coastal Region
Index=13	Ocean
Index=14	Sea
Index=15	River
Index=16	Stream
Index=17	Map Title E-Size
Index=18	Map Title D-Size
Index=19	Map Title C-Size
Index=20	Map Title B-Size
Index=21	Map Title A-Size
Index=22	Subtitle E-Size
Index=23	Subtitle D-Size
Index=24	Subtitle C-Size
Index=25	Subtitle B-Size
Index=26	Subtitle A-Size
Index=27	Data Source
Index=28	Projection
Index=29	Cartographer
Index=30	Subject Title
Index=31	Normal Text
Index=32	U.S. Route HWY
Index=33	U.S. Interstate HWY
Index=34	Banner Text
Index=35	Banner Text, Rounded

Gallery: 2 Shadows - Samples

Grey 20%

Grey 30%

Grey 40%

Grey 50%

Grey 60%

Black

Sienna

Med Sand

Sand

Aqua

Blue

Leaf

Gallery: 6 North Arrows - Samples
ESRI North 1
ESRI North 2
ESRI North 3
ESRI North 4
ESRI North 5
ESRI North 6
ESRI North 7
ESRI North 8
ESRI North 9
ESRI North 10
ESRI North 11
ESRI North 12
ESRI North 13
ESRI North 14
ESRI North 15
ESRI North 16
ESRI North 17
ESRI North 18
ESRI North 19
ESRI North 20
ESRI North 21
ESRI North 22
ESRI North 23
ESRI North 24
ESRI North 25
ESRI North 26
ESRI North 27
ESRI North 28
ESRI North 29
ESRI North 30
ESRI North 31
ESRI North 32
ESRI North 33
ESRI North 34
ESRI North 35
ESRI North 36
ESRI North 37
ESRI North 38
ESRI North 39
ESRI North 40
ESRI North 41
ESRI North 42
ESRI North 43
ESRI North 44
ESRI North 45
ESRI North 46
ESRI North 47
ESRI North 48
ESRI North 49
ESRI North 50
NORTH
ESRI North 51
ESRI North 52
ESRI North 53
ESRI North 54

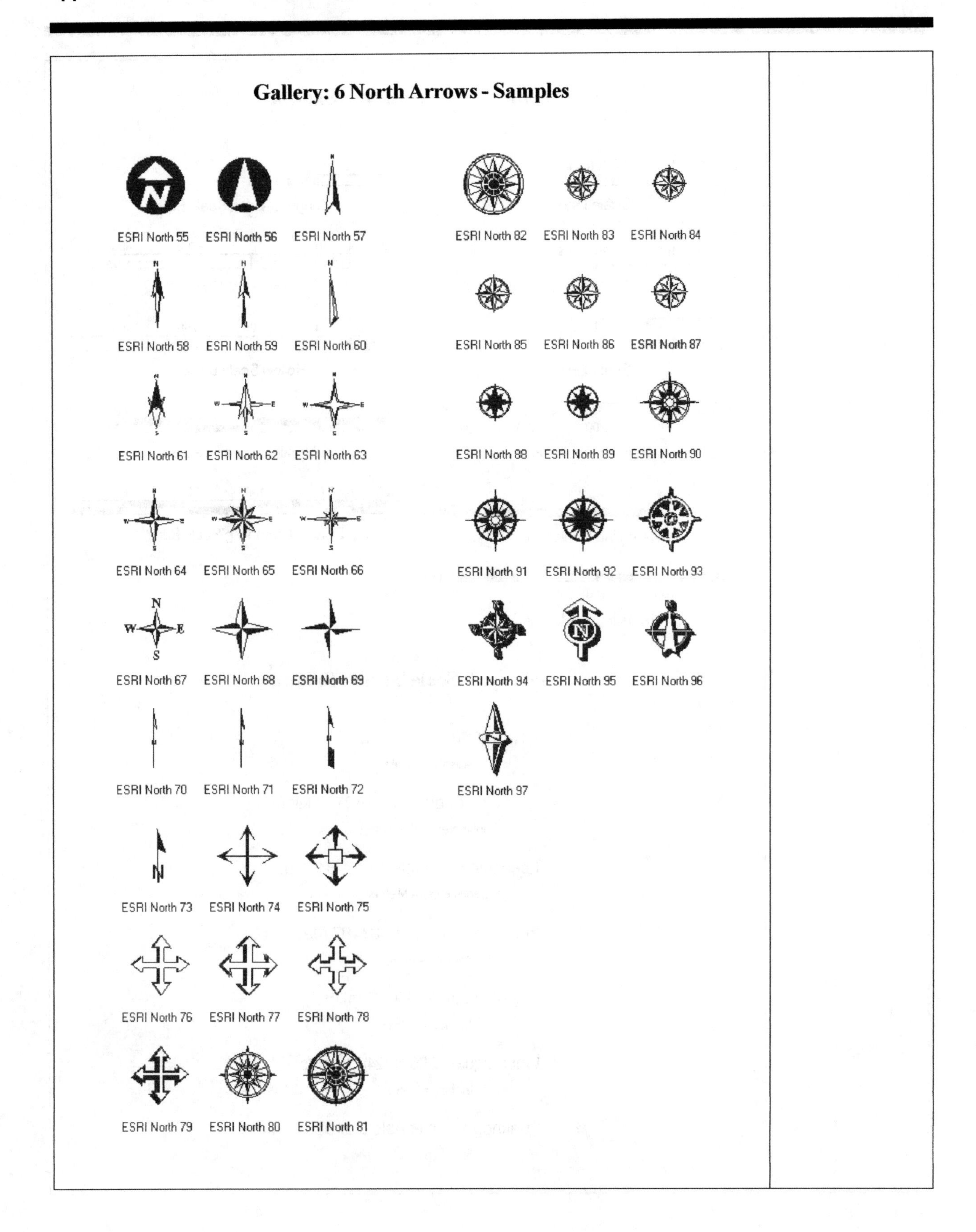
Gallery: 6 North Arrows - Samples
ESRI North 55
ESRI North 56
ESRI North 57
ESRI North 58
ESRI North 59
ESRI North 60
ESRI North 61
ESRI North 62
ESRI North 63
ESRI North 64
ESRI North 65
ESRI North 66
ESRI North 67
ESRI North 68
ESRI North 69
ESRI North 70
ESRI North 71
ESRI North 72
ESRI North 73
ESRI North 74
ESRI North 75
ESRI North 76
ESRI North 77
ESRI North 78
ESRI North 79
ESRI North 80
ESRI North 81
ESRI North 82
ESRI North 83
ESRI North 84
ESRI North 85
ESRI North 86
ESRI North 87
ESRI North 88
ESRI North 89
ESRI North 90
ESRI North 91
ESRI North 92
ESRI North 93
ESRI North 94
ESRI North 95
ESRI North 96
ESRI North 97

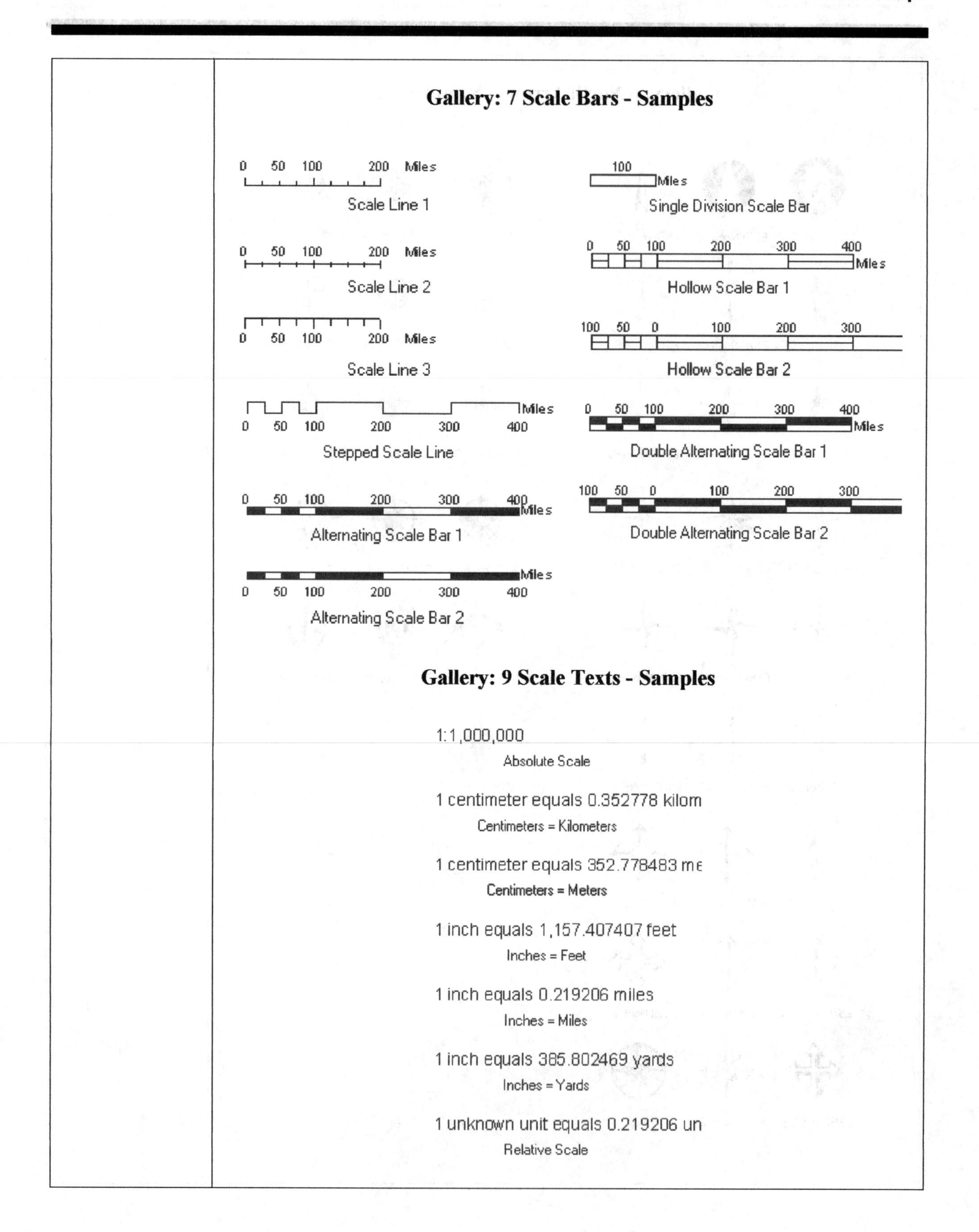
Gallery: 7 Scale Bars - Samples
0 50 100 200 Miles
Scale Line 1
100
Miles
Single Division Scale Bar
0 50 100 200 Miles
Scale Line 2
0 50 100 200 300 400
Miles
Hollow Scale Bar 1
0 50 100 200 Miles
Scale Line 3
100 50 0 100 200 300
Hollow Scale Bar 2
Miles
0 50 100 200 300 400
Stepped Scale Line
0 50 100 200 300 400
Miles
Double Alternating Scale Bar 1
0 50 100 200 300 400
Miles
Alternating Scale Bar 1
100 50 0 100 200 300
Double Alternating Scale Bar 2
Miles
0 50 100 200 300 400
Alternating Scale Bar 2
Gallery: 9 Scale Texts - Samples
1:1,000,000
Absolute Scale
1 centimeter equals 0.352778 kilom
Centimeters = Kilometers
1 centimeter equals 352.778483 me
Centimeters = Meters
1 inch equals 1,157.407407 feet
Inches = Feet
1 inch equals 0.219206 miles
Inches = Miles
1 inch equals 385.802469 yards
Inches = Yards
1 unknown unit equals 0.219206 un
Relative Scale

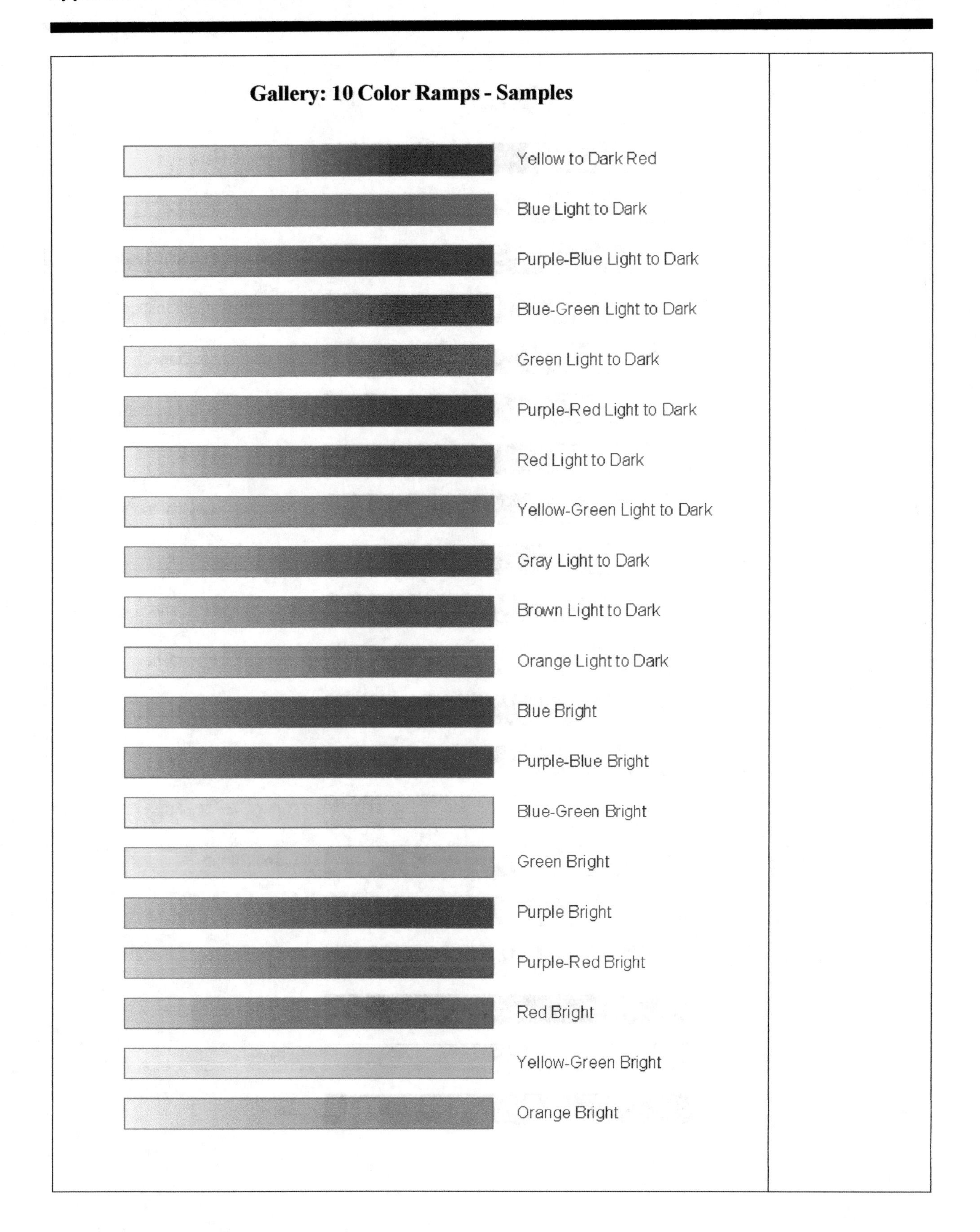
Gallery: 10 Color Ramps - Samples
Yellow to Dark Red
Blue Light to Dark
Purple-Blue Light to Dark
Blue-Green Light to Dark
Green Light to Dark
Purple-Red Light to Dark
Red Light to Dark
Yellow-Green Light to Dark
Gray Light to Dark
Brown Light to Dark
Orange Light to Dark
Blue Bright
Purple-Blue Bright
Blue-Green Bright
Green Bright
Purple Bright
Purple-Red Bright
Red Bright
Yellow-Green Bright
Orange Bright

Gallery: 10 Color Ramps - Samples
Yellow to Dark Red
Blue Light to Dark
Purple-Blue Light to Dark
Blue-Green Light to Dark
Green Light to Dark
Purple-Red Light to Dark
Red Light to Dark
Yellow-Green Light to Dark
Gray Light to Dark
Brown Light to Dark
Orange Light to Dark
Blue Bright
Purple-Blue Bright
Blue-Green Bright
Green Bright
Purple Bright
Purple-Red Bright
Distance
Surface
Slope

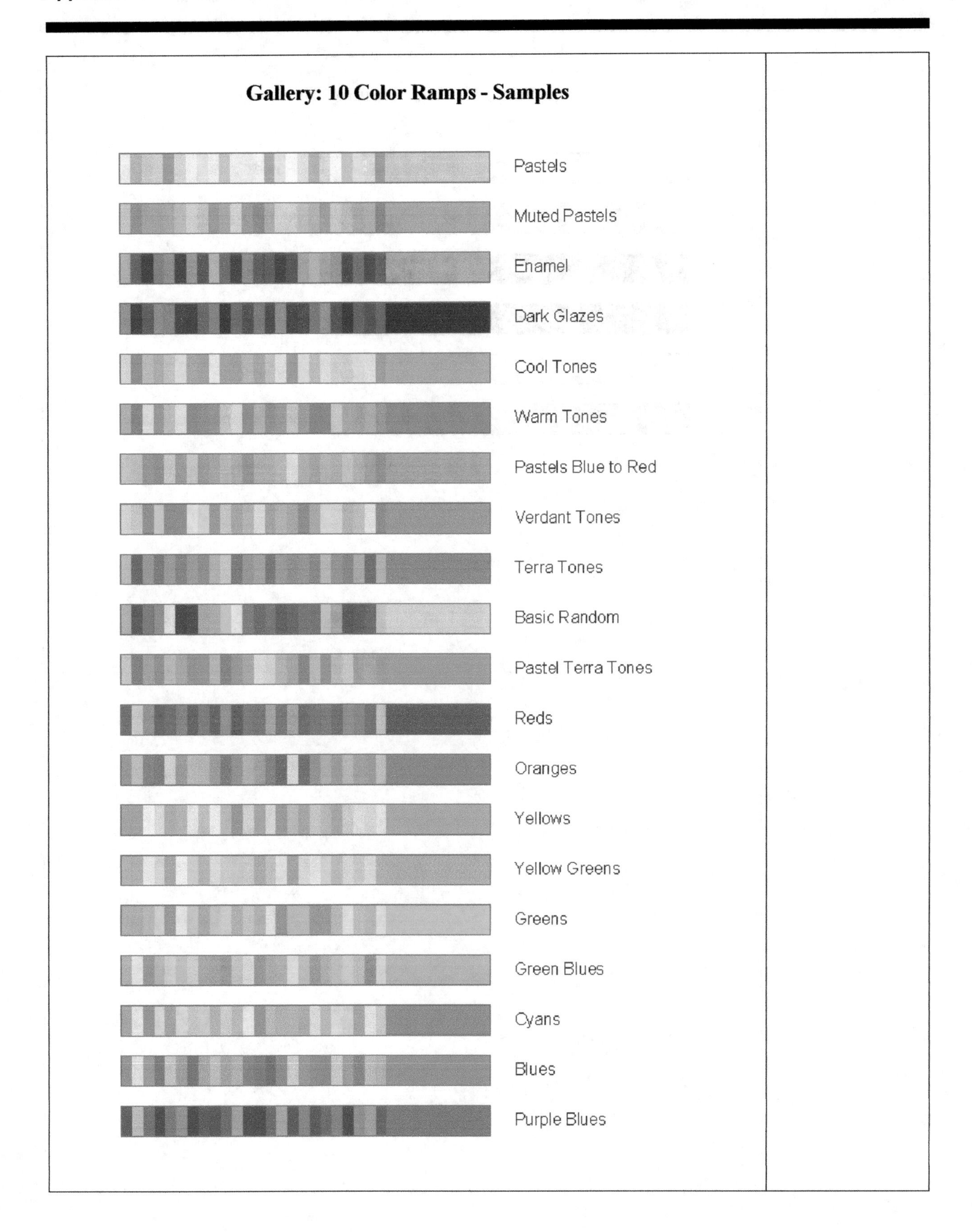
Gallery: 10 Color Ramps - Samples
Pastels
Muted Pastels
Enamel
Dark Glazes
Cool Tones
Warm Tones
Pastels Blue to Red
Verdant Tones
Terra Tones
Basic Random
Pastel Terra Tones
Reds
Oranges
Yellows
Yellow Greens
Greens
Green Blues
Cyans
Blues
Purple Blues

Gallery: 10 Color Ramps - Samples
Pastels
Muted Pastels
Enamel
Dark Glazes
Cool Tones
Warm Tones

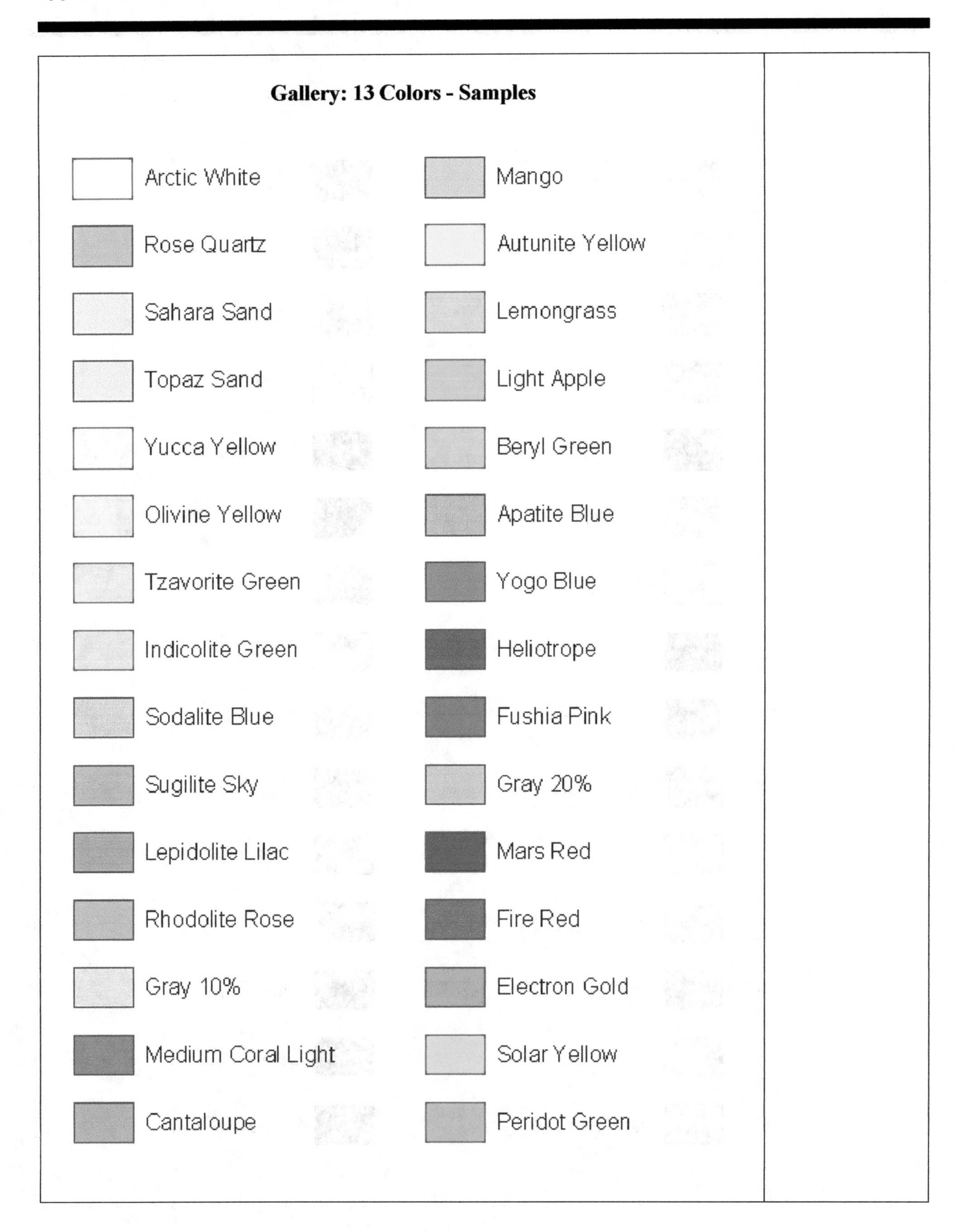
Gallery: 13 Colors - Samples
Arctic White
Rose Quartz
Sahara Sand
Topaz Sand
Yucca Yellow
Olivine Yellow
Tzavorite Green
Indicolite Green
Sodalite Blue
Sugilite Sky
Lepidolite Lilac
Rhodolite Rose
Gray 10%
Medium Coral Light
Cantaloupe
Mango
Autunite Yellow
Lemongrass
Light Apple
Beryl Green
Apatite Blue
Yogo Blue
Heliotrope
Fushia Pink
Gray 20%
Mars Red
Fire Red
Electron Gold
Solar Yellow
Peridot Green

Gallery: 13 Colors - Samples
Medium Apple
Tourmaline Green
Big Sky Blue
Cretean Blue
Amethyst
Ginger Pink
Gray 30%
Poinsettia Red
Flame Red
Seville Orange
Citroen Yellow
Macaw Green
Quetzel Green
Chrysophase
Moorea Blue
Lapis Lazuli
Anemone Violet
Peony Pink
Gray 40%
Tuscan Red
Cherry Cola
Raw Umber
Olivenite Green
Tarragon Green
Leaf Green
Malachite Green
Delft Blue
Ultra Blue
Dark Amethyst
Cattleya Orchid

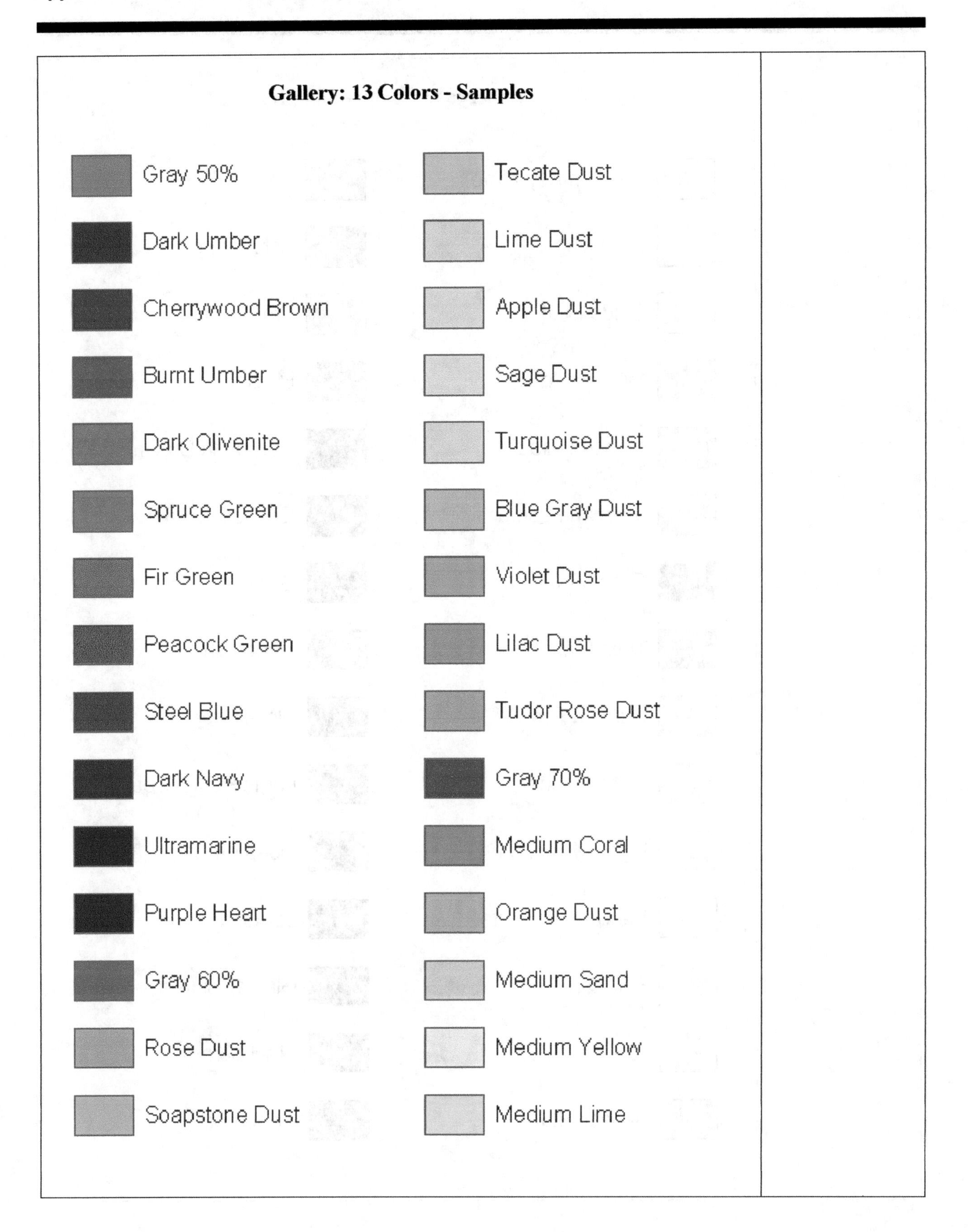
Gallery: 13 Colors - Samples
Gray 50%
Dark Umber
Cherrywood Brown
Burnt Umber
Dark Olivenite
Spruce Green
Fir Green
Peacock Green
Steel Blue
Dark Navy
Ultramarine
Purple Heart
Gray 60%
Rose Dust
Soapstone Dust
Tecate Dust
Lime Dust
Apple Dust
Sage Dust
Turquoise Dust
Blue Gray Dust
Violet Dust
Lilac Dust
Tudor Rose Dust
Gray 70%
Medium Coral
Orange Dust
Medium Sand
Medium Yellow
Medium Lime

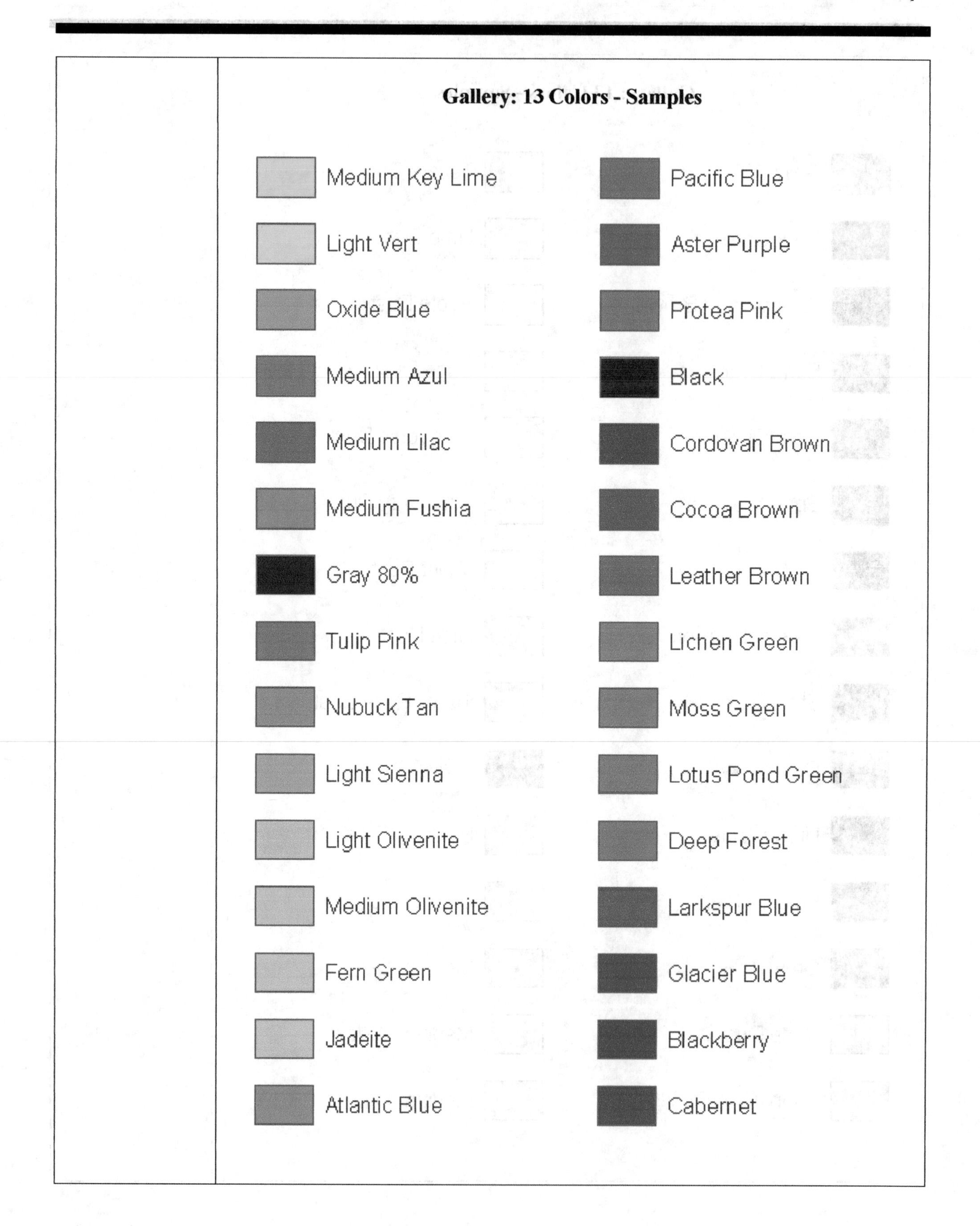
Gallery: 13 Colors - Samples
Medium Key Lime
Light Vert
Oxide Blue
Medium Azul
Medium Lilac
Medium Fushia
Gray 80%
Tulip Pink
Nubuck Tan
Light Sienna
Light Olivenite
Medium Olivenite
Fern Green
Jadeite
Atlantic Blue
Pacific Blue
Aster Purple
Protea Pink
Black
Cordovan Brown
Cocoa Brown
Leather Brown
Lichen Green
Moss Green
Lotus Pond Green
Deep Forest
Larkspur Blue
Glacier Blue
Blackberry
Cabernet

Gallery: 14 Fill Symbols - Samples
Green
Blue
Sun
Hollow
Lake
Rose
Beige
Yellow
Olive
Green
Jade
Blue
Med Blue
Lilac
Violet
Grey
Orange
Coral
Pink
Tan
Lt Orange
Med Green
Med Yellow
100 Year Flood Overlay
500 Year Flood Overlay
Potential Flood Overlay
Biohazard Overlay
Chemical Overlay
Radiation Overlay
Poison Overlay

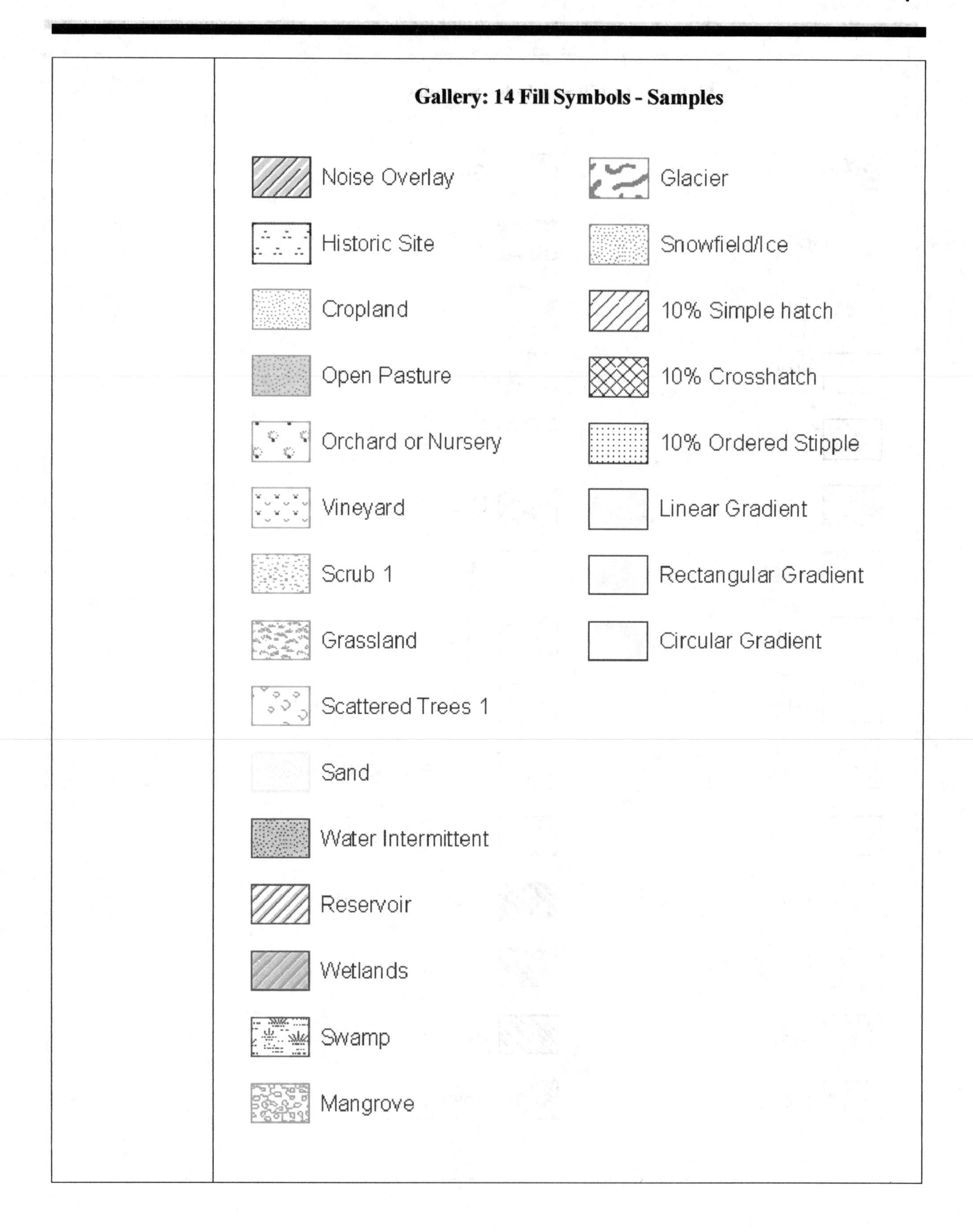
Gallery: 14 Fill Symbols - Samples
Noise Overlay
Historic Site
Cropland
Open Pasture
Orchard or Nursery
Vineyard
Scrub 1
Grassland
Scattered Trees 1
Sand
Water Intermittent
Reservoir
Wetlands
Swamp
Mangrove
Glacier
Snowfield/Ice
10% Simple hatch
10% Crosshatch
10% Ordered Stipple
Linear Gradient
Rectangular Gradient
Circular Gradient

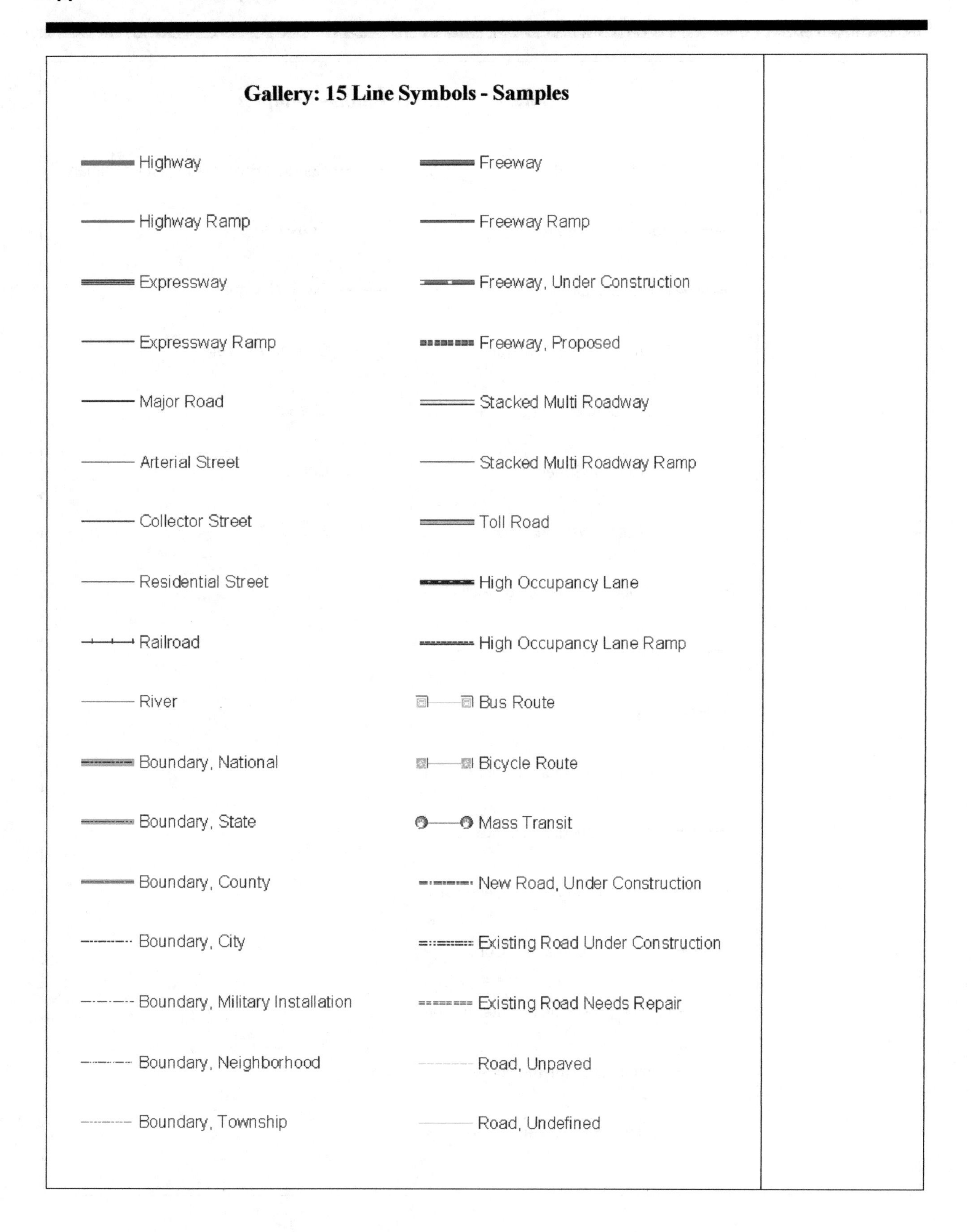
Gallery: 15 Line Symbols - Samples
Highway
Freeway
Highway Ramp
Freeway Ramp
Expressway
Freeway, Under Construction
Expressway Ramp
Freeway, Proposed
Major Road
Stacked Multi Roadway
Arterial Street
Stacked Multi Roadway Ramp
Collector Street
Toll Road
Residential Street
High Occupancy Lane
Railroad
High Occupancy Lane Ramp
River
Bus Route
Boundary, National
Bicycle Route
Boundary, State
Mass Transit
Boundary, County
New Road, Under Construction
Boundary, City
Existing Road Under Construction
Boundary, Military Installation
Existing Road Needs Repair
Boundary, Neighborhood
Road, Unpaved
Boundary, Township
Road, Undefined

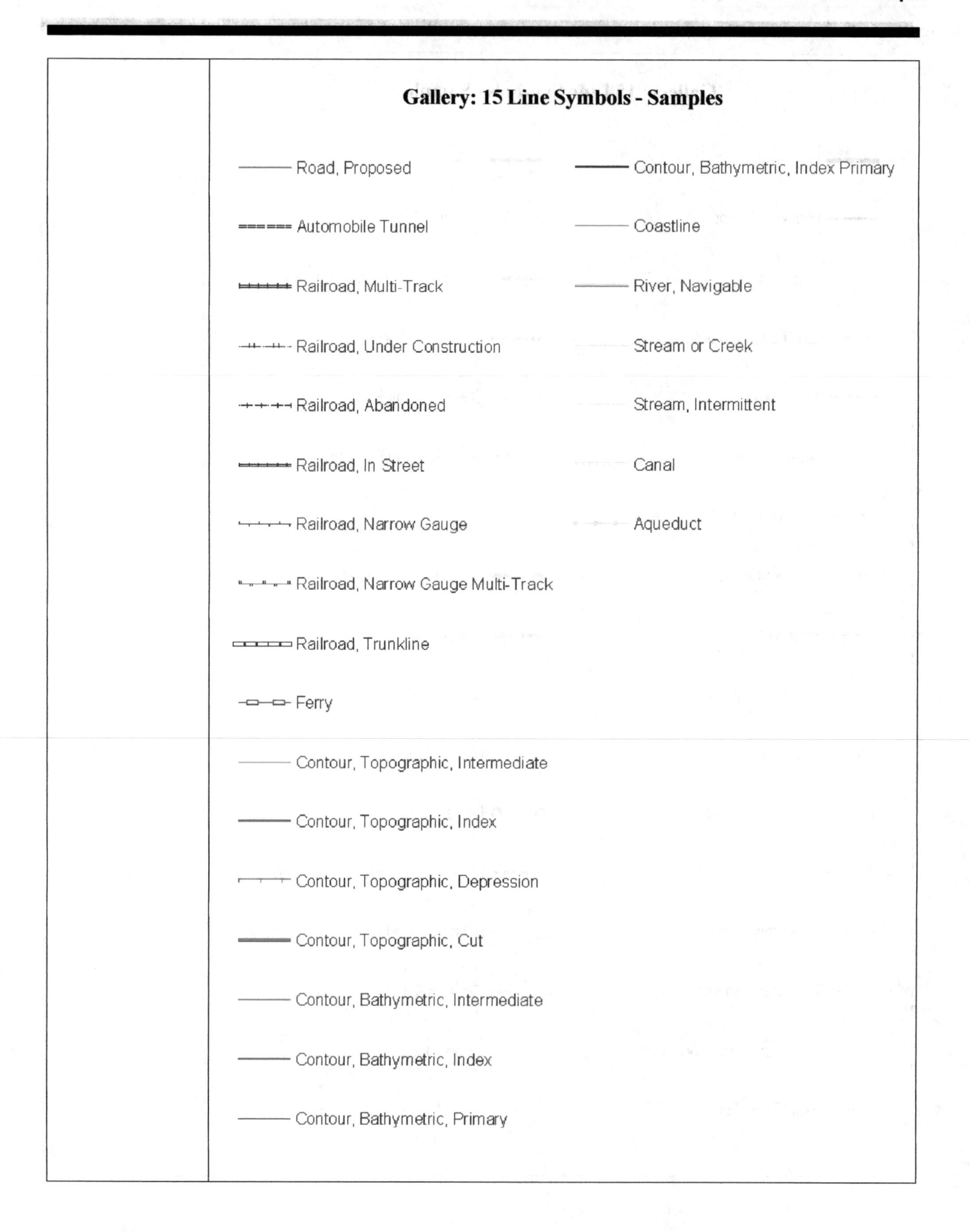
Gallery: 15 Line Symbols - Samples
Road, Proposed
Automobile Tunnel
Railroad, Multi-Track
Railroad, Under Construction
Railroad, Abandoned
Railroad, In Street
Railroad, Narrow Gauge
Railroad, Narrow Gauge Multi-Track
Railroad, Trunkline
Ferry
Contour, Bathymetric, Index Primary
Coastline
River, Navigable
Stream or Creek
Stream, Intermittent
Canal
Aqueduct
Contour, Topographic, Intermediate
Contour, Topographic, Index
Contour, Topographic, Depression
Contour, Topographic, Cut
Contour, Bathymetric, Intermediate
Contour, Bathymetric, Index
Contour, Bathymetric, Primary

Gallery: 15 Line Symbols - Samples

Contour, Topographic, Supplementary	Single, Narrow
Contour, Bathymetric, Supplementary	Single, Wide
Dashed 6:1	Single, Nautical Dashed
Dashed 4:1	Double, Plain
Dashed 2:1	Double, Graded
Dashed 6:6	Double, Nautical Dashed
Dashed 4:4	Triple, Plain
Dashed 2:2	Triple, Wide Center
Dashed 1 Long 1 Short	Triple, Graded
Dashed 1 Long 2 Short	Single, Nautical Dashed 2
Dashed 1 Long 3 Short	Arrow at End
Dashed with 1 Dot	Arrow at Start
Dashed with 2 Dots	Arrows at Start and End
Dashed with 3 Dots	Dam

Gallery: 16 Marker Symbols - Samples

Circle 1	Star 6	Airplane
Square 1	Diamond 1	Airfield
Triangle 1	Diamond 2	Airport
Pentagon 1	Diamond 3	Handicapped 1
Hexagon 1	Diamond 4	Handicapped 2
Octagon 1	Diamond 5	Hospital 1
Rnd Square 1	X	Hospital 2
Circle 2	Cross 1	Interstate HWY 1
Square 2	Cross 2	Interstate HWY 2
Triangle 2	Cross 3	U.S. Route 1
Pentagon 2	Cross 4	U.S. Route 2
Hexagon 2	Pushpin 1	Trans Canada HWY
Octagon 2	Pushpin 2	Circle 4
Rnd Square 2	Pushpin 3	Circle 5
Circle 3	Bolt	Circle 6
Square 3	Identify	Circle 7
Triangle 3	Question	Circle 8
Pentagon 3	Inquiry	Circle 9
Hexagon 3	Check 1	Circle 10
Octagon 3	Check 2	Circle 11
Rnd Square 3	Asterisk 1	Circle 12
Star 1	Asterisk 2	Circle 13
Star 2	Asterisk 3	Circle 14
Star 3	Asterisk 4	Circle 15
Star 4	School 1	Circle 16
Star 5	School 2	Circle 17

Gallery: 16 Marker Symbols - Samples

Circle 18
Circle 19
Circle 20
Circle 21
Circle 22
Circle 23
Circle 24
Square 6
Square 7
Square 8
Square 9
Square 10
Square 11
Square 12
Triangle 4
Triangle 5
Triangle 6
Triangle 7
Triangle 8
Pentagon 4
Pentagon 5
Pentagon 6
Pentagon 7
Hexagon 4
Hexagon 5
Hexagon 6
Hexagon 7
Octagon 4
Octagon 5
Octagon 6
Octagon 7
Rnd Square 4
Rnd Square 5
Rnd Square 6
Rnd Square 7
Dam Lock

Gallery: 17 Text Symbols - Samples

Sample	Symbol	Sample	Symbol
AaBbYyZz	Country 1		
AaBbYyZz	Country 2		
AaBbYyZz	Country 3		
AaBbYyZz	Capital	AaBbYyZz	Subtitle C-Size
AaBbYyZz	County	AaBbYyZz	Subtitle B-Size
AaBbYyZz	Large City	AaBbYyZz	Subtitle A-Size
AaBbYyZz	City	AaBbYyZz	Data Source
AaBbYyZz	Town	AaBbYyZz	Projection
AaBbYyZz	Street	AaBbYyZz	Cartographer
A a B b Y y Z z	Physical Reg	AaBbYyZz	Subject Title
AaBbYyZz	Historic Region	AaBbYyZz	Normal Text
AaBbYyZz	Coastal Region	AaBbYyZz	U.S. Route HWY
A a B b Y y Z z	Ocean		U.S. Interstate HWY
A a B b Y y Z z	Sea		
AaBbYyZz	River	AaBbYyZz	Banner Text
AaBbYyZz	Stream	AaBbYyZz	Banner Text, Rounded

APPENDIX B

COLOR CALIBRATION CHARTS

This appendix contains the calibration charts for the gray scales and for the three color models, the CMYK color model, the HSV color model and the RGB color model. These charts are referenced by various Avenue Wrap subroutines and/or functions contained in Appendix D of this publication. The following are noted for these charts:

Gray Scale The various shades of gray shown range from zero to 255 at increments of two.

CMYK Model For this model, six charts are given for six settings of yellow from zero to 100 at increments of 20. For each of these charts, there is:

- A center grouping of 36 squares (6x6) with the value of zero for black. As the values of cyan and magenta range from zero to 100, the colors of the center group change from the upper left corner square to the low right square of the center grouping of squares.
- Four perimeter or border rectangles, each comprised of six squares with a value of 50 for black. As the value of magenta changes from zero to 100, the color of the top and bottom border squares changes, and as the value of cyan changes, the color of the left and right border squares changes. Thus, the colors of the two extreme horizontal and vertical border squares at each corner of the center grouping are the same.

HSV Model For this model, there is a horizontal color bar with a value of 100 for each of the saturation and lightness. There are also six charts of 36 squares (6x6) with hue values ranging from zero to 100 at increments of 20.

RGB Model For this model, there are also six charts of 36 squares (6x6) with blue values of zero, 51, 102, 153, 204 and 255, and with values of red and green ranging from zero to 100 at increments of 20.

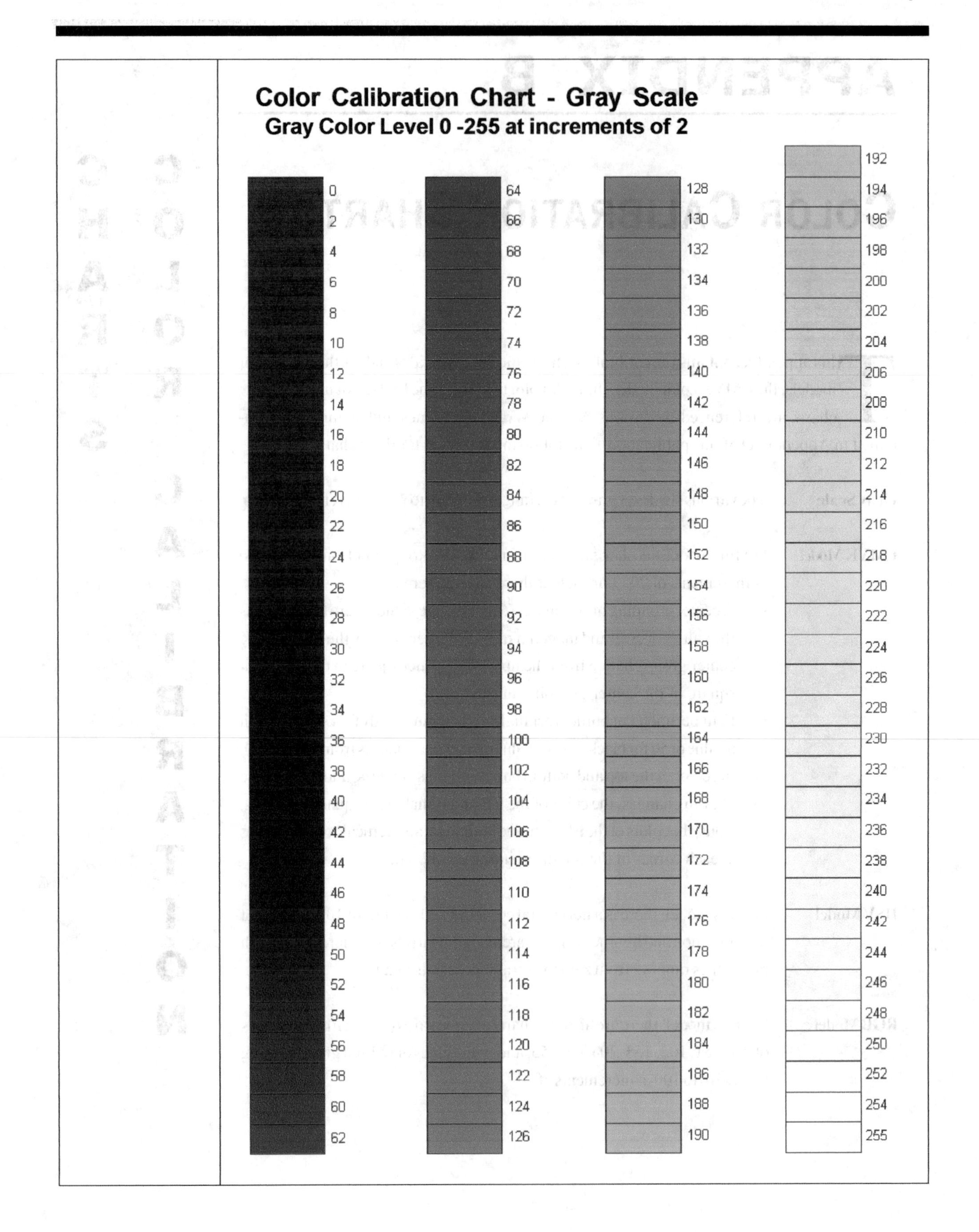
Color Calibration Chart - Gray Scale
Gray Color Level 0 -255 at increments of 2
0
2
4
6
8
10
12
14
16
18
20
22
24
26
28
30
32
34
36
38
40
42
44
46
48
50
52
54
56
58
60
62
64
66
68
70
72
74
76
78
80
82
84
86
88
90
92
94
96
98
100
102
104
106
108
110
112
114
116
118
120
122
124
126
128
130
132
134
136
138
140
142
144
146
148
150
152
154
156
158
160
162
164
166
168
170
172
174
176
178
180
182
184
186
188
190
192
194
196
198
200
202
204
206
208
210
212
214
216
218
220
222
224
226
228
230
232
234
236
238
240
242
244
246
248
250
252
254
255

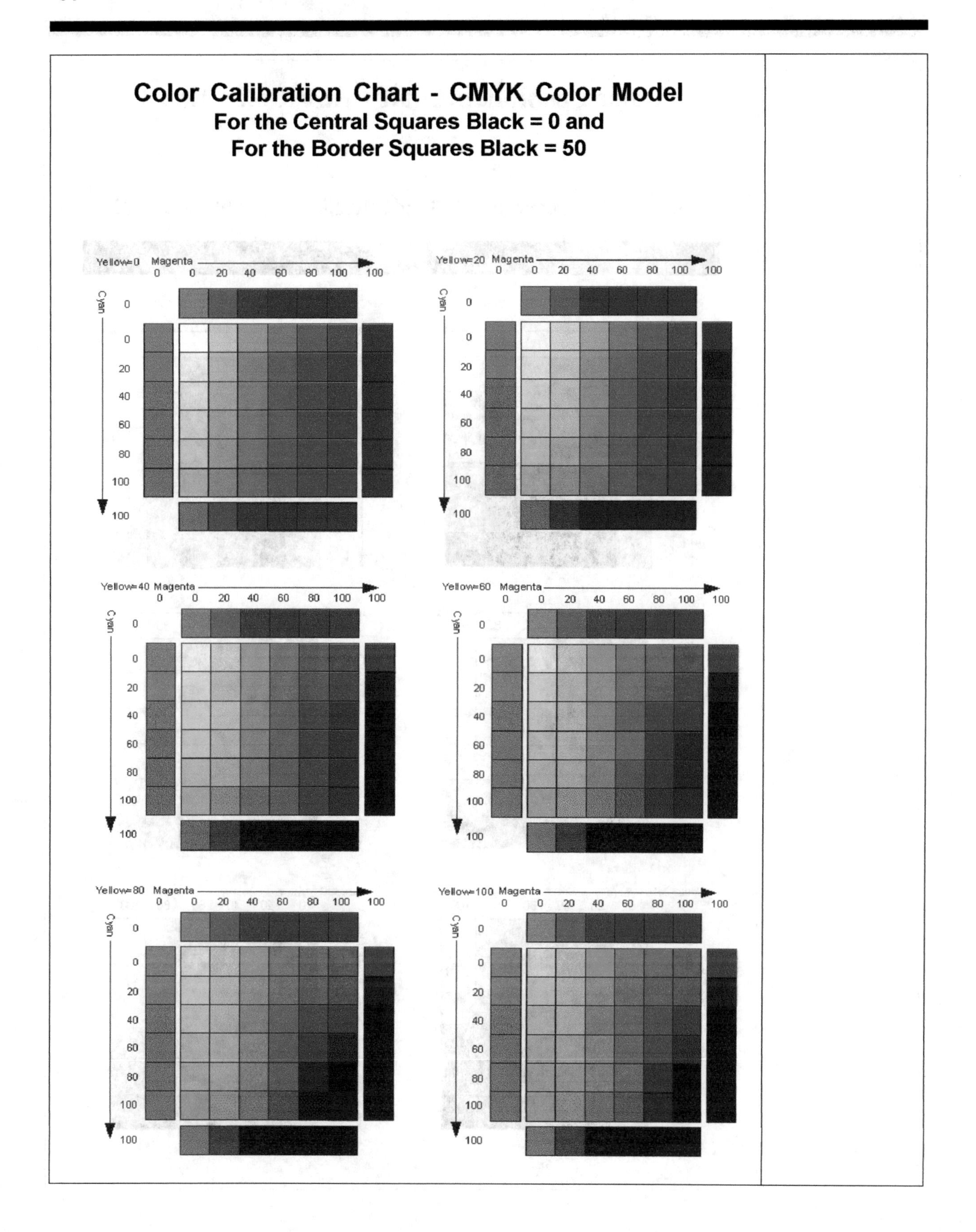
Color Calibration Chart - CMYK Color Model
For the Central Squares Black = 0 and
For the Border Squares Black = 50
Yellow=0 Magenta
Yellow=20 Magenta
Yellow=40 Magenta
Yellow=60 Magenta
Yellow=80 Magenta
Yellow=100 Magenta
Cyan
0 0 20 40 60 80 100 100
0 0 20 40 60 80 100 100

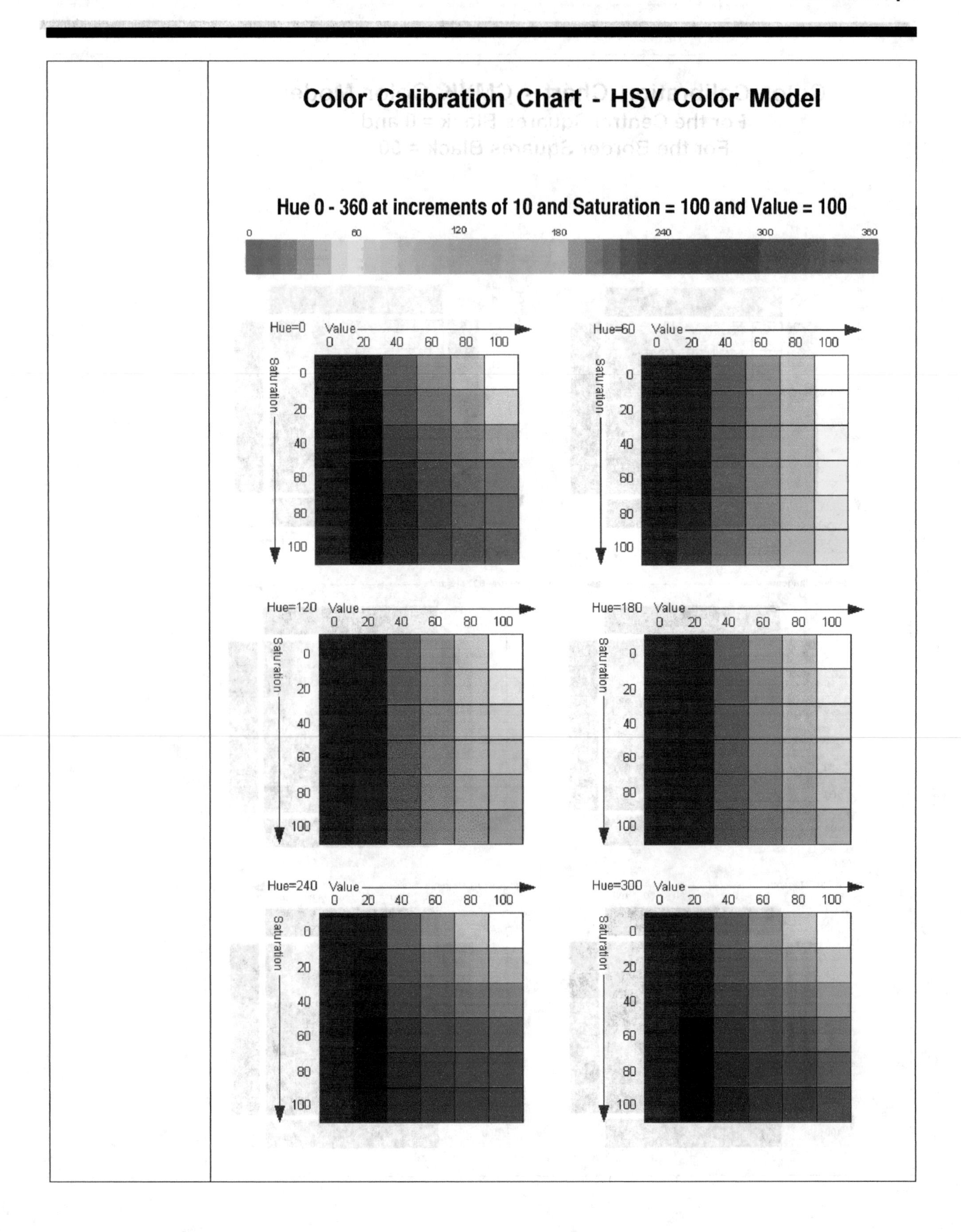

Color Calibration Chart - HSV Color Model
Hue 0 - 360 at increments of 10 and Saturation = 100 and Value = 100
0
60
120
180
240
300
360
Hue=0
Hue=60
Hue=120
Hue=180
Hue=240
Hue=300
Value
Saturation
0
20
40
60
80
100

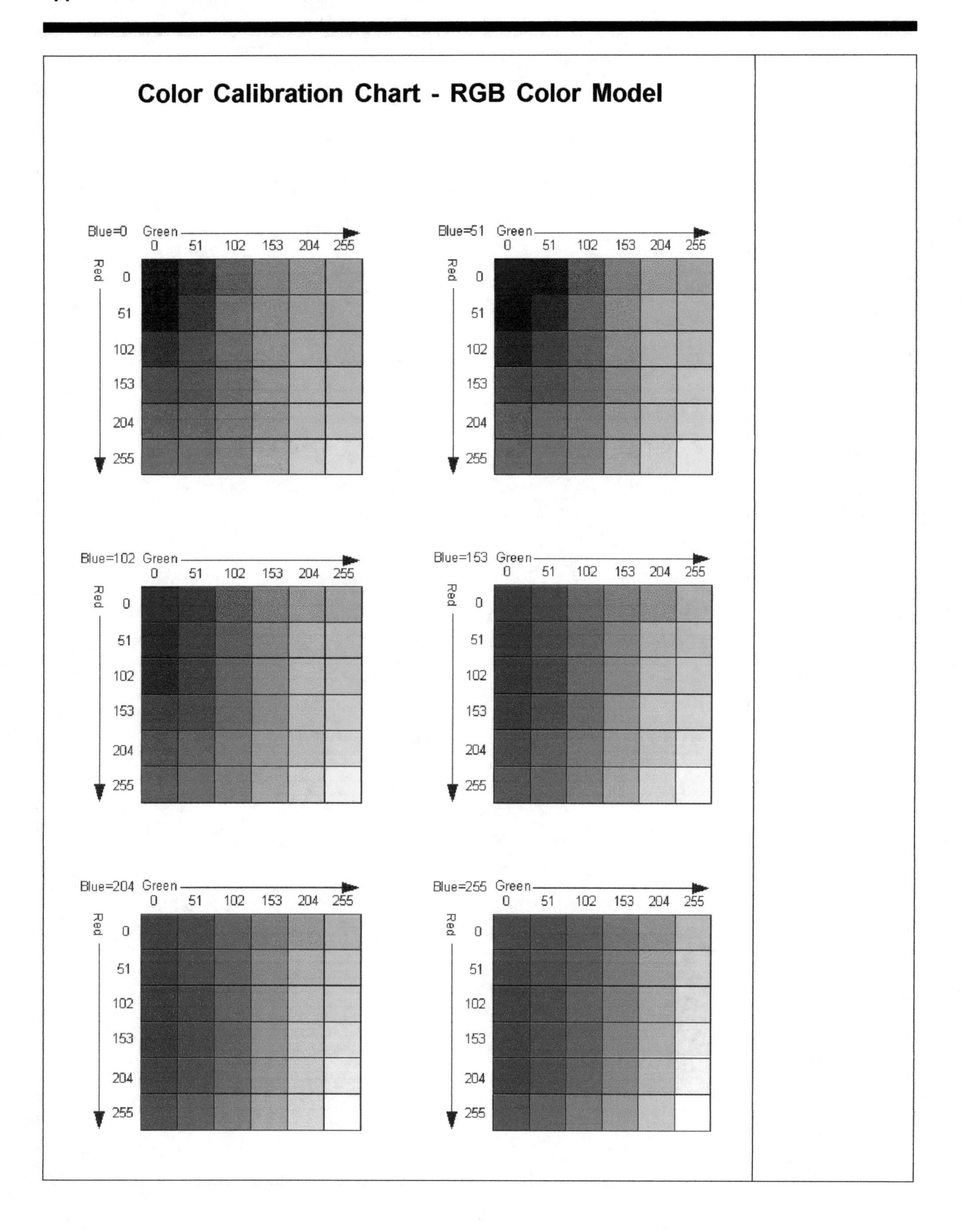
Color Calibration Chart - RGB Color Model
Blue=0 Green
0 51 102 153 204 255
Red
0
51
102
153
204
255
Blue=51 Green
0 51 102 153 204 255
Red
0
51
102
153
204
255
Blue=102 Green
0 51 102 153 204 255
Red
0
51
102
153
204
255
Blue=153 Green
0 51 102 153 204 255
Red
0
51
102
153
204
255
Blue=204 Green
0 51 102 153 204 255
Red
0
51
102
153
204
255
Blue=255 Green
0 51 102 153 204 255
Red
0
51
102
153
204
255

APPENDIX C

AVENUE REQUESTS TO AVENUE WRAPS MAPPING

AVENUE REQUESTS TO AVENUE WRAPS MAPPING

This appendix is comprised of two tables. Table C-1, which contains the various Avenue requests for which Avenue Wraps™ have been written, or there is an equivalent counterpart, and Table C-2, which contains the various subroutines and functions for which there is no direct Avenue request. Table C-1 is organized in alphabetical order by Avenue request, while Table C-2 is organized in alphabetical order by Avenue Wrap™. Shown in both tables is the Chapter and Section, where a description, for each Avenue Wrap™ or equivalent counterpart, may be found.

In Table C-1, there are certain Avenue requests for which Avenue Wraps™ have not been written, but for which there is either (a) a direct equivalent VB statement, (b) sample VB code consisting of several statements presented in the text, or (c) a direct ArcObjects™ counterpart. In such instances, under the Avenue Wrap™ column, the letters VB enclosed in parentheses (VB) accompany the statement corresponding to the Avenue request, or the name of the ArcObject™ counterpart, will appear. If there is a direct ArcObjects™ counterpart for an Avenue request, N/A will appear under the Chapter and Section column denoting that no discussion is given, since the behavior of the request and the ArcObjects™ counterpart are the same.

TABLE C-1
AVENUE REQUESTS AND CORRESPONDING AVENUE WRAPS™

Avenue Request	Avenue Wrap	Chapter/Section
Abs	Abs (VB)	Table 2.1
Add	Add (VB)	Table 2.5
AddDoc	avAddDoc	3.1.1
AddFields	avAddFields	5.2.1
AddRecord	avAddRecord	5.3.1
AsDegrees	icraddeg	12.2.11
Asin	icasinan	12.2.4
AsList	avAsList	8.1.2
AsList	avPlAsList	8.1.5
AsNumber	CDbl (VB)	2.2.3
AsNumber	CInt (VB)	2.2.3
AsNumber	CLng (VB)	2.2.3

TABLE C-1 (cont.)
AVENUE REQUESTS AND CORRESPONDING AVENUE WRAPS™

Avenue Request	Avenue Wrap	Chapter/Section
AsNumber	CSng (VB)	2.2.3
AsNumber	CVar (VB)	2.2.3
AsRadians	icdegrad	12.2.8
AsString	CStr (VB)	2.1.5
AsTokens	avAsTokens	4.5.6
Atan	Atn (VB)	Table 2.1
Atan	icatan	12.2.5
BasicTrim	avBasicTrim	2.6.1
Calculate	avCalculate	5.5.1
Ceiling	Int (VB)	Table 2.2
Ceiling	Fix (VB)	Table 2.2
Choice	avMsgBoxChoice	7.3.1
Clean	avClean	8.4.1
Clear	avBitmapClear	6.2.1
ClearMsg	avClearStatus	3.4.1
ClearSelection	avClearSelection	6.2.2
ClearStatus	avClearStatus	3.4.1
Clone	avClone	2.6.2
Clone	Set (VB)	Table 2.5
Close	avLineFileClose	4.4.1
Convert	avConvertDistance	12.2.2
ConvertArea	avConvertArea	12.2.1
Cos	Cos (VB)	Table 2.1
Count	Len (VB)	2.2.1
Count	Count (VB)	Table 2.5
Delete	avFileDelete	4.2.2
Empty	CreateList	Table 2.5
Execute	avExecute	2.6.3
Exists	avFileExists	4.2.3
FindAllByClass	sample code (VB)	Appendix D
FindDoc	avFindDoc	3.1.2
FindField	FindField	N/A
FindTheme	FindLayer	5.1.17
FindTheme	FindTheme	5.1.18
Floor	Int (VB)	Table 2.2
Floor	Fix (VB)	Table 2.2
Flush	avGetDisplayFlush	3.3.3
Get	Item	Table 2.5
GetActiveDoc	avGetActiveDoc	3.1.3
GetActiveThemes	avGetActiveThemes	3.1.4
GetAngle	avGraphicTextGetAngle	10.3.1
GetAngle	avSymbolGetAngle	10.2.6
GetBaseName	avGetBaseName	4.3.1
GetBounds	GetTextRect	12.2.3

TABLE C-1 (cont.)
AVENUE REQUESTS AND CORRESPONDING AVENUE WRAPS™

Avenue Request	Avenue Wrap	Chapter/Section
GetClassifications	avGetClassifications	11.2.1
GetClassType	avGetClassType	11.2.2
GetColor	avSymbolGetColor	10.2.7
GetDisplay	avGetDisplay	3.3.2
GetEnvVar	avGetEnvVar	2.6.4
GetExtension	avGetExtension	4.3.3
GetFieldNames	avGetFieldNames	11.2.3
GetFields	avGetFields	5.2.4
GetFileName	avGetProjectName	3.1.6
GetFTab	avGetFTab	5.3.2
GetGraphics	avViewGetGraphics	10.1.9
GetLegend	avGetLegend	11.2.4
GetLegendType	avGetLegendType	11.2.5
GetList	avPaletteGetList	11.3.2
GetName	avGetName	3.1.5
GetName	avObjGetName	3.1.11
GetNumClasses	avGetNumClasses	11.2.6
GetNumRecords	avGetNumRecords	5.3.4
GetOLColor	avSymbolGetOLColor	10.2.8
GetOLWidth	avSymbolGetOLWidth	10.2.9
GetPalette	avGetPalette	11.3.1
GetPrecision	avGetPrecision	5.2.5
GetSelected	avGetSelected	10.1.2
GetSelectedExtent	avGetSelectedExtent	3.2.3
GetSelection	avGetSelection	6.1.2
GetShape	avGraphicGetShape	10.2.1
GetShapeClass	avGetShapeType	5.1.3
GetSize	avSymbolGetSize	10.2.10
GetStipple	avSymbolGetStipple	10.2.11
GetStyle	avSymbolGetStyle	10.2.12
GetSymbol	avGraphicTextGetSymbol	10.3.3
GetSymbol	avGraphicGetSymbol	10.2.2
GetSymbols	avLegendGetSymbols	11.2.7
GetText	avGraphicTextGetText	10.3.4
GetThemes	avGetThemes	3.2.4
GetType	avFieldGetType	5.2.2
GetVisibleThemes	avGetVisibleThemes	3.2.6
GetVTab	avGetVTab	5.3.5
GetWorkDir	avGetWorkDir	3.1.8
Info	avMsgBoxInfo	7.1.1
Input	avMsgBoxInput	7.2.1
Insert	Add (VB)	Table 2.5
Intersects	avIntersects	8.4.2
Interval	avInterval	11.1.1

TABLE C-1 (cont.)
AVENUE REQUESTS AND CORRESPONDING AVENUE WRAPS™

Avenue Request	Avenue Wrap	Chapter/Section
Invalidate	avDisplayInvalidate	3.3.1
Invalidate	avGraphicInvalidate	10.1.3
Invalidate	avThemeInvalidate	5.1.15
InvalidateTOC	avInvalidateTOC	5.1.5
Is	avIsFTheme	5.1.7
IsEditable	avIsEditable	5.1.6
IsEmpty	IsEmpty	N/A
IsNoDataClassDisplayed	sample code (VB)	Appendix D
IsNull	IsNull (VB)	Table 2.4
IsNumber	IsNumeric (VB)	Table 2.4
IsVisible	avIsVisible	5.1.8
LCase	LCase (VB)	Table 2.3
Left	Left (VB)	Table 2.3
List	avMsgBoxList	7.1.2
Ln	Log (VB)	Table 2.1
Make	avFieldMake	5.2.3
Make	avGraphicShapeMake	10.2.5
Make	avGraphicTextMake	10.3.5
Make	avLineFileMake	4.4.2
Make	avPointMake	8.2.3
Make	avPolygonMake	8.2.4
Make	avPolylineMake	8.2.6
Make	avSymbolMake	10.2.13
MakeNew	avFTabMakeNew	5.1.2
MakeNew	avVTabMakeNew	5.1.16
MakeXY	avRectMakeXY	8.2.9
Middle	Mid (VB)	Table 2.3
MoveTo	avMoveTo	3.1.10
MultiInput	avMsgBoxMultiInput	7.2.2
MultiList	avMsgBoxMultiList	7.1.3
Natural	avNatural	11.1.2
PanTo	avPanTo	3.3.4
Put	avFileDialogPut	4.6.1
Quantile	avQuantile	11.1.3
Query	avQuery	5.5.2
ReadElt	ReadLine (VB)	4.5.2
Remove	Remove (VB)	Table 2.5
RemoveDoc	avRemoveDoc	3.1.13
RemoveDuplicates	avRemoveDupStrings	2.6.5
RemoveFields	avRemoveFields	5.2.6
RemoveGraphic	avRemoveGraphic	10.1.6
RemoveRecord	avRemoveRecord	6.2.4
Resize	avResize	3.1.14
ReturnArea	avReturnArea	8.3.1

TABLE C-1 (cont.)
AVENUE REQUESTS AND CORRESPONDING AVENUE WRAPS™

Avenue Request	Avenue Wrap	Chapter/Section
ReturnCenter	avReturnCenter	8.3.2
ReturnFamilies	avGetWinFonts	3.1.7
ReturnFiles	avFileDialogReturnFiles	4.6.2
ReturnIntersection	avReturnIntersection	8.4.3
ReturnLength	avReturnLength	8.3.3
ReturnMerged	avReturnMerged	8.4.4
ReturnProjected	sample code (VB)	Appendix D
ReturnUnion	avReturnUnion	8.4.5
ReturnUnProjected	sample code (VB)	Appendix D
ReturnUserCircle	sample code (VB)	9.2.3
ReturnUserPoint	CurrentLocation (VB)	9.2.1
ReturnUserPolygon	sample code (VB)	9.2.4
ReturnUserPolyline	sample code (VB)	9.2.2
ReturnUserRect	sample code (VB)	9.2.5
ReturnValue	avGetFeature	5.4.2
ReturnVisExtent	avReturnVisExtent	3.3.5
Right	Right (VB)	Table 2.3
Round	Fix (VB)	Table 2.2
Run	Sub/Function (VB)	2.5.4
SelectByFTab	avSelectByFTab	6.1.6
SelectByPoint	avSelectByPoint	6.1.7
SelectByPolygon	avSelectByPolygon	6.1.8
Set	avBitmapSet	6.1.1
SetActive	avSetActive	6.1.9
SetAll	avSetAll	6.1.10
SetAngle	avSymbolSetAngle	10.2.14
SetAngle	avGraphicTextSetAngle	10.3.6
SetColor	avSymbolSetColor	10.2.15
SetEditable	avSetEditable	5.1.9
SetEditableTheme	avSetEditableTheme	5.1.10
SetExtension	avSetExtension	4.3.5
SetExtent	avSetExtent	3.3.6
SetName	avObjSetName	3.1.12
SetName	avSetName	3.1.15
SetOlColor	avSymbolSetOLColor	10.2.16
SetOlWidth	avSymbolSetOLWidth	10.2.17
SetSelection	avSetSelection	6.1.11
SetShape	avGraphicSetShape	10.2.3
SetSize	avSymbolSetSize	10.2.18
SetStipple	avSymbolSetStipple	10.2.19
SetStyle	avGraphicTextSetStyle	10.3.7
SetStyle	avSymbolSetStyle	10.2.20
SetSymbol	avGraphicSetSymbol	10.2.4
SetSymbol	avGraphicTextSetSymbol	10.3.8

TABLE C-1 (cont.)
AVENUE REQUESTS AND CORRESPONDING AVENUE WRAPS™

Avenue Request	Avenue Wrap	Chapter/Section
SetSymbols	avLegendSetSymbols	11.2.8
SetText	avGraphicTextSetText	10.3.9
SetValue	avSetValue	5.3.7
SetVisible	avSetVisible	5.1.11
SetWorkDir	avSetWorkDir	3.1.16
ShowMsg	avShowMsg	7.1.5
ShowStopButton	avShowStopButton	3.4.2
Shuffle	(VB)	Table 2.5
Sin	Sin	Table 2.1
SingleSymbol	avSingleSymbol	11.1.5
Split	avSplit	8.4.6
Sort	SortTwoLists	2.6.10
Sqrt	Sqr	Table 2.1
StopEditing	avStopEditing	5.1.13
Substitute	Replace (VB)	Table 2.3
Summarize	avSummarize	5.5.3
Tan	Tan	Table 2.1
Translate	Replace (VB)	Table 2.3
Trim	Trim (VB)	Table 2.3
Truncate	Fix	Table 2.2
UCase	UCase (VB)	Table 2.3
Unique	avUnique	11.1.6
UpdateLegend	avUpdateLegend	11.2.9
UpdateSelection	avUpdateSelection	6.1.13
Warning	avMsgBoxWarning	7.1.4
WriteElt	WriteLine (VB)	4.5.1
XOr	avXORSelection	6.1.14
YesNo	avMsgBoxYesNo	7.3.2
YesNoCancel	avMsgBoxYesNoCancel	7.3.3

TABLE C-2
AVENUE WRAPS™ w/o CORRESPONDING AVENUE REQUESTS

Avenue Request	Avenue Wrap	Chapter/Section
N/A	avaClassMake	8.1.1
N/A	avAsList2	8.1.3
N/A	avAsList3	8.1.4
N/A	avAsPolygon	8.2.1
N/A	avCheckEdits	5.1.1
N/A	avCircleMakeXY	8.2.2
N/A	avDeleteDS	4.2.1
N/A	avDirExists	4.1.1
N/A	avFeatureInvalidate	5.4.1
N/A	avFieldGetType	5.2.2
N/A	avGetBaseName2	4.3.2
N/A	avGetFeatData	5.4.3
N/A	avGetFTabIDs	5.3.3
N/A	avGetGeometry	5.4.4
N/A	avGetGraphicList	10.1.1
N/A	avGetLayerIndx	3.2.1
N/A	avGetLayerType	3.2.2
N/A	avGetPathName	4.3.4
N/A	avGetSelectionClear	6.2.3
N/A	avGetSelectionIDs	6.1.3
N/A	avGetSelFeatures	6.1.4
N/A	avGetThemeExtent	5.1.4
N/A	avGetVisibleCADLayers	3.2.5
N/A	avGetVTabIDs	5.3.6
N/A	avGraphicListDelete	10.1.4
N/A	avGraphicListEmpty	10.1.5
N/A	avGraphicTextSetStyle	10.3.7
N/A	avGraphicTextSetSymbol	10.3.8
N/A	avInit	3.1.9
N/A	avInvSelFeatures	6.1.5
N/A	avListFiles	4.1.2
N/A	avMsgBoxMultiInput2	7.2.3
N/A	avOpenFeatClass	5.6.1
N/A	avOpenWorkspace	5.6.2
N/A	avPlAsList2	8.1.6
N/A	avPlFindVertex	8.1.7
N/A	avPlGet3Pt	8.1.8
N/A	avPlModify	8.1.9
N/A	avPolygonMake2	8.2.5
N/A	avPolyline2Pt	8.2.8
N/A	avPolylineMake2	8.2.7
N/A	avQuantileX	11.1.4
N/A	avSetAll	6.1.10

TABLE C-2 (cont.)
AVENUE WRAPS™ w/o CORRESPONDING AVENUE REQUESTS

Avenue Request	Avenue Wrap	Chapter/Section
N/A	avSetGraphicsLayer	10.1.7
N/A	avSetSelFeatures	6.1.12
N/A	avSetValueG	5.3.8
N/A	avStartOperation	5.1.12
N/A	avStopOperation	5.1.14
N/A	avUpdateAnno	5.4.5
N/A	avViewAddGraphic	10.1.8
N/A	avZoomToSelected	3.3.7
N/A	avZoomToTheme	3.3.8
N/A	avZoomToThemes	3.3.9
N/A	Calc_Callback	3.4.3
N/A	ChangeView	3.3.10
N/A	CopyList	2.6.6
N/A	CopyList2	2.6.7
N/A	CreateAccessDB	5.6.3
N/A	CreateAnnoClass	5.6.4
N/A	CreateFeatClass	5.6.5
N/A	CreateList	2.6.8
N/A	CreateNewGeoDB	5.6.6
N/A	CreateNewShapeFile	5.6.7
N/A	CreateShapeFile	5.6.8
N/A	Dformat	2.6.9
N/A	ExportVBAcode	12.1.1
N/A	GetTextFont	10.3.10
N/A	iccomdis	12.2.6
N/A	iccomppt	12.2.7
N/A	icforce	12.2.9
N/A	icmakdir	12.2.10
N/A	LoadVBAcode	12.1.2
N/A	MakeTextElement	10.3.11
N/A	MakeTextSymbol	10.3.12
N/A	SetViewSnapTol	12.2.12
N/A	HDBbuild	7.2.4
N/A	VDBbuild	7.2.5
N/A	VDBbuild2	7.2.6

APPENDIX D

LISTINGS OF AVENUE WRAPS

Presented in this appendix is the listing for each of the Avenue Wraps™ that were discussed in the previous chapters. To eliminate the retyping of any code, the attached CD includes an ArcMap™ document file, that contains all of the Avenue Wraps™ along with samples which demonstrate the use of the Avenue Wraps™. In fact, the ArcMap document file can be used as a starting point in creating an application. That is, the developer can immediately begin to add new VBA macros and tools to the document file. This appendix, however, does contain the header information for each of the Avenue Wraps™.

Although there are more than 2,000 Avenue requests, 90% of all Avenue programming will probably use only 10% of the Avenue requests. This book presents more than 260 Avenue Wraps which is approximately 10% of the total number of Avenue requests. As such, it is the authors feelings that we have captured the most prevalent requests, which an Avenue programmer would use. Those which we did not include could probably be written using the techniques that are shown in this Appendix. That is one of the reasons for including this Appendix, is that, the reader can review the Avenue Wraps™, see how they were written and perhaps apply this approach in converting other requests that were not included in this book.

One of the issues that became apparent at the outset of our working with ArcObjects™ is that the vocabulary is different from that of Avenue. This, as can be expected, made converting Avenue code into ArcObjects™ somewhat challenging. For example, legends in Avenue are referred to as renderers in ArcObjects™, while themes are now called layers, and so forth. It is hoped that by providing the Avenue programmer with these Avenue Wraps™ we can decrease the ArcObjects™ learning curve, since the listings should help in picking up the vocabulary. The reason being, the reader can find an Avenue request, review the listing and find the corresponding ArcObjects™ terminology.

Following this introduction, there is a section containing various notes, which describe key points concerning the development of the Avenue Wraps™. It is recommended that these notes be reviewed prior to examining the listings. In so doing, the methodology presented in the listings should be a little easier to understand.

As a final note, every programmer has their own style of writing code. In reviewing the listings, the reader may feel that a different approach should have been taken in writing some of the wraps. This very well could be true. As the saying goes, "there is more than one way to skin a cat". What the authors have tried to do in developing the Avenue Wraps™, is not to develop the least amount of code, which more than often is confusing, but rather to present wraparounds that are functional, straightforward and well documented.

Notes regarding the Listings

- ArcMap™ employs a single-document interface, while ArcView® employs a multi-document interface, which changes the meaning of the GetActiveDoc request greatly.

- In Avenue, the variable *theView* was widely used. This variable has been replaced by the variable *pmxDoc* within the Avenue Wraps™.

- Rather than passing objects for layers or tables in the argument list for subroutines or functions, the name of the layer or table is passed as a variable of Variant type.

- When setting a field value for either a feature or table record, use the ***Store*** method to write the value to disk, see the Avenue Wraps™ **avSetValue** and **avSetValueG**.

- In Avenue, the request ReturnValue was applied to a FTab or VTab in order to extract a value out of a table. In ArcObjects™, the ***Value*** property is used to extract the value from an IFeature or IRow object. The following example should explain:

  With Avenue

```
aFTab = aTheme.GetFTab
colS = aFTab.FindField("shape")
aShape = aFTab.ReturnValue(colS, rec)
colA = aFTab.FindField("area")
theArea = aFTab.ReturnValue(colA, rec)
```

  With Avenue Wraps

```
Dim pmxDoc As esricore.IMxDocument
Dim aTheme As Variant
Dim aFTab As esricore.IFields
Dim aFeatClass As esricore.IFeatureClass
Dim aLayer As esricore.IFeatureLayer
Dim rec, colA As Long
Dim aShape As esricore.IFeature
Dim theArea As Double
Call avGetFTab(pmxDoc, aTheme, _
               aFTab, aFeatClass, aLayer)
Call avGetFeature(pmxDoc, aTheme, rec, aShape)
colA = aFTab.FindField("area")
theArea = aShape.Value(colA)
```

- The Avenue requests ChoiceAsString and ListAsString should be substituted with the Avenue Wraps™ **avMsgBoxChoice** and **avMsgBoxList**, respectively.

- An example of converted Avenue code, which cycles through a selected set of features to compute a total value, is shown below:

With Avenue

```
theFTab = theTheme.GetFTab
theSel = theFTab.GetSelection
theField = theFTab.FindField("Deposits")
total = 0.0
for each rec in theSel
   deposit = theFTab.ReturnValue(theField, rec)
   total = total + deposit
end
```

With Avenue Wraps

```
Dim pmxDoc As esricore.IMxDocument
Dim theTheme As Variant
Dim theFTab As esricore.IFields
Dim aFeatClass As esricore.IFeatureClass
Dim aLayer As esricore.IFeatureLayer
Dim theSel As esricore.ISelectionSet
Dim theField, iRec, rec, colA As Long
Dim total, deposit As Double
Dim theSelList As New Collection
Dim pFeat As esricore.IFeature
Call avGetFTab(pmxDoc, theTheme, _
               theFTab, aFeatClass, aLayer)
Call avGetSelection(pmxDoc, theTheme, theSel)
theField = theFTab.FindField("Deposits")
total = 0.0
Call avGetSelectionIDs(theSel, theSelList)
For iRec = 1 to theSelList.Count
   rec = theSelList.Item(iRec)
   Set pFeat = aFeatClass.GetFeature(rec)
   deposit = pFeat.Value(theField)
   total = total + deposit
Next
```

- In Avenue the @ character could be used to create a point. Use the Avenue Wrap™, **avPointMake** to create the point. For example:

With Avenue

```
aPoint = 5000.0 @ 5000.0
```

With Avenue Wraps

```
Dim aPoint As esricore.IPoint
Set aPoint = avPointMake(5000.0, 5000.0)
```

- In Avenue, the request YesNoCancel was applied to the MsgBox Class to determine a course of action from the user. The value that was passed back by the request was a Boolean, using the Avenue Wraps, the returned value is an integer which will be equal to one of the predefined VB constants. The following example should explain:

With Avenue

```
ians = MsgBox.YesNoCancel(Msg, Heading, Default)
if (ians = Nil) then
   .... do something
end
if (ians.Not) then
   .... do something
end
```

With Avenue Wraps

```
Dim Msg, Heading As Variant
Dim Default As Boolean
Dim ians As Integer
Call avMsgBoxYesNoCancel(Msg, Heading, Default, _
                         ians)
If (ians = vbCancel) then
   .... do something
End If
If (ians = vbNo) then
   .... do something
End If
If (ians = vbYes) then
   .... do something
End If
```

- The Avenue request IsNoDataClassDisplayed should be substituted with the ArcObjects™ property, ***UseDefaultSymbol***.

With Avenue

```
noData = aLegend.IsNoDataClassDisplayed
```

With Avenue Wraps

```
Dim pUniqueRend As IUniqueValueRenderer
Dim noData As Boolean
noData = pUniqueRend.UseDefaultSymbol
```

- The Avenue request ClearMsg can be replaced with the Avenue Wrap **avClearStatus**. The following example should explain:

With Avenue

```
av.ClearMsg
```

With Avenue Wraps

```
Call avClearStatus
```

- In Avenue, the programmer could have an if ... then ... end statement on a single data line. In VB or VBA, if an if ... then ... end statement appears in red (denoting an error condition) the data line can be decomposed into a multi-line statement. The following example should explain:

In Avenue

```
if (a = 2.0) then b = 4.0 end
```

In VB/VBA

```
Dim a, b As Double
If (a = 2.0) then
   b = 4.0
End If
```

- The object id (OID) values for shapefiles start at 0 and increase sequentially by one, while for personal geodatabases they start at 1.

- The Avenue request ReturnFamilies could be used to get a list of the available Windows fonts. The Avenue Wrap **avGetWinFonts** can be substitued in its place. The following example should explain:

With Avenue

```
aFontManager = FontManager.The
aFontList = aFontManager.ReturnFamilies
```

With Avenue Wraps

```
Dim aFontList As New Collection
Call avGetWinFonts(aFontList)
```

- The Avenue request FindAllByClass could be used to find all of the graphic text elements in a view. Using Avenue Wraps the following procedure could be used to accomplish the same task:

With Avenue

```
graphList = theView.GetGraphics
gTextList = graphList.FindAllByClass(GraphicText)
```

With Avenue Wraps

```
Dim graphList As New Collection
Dim i As Long
Dim pElement As IElement
Dim gTextList As New Collection
Call avViewGetGraphics(graphList)
If (graphList.Count > 0) Then
   For i = 1 To graphList.Count
       Set pElement = graphList.Item(i)
       If TypeOf pElement Is ITextElement Then
          gTextList.Add pElement
       End If
   Next
End If
```

- The Avenue request ReturnProjected could be used to project a feature into a specific projection. The ArcObjects™ method, ***Project*** can be applied to an IGeometry object to accomplish the same task. The following example shows how a feature can be projected into the map's (pMap) current projection (pSpatialReference). Note that it is the geometry (aShape) of the feature that is actually processed.

With Avenue

```
newShape = aShape.ReturnProjected(thePrj)
```

With Avenue Wraps

```
Dim pmxDoc As esriCore.IMxDocument
Dim pMap As esriCore.IMap
Dim pSpatialReference As ISpatialReference
Dim aShape As esriCore.IGeometry
Set pmxDoc = Application.Document
Set pMap = pmxDoc.FocusMap
Set pSpatialReference = pMap.SpatialReference
aShape.Project pSpatialReference
```

- The Avenue request ReturnUnProjected could be used to unproject a feature into its own natural projection. The ArcObjects™ method, ***Project*** can be applied to an IGeometry object to accomplish the same task. The following example shows how a feature (aFeature) can be unprojected from the map's (pMap) current projection (pSpatialReference) into the projection of the layer (pFeatureClass) in which the feature resides. Note that it is the geometry (aShape) of the feature that is actually processed.

With Avenue

```
theNewShape = aShape.ReturnUnProjected(thePrj)
```

With Avenue Wraps

```
Dim pmxDoc As esriCore.IMxDocument
Dim pMap As esriCore.IMap
Dim pSpatialReference As ISpatialReference
Dim pObjectClass As esriCore.IObjectClass
Dim aFeature As esriCore.IFeature
Dim pFeatureClass As esriCore.IFeatureClass
Dim pGeoDataSet As esriCore.IGeoDataset
Dim aShape As esriCore.IGeometry
Set pmxDoc = Application.Document
Set pMap = pmxDoc.FocusMap
Set pSpatialReference = pMap.SpatialReference
Set pObjectClass = aFeature.Class
Set pFeatureClass = pObjectClass
Set pGeoDataSet = pFeatureClass
Set aShape.SpatialReference = pSpatialReference
aShape.Project pGeoDataSet.SpatialReference
```

How to Install the CEDRA Avenue Wraps Document File:

Step 1:

The **CEDRA Avenue Wraps™** software requires approximately five (5) megabytes of disk space and will operate under **Windows NT®, Windows 2000® and Windows XP®**. One top level directory, whose default name is **CEDRA**, will be created. Should the user wish to use a different name, the user can do so. Within the **CEDRA** directory a single **ArcMap™** document file called **avwraps.mxd** will be stored. This file contains all of the modules and forms that comprise the **Avenue Wraps**.

It is assumed in this installation discussion, that the **"C:"** partition will be used to contain the software. If this is not the case, it is possible to substitute the appropriate drive identifier when performing the installation. It is also assumed that the individual installing the software is somewhat familiar with **PC** terminology, has a working knowledge of the **PC** and the available text editors that are installed on the **PC**, and is somewhat familiar with **ArcGIS™**.

Note that in order to operate **CEDRA Avenue Wraps**, **ArcGIS™** Version 8.1.2 or higher needs to be installed on the **PC**. The user should verify that this requirement is satisfied this time.

The **CEDRA Avenue Wraps** software consists of a single CD and contains the **ArcMap** document file in a compressed file format. Prior to installing the software, a partition on the **PC** should be found that contains the necessary amount of free disk space, five (5) megabytes.

Step 2:

The contents of the **CEDRA Avenue Wraps** software can now be extracted and stored onto the **PC**. The CD should now be inserted into the appropriate drive.

Step 3:

Select the **Start** button from the task bar followed by selecting the **Run...** menu item.

Step 4:

The **CEDRA** software installation program can be invoked by typing:

```
D:SETUP
```

If the CD drive identifier is something other than D, the user should make the appropriate substitution in the above command.

The program will then pose a series of screens guiding the user through the installation process. Once the final screen has been displayed, the program will decompress the document file, and store the file in the appropriate directory location. After the file has been decompressed, the user can invoke **ArcMap** and open the document file.

How to Invoke the CEDRA Avenue Wraps Document File:

This version of **CEDRA Avenue Wraps™** is available as an **ArcMap™** document file called **avwraps.mxd**. This document file contains only the modules and forms that comprise the **Avenue Wraps**. Using **avwraps.mxd** the developer can create additional forms, modules and controls to establish a custom application. To begin using the **CEDRA Avenue Wraps** perform the following:

➤1 Invoke the **ArcMap™** program.

➤2 **Click** in the radial button to the left of the *An existing map:* label, then **click** the **OK** button.

➤3 **Navigate** to the **\cedra** directory, **click** on the **avwraps.mxd** file name, and then **click** the **OK** button.

➤4 At this point the developer can begin reviewing the **CEDRA Avenue Wraps** and building a custom application.

Sample Data:

In addition to the **ArcMap** document file, **avwraps.mxd**, seven VBA modules and two shapefiles are included containing sample code and data, which illustrate the use of the **CEDRA Avenue Wraps**. These samples include:

Module1.bas Sample illustrating how to process graphics and symbols, the graphics that are created are based upon an arbitrary coordinate system.

Module2.bas Sample illustrating how to process feature geometry. This is done by using the first selected feature in a polygon theme.

Module3.bas Sample illustrating how to create a Shapefile and add a feature to it. The sample will also show how an operation can be defined.

Module4.bas Sample illustrating how to perform various shape editing operations. This sample requires seven polygon features and one polyline feature be selected prior to executing this macro. The first selected polygon and the selected polyline features will be used in a split operation. The remaining selected polygons will be used to demonstrate (a) merging, (b) intersecting and (c) unioning operations. The shapefiles **L_0pg** and **L_0pl**, which are included in the distribution set, can be added to **ArcMap** and used in this sample.

Module5.bas Sample illustrating how to create, add records, populate and summarize a table.

Module6.bas Sample illustrating how to create a new shapefile that has a default spatial reference and three attributes using a name that the user enters in a file dialog box. The shapefile is to contain Polyline features and will be added to the map once it has been created.

Module7.bas Sample illustrating how to create various types of message boxes.

To import and execute a sample perform the following:

➤1 Invoke the **ArcMap™** program.

➤2 **Click** in the radial button to the left of the *An existing map:* label, then **click** the **OK** button.

➤3 **Navigate** to the **\cedra** directory, **click** on the **avwraps.mxd** file name, and then **click** the **OK** button.

➤4 Click at the **Tools** menu and then at the **Macros** and **Visual Basic Editor** sub-menus.

➤5 Click at the plus sign, +, to the left of the **Project (avwraps.mxd)** label in the project window to expand the project document.

➤6 Click at the **File** menu and then at the **Import File...** sub-menu.

➤7 **Navigate** to the **\cedra** directory, **click** on the desired sample module file name, and then **click** the **Open** button.

➤8 Click at the plus sign, +, to the left of the **Modules** label in the project window to expand the module document.

➤9 Scroll down the list of modules and find the sample module file that was imported. Once found, double-click on the name of the module to open the module.

➤10 Click at the Run Sub/UserForm tool (▶) to execute the sample code.

```
'
' * * * * * * * * * * * * * * * * * * * * * * * * * * * * * * * * * * *
' *                                                                    *
' *  Name: avaClassMake                        File Name: avaclsmk.bas *
' *                                                                    *
' *  PURPOSE:  TO CREATE A SPECIAL FEATURE OBJECT FROM A POINT LIST    *
' *                                                                    *
' *  GIVEN:    aClass    = the type of special feature                 *
' *                        11 for PolyLineM                            *
' *                        12 for PolyLineZ                            *
' *                        13 for PolygonM                             *
' *                        14 for PolygonZ                             *
' *                        15 for PointM                               *
' *                        16 for PointZ                               *
' *                        17 for MultiPointM                          *
' *                        18 for MultiPointZ                          *
' *                        31 for PolyLineM and PolyLineZ              *
' *                        32 for PolygonM and PolygonZ                *
' *                        33 for PointM and PointZ                    *
' *                        34 for MultiPointM and MultiPointZ          *
' *                        41 for PolyLine                             *
' *                        42 for Polygon                              *
' *                        43 for Point                                *
' *                        44 for MultiPoint                           *
' *            shapeList = the list of points comprising the feature   *
' *                        structure of shapeList is:                  *
' *                        Item 1: number of parts                     *
' *                        Item 2: number of points in part 1          *
' *                        Item 3: x value of point 1 in part 1        *
' *                        Item 4: y value of point 1 in part 1        *
' *                        Item 5: z value of point 1 in part 1        *
' *                        Item 6: m value of point 1 in part 1        *
' *                        Item 7: id value of point 1 in part 1       *
' *                        Item 8: Repeat Items 3 - 7 for each         *
' *                                point                               *
' *                        Repeat Items 2 - 8 for each part            *
' *                                                                    *
' *  RETURN:   theFeat   = the special feature                         *
' *                                                                    *
' *  Dim aClass As Integer                                             *
' *  Dim shapeList As New Collection                                   *
' *  Dim theFeat As IPoint, IMultiPoint, IPolyline, or IPolygon        *
' *                                                                    *
' * * * * * * * * * * * * * * * * * * * * * * * * * * * * * * * * * * *
'
Public Sub avaClassMake(aclass, shapeList, theFeat)
'
' * * * * * * * * * * * * * * * * * * * * * * * * * * * * * * * * * * *
' *                                                                    *
' *  Name: avAddDoc                            File Name: avadddoc.bas *
' *                                                                    *
' *  PURPOSE:  TO ADD A LAYER OR TABLE TO THE MAP                      *
' *                                                                    *
' *  GIVEN:    aDoc      = the document to be added                    *
' *                                                                    *
' *  RETURN:   avAddDoc = error flag (0 = no error, 1 = error)         *
' *                                                                    *
' *  Dim aDoc As IUnknown                                              *
' *  Dim avAddDoc As Integer                                           *
' *                                                                    *
' * * * * * * * * * * * * * * * * * * * * * * * * * * * * * * * * * * *
'
Public Function avAddDoc(aDoc As IUnknown)
'
' * * * * * * * * * * * * * * * * * * * * * * * * * * * * * * * * * * *
' *                                                                    *
' *  Name: avAddFields                         File Name: avaddfld.bas *
' *                                                                    *
' *  PURPOSE:  TO ADD FIELDS INTO A LAYER OR TABLE                     *
' *                                                                    *
' *  GIVEN:    pmxDoc     = the active view                            *
' *            theTheme   = the theme or table to be processed         *
' *            theFields  = list of fields to be added, the items in   *
```

```
' *                         this list are IFieldEdit objects, not     *
' *                         strings                                   *
' *                                                                   *
' *  RETURN:   avAddFields = error flag (0 = no error, 1 = error)     *
' *                                                                   *
' *  NOTE:     In order to add fields into a layer or table the       *
' *            editor can not be in an edit state, this routine will  *
' *            stop the editor if the editor is in an edit state      *
' *            thereby saving any changes that may have been made,    *
' *            prior to adding the fields                             *
' *                                                                   *
' *  Dim pmxDoc As IMxDocument                                        *
' *  Dim theTheme As Variant                                          *
' *  Dim theFields As New Collection                                  *
' *  Dim avAddFields As Integer                                       *
' *                                                                   *
' * * * * * * * * * * * * * * * * * * * * * * * * * * * * * * * * * * *
'
Public Function avAddFields(pmxDoc As esriCore.IMxDocument, _
                            theTheme, theFieldS)
'
' * * * * * * * * * * * * * * * * * * * * * * * * * * * * * * * * * * *
' *                                                                   *
' *  Name: avAddRecord                      File Name: avaddrec.bas   *
' *                                                                   *
' *  PURPOSE:  TO ADD A RECORD INTO A LAYER OR TABLE                  *
' *                                                                   *
' *  GIVEN:    pmxDoc      = the active view                          *
' *            theTheme    = the theme or table to be processed       *
' *                                                                   *
' *  RETURN:   avAddRecord = the id of the record that was added, if  *
' *                          a record can not be added will be -1     *
' *                                                                   *
' *  Dim pmxDoc As IMxDocument                                        *
' *  Dim theTheme As Variant                                          *
' *  Dim avAddRecord As Long                                          *
' *                                                                   *
' * * * * * * * * * * * * * * * * * * * * * * * * * * * * * * * * * * *
'
Public Function avAddRecord(pmxDoc As esriCore.IMxDocument, _
                            theTheme) As Long
'
' * * * * * * * * * * * * * * * * * * * * * * * * * * * * * * * * * * *
' *                                                                   *
' *  Name: avAsList                         File Name: avaslist.bas   *
' *                                                                   *
' *  PURPOSE:  TO CREATE A LIST CONTAINING THE POINTS THAT COMPRISE   *
' *            A LINE OR POLYGON FEATURE                              *
' *                                                                   *
' *  GIVEN:    theFeature = feature to be processed                   *
' *                                                                   *
' *  RETURN:   shapeList  = the shape's list of points for each part  *
' *                         structure of shapeList is:                *
' *                         Item 1: collection of points in part 1    *
' *                         Repeat Item 1 for each part               *
' *                         So that, shapeList is a list of           *
' *                         collections with each collection          *
' *                         containing points                         *
' *                                                                   *
' *  NOTE:     Use subroutine avPlAsList when an IGeometry object is  *
' *            known and not an IFeature object                       *
' *                                                                   *
' *  Dim theFeature As IFeature                                       *
' *  Dim shapeList As New Collection                                  *
' *                                                                   *
' * * * * * * * * * * * * * * * * * * * * * * * * * * * * * * * * * * *
'
Public Sub avAsList(theFeature As esriCore.IFeature, shapeList)
'
```

```
' * * * * * * * * * * * * * * * * * * * * * * * * * * * * * * * * * * *
' *                                                                     *
' *  Name: avAsList2                            File Name: avaslis2.bas *
' *                                                                     *
' *  PURPOSE:  TO CREATE A LIST CONTAINING THE POINTS THAT COMPRISE     *
' *            A LINE OR POLYGON FEATURE                                *
' *                                                                     *
' *  GIVEN:    theFeature = feature to be processed                     *
' *                                                                     *
' *  RETURN:   shapeList  = the list of points comprising the feature   *
' *                         structure of shapeList is:                  *
' *                         Item 1: number of parts                     *
' *                         Item 2: number of points in part 1          *
' *                         Item 3: point 1 in part 1                   *
' *                         Item 4: point 2 in part 1                   *
' *                         Item 5:        .                            *
' *                         Item 6:        .                            *
' *                         Item n: point n in part 1                   *
' *                         Repeat Items 2 - n for each part            *
' *                                                                     *
' *  NOTE:     Use subroutine avAsList3 when an IGeometry object is     *
' *            known and not an IFeature object                         *
' *                                                                     *
' *  Dim theFeature As IFeature                                         *
' *  Dim shapeList As New Collection                                    *
' *                                                                     *
' * * * * * * * * * * * * * * * * * * * * * * * * * * * * * * * * * * *
'
Public Sub avAsList2(theFeature As esriCore.IFeature, shapeList)
'
' * * * * * * * * * * * * * * * * * * * * * * * * * * * * * * * * * * *
' *                                                                     *
' *  Name: avAsList3                            File Name: avaslis3.bas *
' *                                                                     *
' *  PURPOSE:  TO CREATE A LIST CONTAINING THE POINTS THAT COMPRISE     *
' *            A LINE OR POLYGON                                        *
' *                                                                     *
' *  GIVEN:    theShape  = feature to be processed                      *
' *                                                                     *
' *  RETURN:   shapeList = the list of points comprising the feature    *
' *                        structure of shapeList is:                   *
' *                        Item 1: number of parts                      *
' *                        Item 2: number of points in part 1           *
' *                        Item 3: point 1 in part 1                    *
' *                        Item 4: point 2 in part 1                    *
' *                        Item 5:        .                             *
' *                        Item 6:        .                             *
' *                        Item n: point n in part 1                    *
' *                        Repeat Items 2 - n for each part             *
' *                                                                     *
' *  NOTE:     Use subroutine avAsList2 when an IFeature object is      *
' *            known and not an IGeometry object                        *
' *                                                                     *
' *  Dim theShape As IGeometry                                          *
' *  Dim shapeList As New Collection                                    *
' *                                                                     *
' * * * * * * * * * * * * * * * * * * * * * * * * * * * * * * * * * * *
'
Public Sub avAsList3(theShape As esriCore.IGeometry, shapeList)
'
' * * * * * * * * * * * * * * * * * * * * * * * * * * * * * * * * * * *
' *                                                                     *
' *  Name: avAsPolygon                          File Name: avaspoly.bas *
' *                                                                     *
' *  PURPOSE:  CONVERT INPUT INTO POLYGON GEOMETRY                      *
' *                                                                     *
' *  GIVEN:    pInput      = the input to be converted                  *
' *                                                                     *
' *  RETURN:   avAsPolygon = polygon geometry                           *
' *                                                                     *
' *  Dim pInput As IUnknown                                             *
' *  Dim avAsPolygon As IGeometry                                       *
' *                                                                     *
```

```
' * * * * * * * * * * * * * * * * * * * * * * * * * * * * * * * * * * * *
'
Public Function avAsPolygon(pInput As IUnknown) As esriCore.IGeometry
'
' * * * * * * * * * * * * * * * * * * * * * * * * * * * * * * * * * * * *
' *                                                                      *
' *  Name: avAsTokens                          File Name: avastokn.bas  *
' *                                                                      *
' *  PURPOSE:  Read a text string, a word delineator and an indicator   *
' *            whether to change all characters to upper or lower       *
' *            case characters, and                                      *
' *            1. Remove the leading and trailing blank spaces; and     *
' *            2. Create a list of  words from the contents of the      *
' *               read text string excluding therefrom any blank        *
' *               spaces that may be present between the words; and     *
' *            3. Compute the number of words returned list.            *
' *                                                                      *
' *  GIVEN:     theString  = the input string                           *
' *             delString  = the word delineator                        *
' *             UpperLower = U - change all characters to upper case    *
' *                        = L - change all characters to lower case    *
' *                        = X - no change in terms of case and do not  *
' *                              trim leading/trailing characters       *
' *                        = any other character - no change            *
' *                                                                      *
' *  RETURN:    theList    = the returned list of words                 *
' *             nWords     = number of words extracted                  *
' *                                                                      *
' *  Dim theString, delString, UpperLower As String                     *
' *  Dim theList As New Collection                                      *
' *  Dim nWords As Integer                                              *
' *                                                                      *
' * * * * * * * * * * * * * * * * * * * * * * * * * * * * * * * * * * * *
'
Public Sub avAsTokens(theString, delString, UpperLower, theList, nWords)
'
' * * * * * * * * * * * * * * * * * * * * * * * * * * * * * * * * * * * *
' *                                                                      *
' *  Name: avBasicTrim                         File Name:  avtrmchr.bas *
' *                                                                      *
' *  PURPOSE:  Remove from a given string the specified leading         *
' *            and/or trailing characters                               *
' *                                                                      *
' *  GIVEN:    theString   = the given string to be processed           *
' *            LeadChar    = the characters to be removed at the        *
' *                          start of the given string                  *
' *            TrailChar   = the characters to be removed at the        *
' *                          end of the given string                    *
' *                                                                      *
' *  RETURN:   avBasicTrim = the resultant string                       *
' *                                                                      *
' *  NOTE:     Blank characters are not removed from theString by       *
' *            this function                                            *
' *                                                                      *
' *  Dim theString, LeadChar, TrailChar As String                       *
' *  Dim avBasicTrim As String                                          *
' *                                                                      *
' * * * * * * * * * * * * * * * * * * * * * * * * * * * * * * * * * * * *
'
Public Function avBasicTrim(theString, LeadChar, TrailChar)
'
' * * * * * * * * * * * * * * * * * * * * * * * * * * * * * * * * * * * *
' *                                                                      *
' *  Name: avBitmapClear                       File Name: avbitclr.bas  *
' *                                                                      *
' *  PURPOSE:  REMOVE A RECORD FROM THE SELECTED SET FOR A LAYER OR     *
' *            TABLE                                                    *
' *                                                                      *
' *  GIVEN:    psTableSel = selection set for a theme or table          *
' *            theRcrd    = record to be removed from the selection     *
' *                                                                      *
' *  RETURN:   nothing                                                  *
' *                                                                      *
```

```
' *  Dim psTableSel As ISelectionSet                                     *
' *  Dim theRcrd As Long                                                 *
' *                                                                      *
' * * * * * * * * * * * * * * * * * * * * * * * * * * * * * * * * * * * *
'
Public Sub avBitmapClear(psTableSel As esriCore.ISelectionSet, theRcrd)
'
' * * * * * * * * * * * * * * * * * * * * * * * * * * * * * * * * * * * *
' *                                                                      *
' *  Name: avBitmapSet                         File Name: avbitset.bas  *
' *                                                                      *
' *  PURPOSE:  ADD A RECORD TO THE SELECTED SET FOR A LAYER OR TABLE    *
' *                                                                      *
' *  GIVEN:     pmxDoc   = the active view                               *
' *            theTheme = the theme or table to be processed             *
' *            theRcrd  = the record to be added to the selection        *
' *                                                                      *
' *  RETURN:    nothing                                                  *
' *                                                                      *
' *  Dim pmxDoc As IMxDocument                                           *
' *  Dim theTheme As Variant                                             *
' *  Dim theRcrd As Long                                                 *
' *                                                                      *
' * * * * * * * * * * * * * * * * * * * * * * * * * * * * * * * * * * * *
'
Public Sub avBitmapSet(pmxDoc As esriCore.IMxDocument, theTheme, _
                       theRcrd)
'
' * * * * * * * * * * * * * * * * * * * * * * * * * * * * * * * * * * * *
' *                                                                      *
' *  Name: avCalculate                         File Name: avcalcul.bas  *
' *                                                                      *
' *  PURPOSE:  TO APPLY A CALCULATION TO A FIELD IN A LAYER OR TABLE    *
' *            THE RECORDS PROCESSED ARE THOSE THAT ARE SELECTED         *
' *                                                                      *
' *  GIVEN:     pmxDoc      = the active view                            *
' *            theTheme    = name of theme or table to be processed      *
' *            aCalcString = calculation string to be applied            *
' *                          sample string field equation                *
' *                          aCalcStr = """abcd"""                       *
' *                          sample numeric field equation               *
' *                          aCalcStr = "([SLN] - " + CStr(ii) + ")"     *
' *            aField      = field to be populated (index value)         *
' *                                                                      *
' *  RETURN:    avCalculate = error flag as noted below                  *
' *                           0 : no error                               *
' *                           1 : theme or table not found               *
' *                           2 : error in performing calculation        *
' *                           3 : no records selected                    *
' *                           4 : an edit session has not been started   *
' *                                                                      *
' *  NOTE:      (a) If the table contains selected records, then only    *
' *                 the selected records will be processed, if there     *
' *                 are no selected records, then the entire table       *
' *                 will be processed                                    *
' *             (b) The syntax for aCalcString shown above works for     *
' *                 both shapefiles and personal geodatabases            *
' *                                                                      *
' *  Dim pmxDoc As IMxDocument                                           *
' *  Dim theTheme As Variant                                             *
' *  Dim aCalcString As String                                           *
' *  Dim aField As Long                                                  *
' *  Dim avCalculate As Integer                                          *
' *                                                                      *
' * * * * * * * * * * * * * * * * * * * * * * * * * * * * * * * * * * * *
'
Public Function avCalculate(pmxDoc As esriCore.IMxDocument, theTheme, _
                            aCalcString, aField) As Integer
'
```

```
' * * * * * * * * * * * * * * * * * * * * * * * * * * * * * * * * * * *
' *                                                                   *
' *  Name: avCheckEdits                       File Name: avchkedt.bas *
' *                                                                   *
' *  PURPOSE:  TO PERFORM CHECKS ON THE EDITING OF DATA               *
' *                                                                   *
' *  GIVEN:    pEditor  = the ArcMap Editor extension                 *
' *           pDataSet = the dataset to be processed, if NOTHING is   *
' *                      specified and if the editor is in an edit    *
' *                      state, the editor is stopped saving any      *
' *                      edits that may have been made                *
' *                                                                   *
' *  RETURN:   nothing                                                *
' *                                                                   *
' *  NOTE:     This routine first checks if the editor is in an edit  *
' *            state, if not, this routine does nothing, if it is in  *
' *            an edit state it will check if the dataset passed in   *
' *            is currently being edited, if not the routine saves    *
' *            the edits on the dataset currently being edited and    *
' *            starts the editor on the dataset that is passed in     *
' *                                                                   *
' *  Dim pEditor As IEditor                                           *
' *  Dim pDataSet As IDataset                                         *
' *                                                                   *
' * * * * * * * * * * * * * * * * * * * * * * * * * * * * * * * * * * *
'
Public Sub avCheckEdits(pEditor As esriCore.IEditor, _
                        pDataSet As esriCore.IDataset)
'
' * * * * * * * * * * * * * * * * * * * * * * * * * * * * * * * * * * *
' *                                                                   *
' *  Name: avCircleMakeXY                     File Name: avcccrmk.bas *
' *                                                                   *
' *  PURPOSE:  TO CREATE A CIRCLE FROM COORDINATES AND A RADIUS       *
' *                                                                   *
' *  GIVEN:    xPt           = x coordinate of circle center          *
' *            yPt           = y coordinate of circle center          *
' *            rad           = radius of circle                       *
' *                                                                   *
' *  RETURN:   avCircleMakeXY = the curve feature                     *
' *                                                                   *
' *  Dim xPt, yPt, rad As Double                                      *
' *  Dim avCircleMakeXY As ICurve                                     *
' *                                                                   *
' * * * * * * * * * * * * * * * * * * * * * * * * * * * * * * * * * * *
'
Public Function avCircleMakeXY(xPt, yPt, rad) As esriCore.ICurve
'
' * * * * * * * * * * * * * * * * * * * * * * * * * * * * * * * * * * *
' *                                                                   *
' *  Name: avClean                             File Name: avclean.bas *
' *                                                                   *
' *  PURPOSE:  TO VERIFY AND ENFORCE THE CORRECTNESS OF A SHAPE       *
' *                                                                   *
' *  GIVEN:    aShape1 = shape to be cleaned                          *
' *                                                                   *
' *  RETURN:   avClean = new shape reflecting the cleaning            *
' *                                                                   *
' *  Dim aShape1 As IGeometry                                         *
' *  Dim avClean As IGeometry                                         *
' *                                                                   *
' * * * * * * * * * * * * * * * * * * * * * * * * * * * * * * * * * * *
'
Public Function avClean(aShape1 As esriCore.IGeometry) _
                                           As esriCore.IGeometry
'
```

```
' * * * * * * * * * * * * * * * * * * * * * * * * * * * * * * * * * * *
' *                                                                   *
' *  Name: avClearSelection                   File Name: avclselt.bas  *
' *                                                                   *
' *  PURPOSE:  CLEAR THE SELECTION SET FOR A LAYER OR TABLE           *
' *                                                                   *
' *  GIVEN:     pmxDoc   = the active view                            *
' *            theTheme = the theme or table to be processed, if NULL *
' *                       is specified all selected features in all   *
' *                       themes will be deselected                   *
' *                                                                   *
' *  RETURN:    nothing                                               *
' *                                                                   *
' *  Dim pmxDoc As IMxDocument                                        *
' *  Dim theTheme As Variant                                          *
' *                                                                   *
' * * * * * * * * * * * * * * * * * * * * * * * * * * * * * * * * * * *
'
Public Sub avClearSelection(pmxDoc As esriCore.IMxDocument, theTheme)
'
' * * * * * * * * * * * * * * * * * * * * * * * * * * * * * * * * * * *
' *                                                                   *
' *  Name: avClearStatus                      File Name: avclrsta.bas  *
' *                                                                   *
' *  PURPOSE:  CLEAR THE STATUS BAR AREA                              *
' *                                                                   *
' *  GIVEN:     nothing                                               *
' *                                                                   *
' *  RETURN:    nothing                                               *
' *                                                                   *
' * * * * * * * * * * * * * * * * * * * * * * * * * * * * * * * * * * *
'
Public Sub avClearStatus()
'
' * * * * * * * * * * * * * * * * * * * * * * * * * * * * * * * * * * *
' *                                                                   *
' *  Name: avClone                              File Name: avclone.bas *
' *                                                                   *
' *  PURPOSE:  MAKE A NEW OBJECT BY COPYING AN EXISTING OBJECT        *
' *                                                                   *
' *  GIVEN:     theObject = object which is to be copied              *
' *                                                                   *
' *  RETURN:    avClone   = copy of the object                        *
' *                                                                   *
' *  Dim theObject As IUnknown                                        *
' *  Dim avClone As IClone                                            *
' *                                                                   *
' * * * * * * * * * * * * * * * * * * * * * * * * * * * * * * * * * * *
'
Public Function avClone(theObject As IUnknown) As esriCore.IClone
'
' * * * * * * * * * * * * * * * * * * * * * * * * * * * * * * * * * * *
' *                                                                   *
' *  Name: avConvertArea                      File Name: avconara.bas  *
' *                                                                   *
' *  PURPOSE:  CONVERT AN AREA VALUE FROM ONE UNIT INTO ANOTHER       *
' *                                                                   *
' *  GIVEN:     theValue = the value to be converted                  *
' *             fromUnit = the from unit of measure                   *
' *             toUnit   = the to unit of measure                     *
' *                                                                   *
' *  RETURN:    nothing                                               *
' *                                                                   *
' *  NOTE:      The argument theValue is modified by this procedure   *
' *                                                                   *
' *  Dim theValue As Variant                                          *
' *  Dim fromUnit As esriUnits                                        *
' *  Dim toUnit As esriUnits                                          *
' *                                                                   *
' * * * * * * * * * * * * * * * * * * * * * * * * * * * * * * * * * * *
'
Public Sub avConvertArea(theValue, fromUnit, toUnit)
'
```

```
'**********************************************************************
'*                                                                    *
'*  Name: avConvertDistance                File Name: avcondis.bas   *
'*                                                                    *
'*  PURPOSE:  CONVERT A DISTANCE VALUE FROM ONE UNIT INTO ANOTHER     *
'*                                                                    *
'*  GIVEN:    theValue = the value to be converted                    *
'*            fromUnit = the from unit of measure                     *
'*            toUnit   = the to unit of measure                       *
'*                                                                    *
'*  RETURN:   nothing                                                 *
'*                                                                    *
'*  NOTE:     The argument theValue is modified by this procedure     *
'*                                                                    *
'*  Dim theValue As Variant                                           *
'*  Dim fromUnit As esriUnits                                         *
'*  Dim toUnit As esriUnits                                           *
'*                                                                    *
'**********************************************************************
'
Public Sub avConvertDistance(theValue, fromUnit, toUnit)
'
'**********************************************************************
'*                                                                    *
'*  Name: avDeleteDS                       File Name: avdeleds.bas   *
'*                                                                    *
'*  PURPOSE:  DELETE A DATASET SUCH AS A SHAPEFILE OR DBASE FILE      *
'*                                                                    *
'*  GIVEN:    name        = name of the dataset to be deleted, if the *
'*                          name does not contain a complete pathname *
'*                          the current working directory will be     *
'*                          used, some examples of name for a         *
'*                          shapefile:                                *
'*                             c:\project\test\l_01n                  *
'*                          access database:                          *
'*                             c:\project\test\montgomery             *
'*                          dBase file:                               *
'*                             c:\project\test\table                  *
'*                                                                    *
'*  RETURN:   avDeleteDS = error flag (0 = no error, 1 = error)       *
'*                                                                    *
'*  NOTE:     (a) The dataset must not appear in the Table of         *
'*                Contents if it does an error will be generated,     *
'*                use the subroutine avRemoveDoc to remove the        *
'*                dataset from the Table of Contents before calling   *
'*                this function                                       *
'*            (b) If the name passed in contains an extension such    *
'*                as .shp, .mdb, .dbf, etc., it will be stripped off  *
'*                and no error will be generated                      *
'*                                                                    *
'*  Dim name As String                                                *
'*  Dim avDeleteDS As Integer                                         *
'*                                                                    *
'**********************************************************************
'
Public Function avDeleteDS(name As String)
'
'**********************************************************************
'*                                                                    *
'*  Name: avDirExists                      File Name: avdirext.bas   *
'*                                                                    *
'*  PURPOSE:  DETERMINE IF A DIRECTORY EXISTS OR NOT                  *
'*                                                                    *
'*  GIVEN:    name        = name of directory to be checked, if the   *
'*                          directory is not in the current folder a  *
'*                          complete pathname must be specified       *
'*                                                                    *
'*  RETURN:   avDirExists = existence flag (true = yes, false = no)   *
'*                                                                    *
'*  Dim name As String                                                *
'*  Dim avDirExists As Boolean                                        *
'*                                                                    *
'**********************************************************************
```

```
'
Public Function avDirExists(name) As Boolean
'
' * * * * * * * * * * * * * * * * * * * * * * * * * * * * * * * * * * *
' *                                                                   *
' *  Name: avDisplayInvalidate              File Name: avdisinv.bas   *
' *                                                                   *
' *  PURPOSE:  TO REFRESH OR REDRAW THE ACTIVE DISPLAY                *
' *                                                                   *
' *  GIVEN:    aFlag   = when to redraw (true = at the next refresh,  *
' *                      false = immediately)                         *
' *                                                                   *
' *  RETURN:   nothing                                                *
' *                                                                   *
' *  Dim aFlag As Boolean                                             *
' *                                                                   *
' * * * * * * * * * * * * * * * * * * * * * * * * * * * * * * * * * * *
'
Public Sub avDisplayInvalidate(aFlag)
'
' * * * * * * * * * * * * * * * * * * * * * * * * * * * * * * * * * * *
' *                                                                   *
' *  Name: avExecute                        File Name: avexecut.bas   *
' *                                                                   *
' *  PURPOSE:  TO EXECUTE A SYSTEM LEVEL COMMAND                      *
' *                                                                   *
' *  GIVEN:    aCommand = the command to be executed                  *
' *                                                                   *
' *  RETURN:   nothing                                                *
' *                                                                   *
' *  NOTE:     Once the command has been issued, the statements that  *
' *            follow the call to avExecute will be immediately       *
' *            executed, to pause ArcMap until the command is done,   *
' *            one possibility is to perform a loop checking for the  *
' *            existence of a file that is created when the command   *
' *            has finished processing                                *
' *                                                                   *
' *  Dim aCommand As String                                           *
' *                                                                   *
' * * * * * * * * * * * * * * * * * * * * * * * * * * * * * * * * * * *
'
Public Sub avExecute(aCommand)
'
' * * * * * * * * * * * * * * * * * * * * * * * * * * * * * * * * * * *
' *                                                                   *
' *  Name: avFTabMakeNew                    File Name: avftabmk.bas   *
' *                                                                   *
' *  PURPOSE:  CREATE A NEW SHAPEFILE                                 *
' *                                                                   *
' *  GIVEN:    aFileName     = name of the shapefile to be created,   *
' *                            if the name does not contain a         *
' *                            complete pathname the current working  *
' *                            directory will be used, some examples  *
' *                            of name include:                       *
' *                               c:\project\test\l_0ln               *
' *                               c:\project\test\l_0ln.shp           *
' *                               l_0ln                               *
' *                               l_0ln.shp                           *
' *                            the name can or can not contain the    *
' *                            extension .shp                         *
' *            aClass        = type of shapefile to be created        *
' *                            POINT                                  *
' *                            MULTIPOINT                             *
' *                            POLYLINE                               *
' *                            POLYGON                                *
' *                            POINTM                                 *
' *                            MULTIPOINTM                            *
' *                            POLYLINEM                              *
' *                            POLYGONM                               *
' *                            POINTZ                                 *
' *                            MULTIPOINTZ                            *
' *                            POLYLINEZ                              *
' *                            POLYGONZ                               *
```

```
' *                                                                       *
' *  RETURN:   avFTabMakeNew = feature layer object that is created      *
' *                                                                       *
' *  NOTE:     (a) Three fields called FID, SHAPE and ID will be         *
' *                created by this routine, the function avAddDoc can *
' *                be used to add the shapefile to the map, if need    *
' *                be                                                   *
' *            (b) If the shapefile to be created exists on disk, the *
' *                routine will abort the existing shapefile will not *
' *                be overwritten                                       *
' *                                                                       *
' *  Dim aFileName, aClass As String                                     *
' *  Dim avFTabMakeNew As IFeatureLayer                                  *
' *                                                                       *
' * * * * * * * * * * * * * * * * * * * * * * * * * * * * * * * * * * * *
'
Public Function avFTabMakeNew(aFileName, aclass) _
                                          As esriCore.IFeatureLayer
'
' * * * * * * * * * * * * * * * * * * * * * * * * * * * * * * * * * * * *
' *                                                                       *
' *  Name: avFeatureInvalidate                File Name: avfeainv.bas    *
' *                                                                       *
' *  PURPOSE:  REDRAW A FEATURE                                           *
' *                                                                       *
' *  GIVEN:    pmxDoc     = the active view                               *
' *            theFeature = feature to be redrawn                         *
' *                                                                       *
' *  RETURN:   nothing                                                    *
' *                                                                       *
' *  Dim pmxDoc As IMxDocument                                            *
' *  Dim theFeature As IFeature                                           *
' *                                                                       *
' * * * * * * * * * * * * * * * * * * * * * * * * * * * * * * * * * * * *
'
Public Sub avFeatureInvalidate(pmxDoc As esriCore.IMxDocument, _
                               theFeature As esriCore.IFeature)
'
' * * * * * * * * * * * * * * * * * * * * * * * * * * * * * * * * * * * *
' *                                                                       *
' *  Name: avFieldGetType                     File Name: avfldtyp.bas    *
' *                                                                       *
' *  PURPOSE:  DETERMINE THE TYPE OF FIELD THAT A FIELD OBJECT IS        *
' *                                                                       *
' *  GIVEN:    pField         = field object to be processed             *
' *                                                                       *
' *  RETURN:   avFieldGetType = numeric value denoting type of field     *
' *                             0 : Small Integer                         *
' *                             1 : Long Integer                          *
' *                             2 : Single-precision float                *
' *                             3 : Double-precision float                *
' *                             4 : String                                *
' *                             5 : Date                                  *
' *                             6 : Long Integer denoting the OID         *
' *                             7 : Geometry                              *
' *                             8 : Blob                                  *
' *                                                                       *
' *  Dim pField As iField                                                 *
' *  Dim avFieldGetType As esriFieldType                                  *
' *                                                                       *
' * * * * * * * * * * * * * * * * * * * * * * * * * * * * * * * * * * * *
'
Public Function avFieldGetType(pField As esriCore.iField) _
                                          As esriCore.esriFieldType
'
' * * * * * * * * * * * * * * * * * * * * * * * * * * * * * * * * * * * *
' *                                                                       *
' *  Name: avFieldMake                        File Name: avfldmak.bas    *
' *                                                                       *
' *  PURPOSE:  CREATE A FIELD THAT CAN BE ADDED TO A LAYER OR TABLE      *
' *                                                                       *
' *  GIVEN:    aName       = name of field to be created                 *
' *            aFieldType  = type of field to be created as denoted      *
```

```
' *                         by the strings shown below on the left,  *
' *                         to the right of these strings are the    *
' *                         field types that are actually created    *
' *                          BYTE            : Small Integer         *
' *                          CHAR            : String                *
' *                          DATE            : Date                  *
' *                          DECIMAL         : Single                *
' *                          DOUBLE          : Double                *
' *                          FLOAT           : Single                *
' *                          ISODATE         : Date                  *
' *                          ISODATETIME     : Date                  *
' *                          ISOTIME         : Date                  *
' *                          LOGICAL         : String                *
' *                          LONG            : Integer               *
' *                          MONEY           : Double                *
' *                          SHORT           : Small Integer         *
' *                          BLOB            : Blob                  *
' *                          VCHAR           : String                *
' *            nchr       = total character width of field including *
' *                         decimal point and negative sign, if they *
' *                          are to appear in the field              *
' *            ndr        = number of digits to the right of the     *
' *                         decimal point, 0 for non-numeric fields  *
' *                                                                  *
' *  RETURN:   avFieldMake = field object that was created           *
' *                                                                  *
' *  NOTE:     This routine can not be used to create a geometry     *
' *             field                                                *
' *                                                                  *
' *  Dim aName, aFieldType As String                                 *
' *  Dim nchr, ndr As Long                                           *
' *  Dim avFieldMake As IFieldEdit                                   *
' *                                                                  *
' * * * * * * * * * * * * * * * * * * * * * * * * * * * * * * * * * *
'
Public Function avFieldMake(aName, aFieldType, nChr, ndr) _
                                               As esriCore.IFieldEdit
'
' * * * * * * * * * * * * * * * * * * * * * * * * * * * * * * * * * *
' *                                                                  *
' *  Name: avFileDelete                     File Name: avfildel.bas  *
' *                                                                  *
' *  PURPOSE:  DELETE A FILE                                         *
' *                                                                  *
' *  GIVEN:    name          = name of file to be deleted, if the file *
' *                            is not in the current folder a complete *
' *                            pathname must be specified            *
' *                                                                  *
' *  RETURN:   avFileDelete = error flag (0 = no error, 1 = error)   *
' *                                                                  *
' *  Dim name As String                                              *
' *  Dim avFileDelete As Integer                                     *
' *                                                                  *
' * * * * * * * * * * * * * * * * * * * * * * * * * * * * * * * * * *
'
Public Function avFileDelete(name)
'
' * * * * * * * * * * * * * * * * * * * * * * * * * * * * * * * * * *
' *                                                                  *
' *  Name: avFileDialogPut                  File Name: avfdpfil.bas  *
' *                                                                  *
' *  PURPOSE:  TO CREATE A FILE USING A NAME THAT THE USER SPECIFIES *
' *                                                                  *
' *  GIVEN:    defName  = default filename to be displayed           *
' *           aPattrn  = defines the pattern for similar files. Use  *
' *                      an asterisk as a wild card character, for   *
' *                      example: *.ave  or *.*                      *
' *            Heading  = message box caption                        *
' *                                                                  *
' *  RETURN:   fileName = name of file to be created, if the user    *
' *                      Cancels the command, fileName will be set   *
' *                      to a blank character (a single space)       *
' *                                                                  *
```

```
' *  Dim defName As String                                              *
' *  Dim aPattrn As String                                              *
' *  Dim Heading As String                                              *
' *  Dim fileName As String                                             *
' *                                                                     *
' * * * * * * * * * * * * * * * * * * * * * * * * * * * * * * * * * * * *
'
Public Sub avFileDialogPut(defname, aPattrn, Heading, fileName)
'
' * * * * * * * * * * * * * * * * * * * * * * * * * * * * * * * * * * * *
' *                                                                     *
' *  Name: avFileDialogReturnFiles          File Name: avfdrfil.bas    *
' *                                                                     *
' *  PURPOSE:  GET A LIST OF FILE NAMES WHICH THE USER SELECTS          *
' *                                                                     *
' *  GIVEN:    patrns   = list of file patterns that can be displayed  *
' *            labels   = list of labels corresponding to the list of  *
' *                       patterns                                      *
' *            Heading  = message box caption                          *
' *            defIndex = index into pattern list denoting the         *
' *                       default pattern to be displayed              *
' *                                                                     *
' *  RETURN:   fileList = list of file names, of string type, that     *
' *                       were selected by the user, if the user       *
' *                       Cancels the command this list will be empty  *
' *                                                                     *
' *  Dim patrns As New Collection                                       *
' *  Dim labels As New Collection                                       *
' *  Dim Heading As String                                              *
' *  Dim defIndex As Long                                               *
' *  Dim fileList As New Collection                                     *
' *                                                                     *
' * * * * * * * * * * * * * * * * * * * * * * * * * * * * * * * * * * * *
'
Public Sub avFileDialogReturnFiles(patrns, Labels, Heading, defIndex, _
                                   fileList)
'
' * * * * * * * * * * * * * * * * * * * * * * * * * * * * * * * * * * * *
' *                                                                     *
' *  Name: avFileExists                      File Name: avfilext.bas   *
' *                                                                     *
' *  PURPOSE:  DETERMINE IF A FILE EXISTS OR NOT                        *
' *                                                                     *
' *  GIVEN:    name         = name of file to be checked, if the file  *
' *                           is not in the current folder a complete  *
' *                           pathname must be specified               *
' *                                                                     *
' *  RETURN:   avFileExists = existence flag (true = yes, false = no)  *
' *                                                                     *
' *  Dim name As String                                                 *
' *  Dim avFileExists As Boolean                                        *
' *                                                                     *
' * * * * * * * * * * * * * * * * * * * * * * * * * * * * * * * * * * * *
'
Public Function avFileExists(name) As Boolean
'
' * * * * * * * * * * * * * * * * * * * * * * * * * * * * * * * * * * * *
' *                                                                     *
' *  Name: avFindDoc                         File Name: avfnddoc.bas   *
' *                                                                     *
' *  PURPOSE:  GET THE INDEX FOR A SPECIFIED LAYER OR TABLE NAME        *
' *                                                                     *
' *  GIVEN:    name      = name of theme or table to be found          *
' *                                                                     *
' *  RETURN:   avFindDoc = index into Table of Contents of the layer   *
' *                        or table, -1 if not found                   *
' *                                                                     *
' *  NOTE:     The index value for a table will have the number of     *
' *            layers added to it, so that we know a table is being    *
' *            processed                                                *
' *                                                                     *
' *  Dim name As Variant                                                *
' *  Dim avFindDoc As Long                                              *
```

```
' *                                                             *
' * * * * * * * * * * * * * * * * * * * * * * * * * * * * * * * *
'
Public Function avFindDoc(name) As Long
'
' * * * * * * * * * * * * * * * * * * * * * * * * * * * * * * * *
' *                                                             *
' *  Name: avGetActiveDoc                File Name: avgtadoc.bas *
' *                                                             *
' *  PURPOSE:  GET THE CURRENT DOCUMENT OR FOCUS MAP             *
' *                                                             *
' *  GIVEN:    nothing                                          *
' *                                                             *
' *  RETURN:   pMxApp      = the application                    *
' *            pmxDoc      = the document                       *
' *            pActiveView = the active view                    *
' *            pMap        = the focus map                      *
' *                                                             *
' *  Dim pMxApp As IMxApplication                               *
' *  Dim pmxDoc As IMxDocument                                  *
' *  Dim pActiveView As IActiveView                             *
' *  Dim pMap As IMap                                           *
' *                                                             *
' * * * * * * * * * * * * * * * * * * * * * * * * * * * * * * * *
'
Public Sub avGetActiveDoc(pMxApp As esriCore.IMxApplication, _
                          pmxDoc As esriCore.IMxDocument, _
                          pActiveView As esriCore.IActiveView, _
                          pMap As esriCore.IMap)
'
' * * * * * * * * * * * * * * * * * * * * * * * * * * * * * * * *
' *                                                             *
' *  Name: avGetActiveThemes             File Name: avgtathm.bas *
' *                                                             *
' *  PURPOSE:  GET A LIST OF THE ACTIVE OR SELECTED THEMES       *
' *                                                             *
' *  GIVEN:    pmxDoc     = the active view                     *
' *                                                             *
' *  RETURN:   ThemesList = list of themes                      *
' *                                                             *
' *  Dim pmxDoc As IMxDocument                                  *
' *  Dim ThemesList As New Collection                           *
' *                                                             *
' * * * * * * * * * * * * * * * * * * * * * * * * * * * * * * * *
'
Public Sub avGetActiveThemes(pmxDoc As esriCore.IMxDocument, ThemesList)
'
' * * * * * * * * * * * * * * * * * * * * * * * * * * * * * * * *
' *                                                             *
' *  Name: avGetBaseName                 File Name: avgbasnm.bas *
' *                                                             *
' *  PURPOSE:  GET THE BASE NAME THAT APPEARS IN A PATH NAME     *
' *                                                             *
' *  GIVEN:    aPath         = the full path name to be processed *
' *                                                             *
' *  RETURN:   avGetBaseName = base name appearing in a path name *
' *                            including the filename extension, if *
' *                            one is present in the base name  *
' *                                given              return    *
' *                            c:\test\vb\aFile.shp   aFile.shp *
' *                            c:\test\vb\aFile       aFile     *
' *                            c:\test\vb\            vb        *
' *                            c:\test\vb             vb        *
' *                            c:\a                   a         *
' *                            c:\                              *
' *                            aFile.txt              aFile.txt *
' *                            Second from last example (c:\) yields *
' *                            an empty string ("")             *
' *                                                             *
' *  Dim aPath As String                                        *
' *  Dim avGetBaseName As String                                *
' *                                                             *
' * * * * * * * * * * * * * * * * * * * * * * * * * * * * * * * *
```

```
'
Public Function avGetBaseName(aPath) As String
'
' * * * * * * * * * * * * * * * * * * * * * * * * * * * * * * * * * * *
' *                                                                   *
' *  Name: avGetBaseName2                   File Name: avgbasn2.bas  *
' *                                                                   *
' *  PURPOSE:  GET THE BASE NAME THAT APPEARS IN A PATH NAME MINUS   *
' *            ANY EXTENSION THAT MAY APPEAR IN THE BASE NAME         *
' *                                                                   *
' *  GIVEN:    aPath          = the full path name to be processed    *
' *                                                                   *
' *  RETURN:   avGetBaseName2 = base name appearing in a path name    *
' *                             without the filename extension        *
' *                                 given                 return      *
' *                             c:\test\vb\aFile.shp      aFile       *
' *                             c:\test\vb\aFile          aFile       *
' *                             c:\test\vb\               vb          *
' *                             c:\test\vb                vb          *
' *                             c:\a                      a           *
' *                             c:\                                   *
' *                             aFile.txt                 aFile       *
' *                             Second from last example (c:\) yields *
' *                             an empty string ("")                  *
' *                                                                   *
' *  Dim aPath As String                                              *
' *  Dim avGetBaseName2 As String                                     *
' *                                                                   *
' * * * * * * * * * * * * * * * * * * * * * * * * * * * * * * * * * * *
'
Public Function avGetBaseName2(aPath) As String
'
' * * * * * * * * * * * * * * * * * * * * * * * * * * * * * * * * * * *
' *                                                                   *
' *  Name: avGetClassType                   File Name: avclstyp.bas  *
' *                                                                   *
' *  PURPOSE:  DETERMINE THE TYPE OF CLASSIFICATION THAT HAS BEEN    *
' *            USED TO CREATE A LEGEND                                *
' *                                                                   *
' *  GIVEN:    theLegend      = legend to be processed                *
' *                                                                   *
' *  RETURN:   avGetClassType = type of classification used in the    *
' *                             generation of a legend (renderer)     *
' *                             Manual (SingleSymbol and Unique)      *
' *                             DefinedInterval                       *
' *                             EqualInterval                         *
' *                             NaturalBreaks                         *
' *                             Quantile                              *
' *                             StandardDeviation                     *
' *                                                                   *
' *  Dim theLegend As IFeatureRenderer                                *
' *  Dim avGetClassType As String                                     *
' *                                                                   *
' * * * * * * * * * * * * * * * * * * * * * * * * * * * * * * * * * * *
'
Public Function avGetClassType(thelegend As esriCore.IFeatureRenderer) _
                                                            As String
'
' * * * * * * * * * * * * * * * * * * * * * * * * * * * * * * * * * * *
' *                                                                   *
' *  Name: avGetClassifications             File Name: avlgtcls.bas  *
' *                                                                   *
' *  PURPOSE:  TO GET A LIST OF THE CLASSES USED IN A CLASSIFICATION *
' *                                                                   *
' *  GIVEN:    theLegend = legend to be processed                     *
' *                                                                   *
' *  RETURN:   classList = list of classifications                    *
' *                                                                   *
' *  Dim theLegend As IFeatureRenderer                                *
' *  Dim classList As New Collection                                  *
' *                                                                   *
' * * * * * * * * * * * * * * * * * * * * * * * * * * * * * * * * * * *
'
```

```
Public Sub avGetClassifications(thelegend _
                          As esriCore.IFeatureRenderer, classList)
'
' * * * * * * * * * * * * * * * * * * * * * * * * * * * * * * * * * * *
' *                                                                   *
' *  Name: avGetDisplayFlush                   File Name: avdflush.bas *
' *                                                                   *
' *  PURPOSE:  SCRIPT TO MAKE SURE THE DISPLAY IS UP TO DATE BY       *
' *            FORCING ANY BUFFERED DRAWS TO BE DISPLAYED             *
' *                                                                   *
' *  GIVEN:    nothing                                                *
' *                                                                   *
' *  RETURN:   nothing                                                *
' *                                                                   *
' * * * * * * * * * * * * * * * * * * * * * * * * * * * * * * * * * * *
'
Public Sub avGetDisplayFlush()
'
' * * * * * * * * * * * * * * * * * * * * * * * * * * * * * * * * * * *
' *                                                                   *
' *  Name: avGetDisplay                        File Name: avgtdspl.bas *
' *                                                                   *
' *  PURPOSE:  GET THE CURRENT FOCUS MAP DISPLAY                      *
' *                                                                   *
' *  GIVEN:    pActiveView    = the focus map active view             *
' *                                                                   *
' *  RETURN:   pScreenDisplay = the screen display                    *
' *            pDT            = the screen display transformation     *
' *                                                                   *
' *  Dim pActiveView As IActiveView                                   *
' *  Dim pScreenDisplay As IScreenDisplay                             *
' *  Dim pDT As IDisplayTransformation                                *
' *                                                                   *
' * * * * * * * * * * * * * * * * * * * * * * * * * * * * * * * * * * *
'
Public Sub avGetDisplay(pActiveView As esriCore.IActiveView, _
                    pScreenDisplay As esriCore.IScreenDisplay, _
                    pDT As esriCore.IDisplayTransformation)
'
' * * * * * * * * * * * * * * * * * * * * * * * * * * * * * * * * * * *
' *                                                                   *
' *  Name: avGetEnvVar                         File Name: avgtenvv.bas *
' *                                                                   *
' *  PURPOSE:  GET THE FULL PATH FOR AN ENVIRONMENT VARIABLE          *
' *                                                                   *
' *  GIVEN:    aPath       = name of the environment variable name    *
' *                          to be processed                          *
' *                                                                   *
' *  RETURN:   avGetEnvVar = full path associated with the variable   *
' *                              given           return               *
' *                             ARCHOME       C:\ARCGIS\ARCEXE81      *
' *                               TMP         C:\WINDOWS\TEMP         *
' *                               ABC                                 *
' *                         The last example yields an empty string   *
' *                         (""), assuming the variable ABC does not  *
' *                          exist                                    *
' *                                                                   *
' *  Dim aPath As String                                              *
' *  Dim avGetEnvVar As String                                        *
' *                                                                   *
' * * * * * * * * * * * * * * * * * * * * * * * * * * * * * * * * * * *
'
Public Function avGetEnvVar(aPath)
'
' * * * * * * * * * * * * * * * * * * * * * * * * * * * * * * * * * * *
' *                                                                   *
' *  Name: avGetExtension                      File Name: avgnmext.bas *
' *                                                                   *
' *  PURPOSE:  GET THE FILE EXTENSION IN A BASE NAME OF A PATH NAME   *
' *                                                                   *
' *  GIVEN:    aPath          = the full path name to be processed    *
' *                                                                   *
' *  RETURN:   avGetExtension = the filename extension                *
```

```
' *                            given                    return    *
' *                          c:\test\vb\aFile.shp        shp      *
' *                          c:\test\vb\aFile                     *
' *                          c:\test\vb\                          *
' *                          c:\test\vb                           *
' *                          c:\a                                 *
' *                          c:\                                  *
' *                          aFile.shp                   shp      *
' *                          Only the first and last examples     *
' *                          yield non-empty strings ("")         *
' *                                                               *
' *  Dim aPath As String                                          *
' *  Dim avGetExtension As String                                 *
' *                                                               *
' * * * * * * * * * * * * * * * * * * * * * * * * * * * * * * * * *
'
Public Function avGetExtension(aPath) As String
'
' * * * * * * * * * * * * * * * * * * * * * * * * * * * * * * * * *
' *                                                               *
' *  Name: avGetFTabIDs                  File Name: avgtftid.bas  *
' *                                                               *
' *  PURPOSE:  GET A LIST OF THE OBJECT IDS IN A LAYER            *
' *                                                               *
' *  GIVEN:    pmxDoc      = the active view                      *
' *            theTheme    = the theme to be processed            *
' *                                                               *
' *  RETURN:   theRecsList = the list of OIDs for the theme, this *
' *                          list will include all OIDs for all of *
' *                          the features in the theme            *
' *                                                               *
' *  Dim pmxDoc As IMxDocument                                    *
' *  Dim theTheme As Variant                                      *
' *  Dim theRecsList as New Collection                            *
' *                                                               *
' * * * * * * * * * * * * * * * * * * * * * * * * * * * * * * * * *
'
Public Sub avGetFTabIDs(pmxDoc As esriCore.IMxDocument, theTheme, _
                        theRecsList)
'
' * * * * * * * * * * * * * * * * * * * * * * * * * * * * * * * * *
' *                                                               *
' *  Name: avGetFTab                     File Name: avgtftab.bas  *
' *                                                               *
' *  PURPOSE:  GET THE ATTRIBUTE TABLE, FEATURE CLASS AND ASSOCIATED *
' *            LAYER FOR A SPECIFIED THEME                        *
' *                                                               *
' *  GIVEN:    pmxDoc           = the active view                 *
' *            theTheme         = the theme to be processed       *
' *                                                               *
' *  RETURN:   theFTab          = the attribute table for the theme *
' *            theFeatureClass = the feature class for the theme  *
' *            theLayer         = the associated layer for the theme *
' *                                                               *
' *  NOTE:     If a table, rather than a theme, is specified the  *
' *            theFeatureClass and theLayer will be set to Nothing *
' *            while theFTab object will reflect the attributes for *
' *            the table                                          *
' *                                                               *
' *  Dim pmxDoc As IMxDocument                                    *
' *  Dim theTheme As Variant                                      *
' *  Dim theFTab As IFields                                       *
' *  Dim theFeatureClass As IFeatureClass                         *
' *  Dim theLayer As IFeatureLayer                                *
' *                                                               *
' * * * * * * * * * * * * * * * * * * * * * * * * * * * * * * * * *
'
Public Sub avGetFTab(pmxDoc As esriCore.IMxDocument, theTheme, _
                     theFTab As esriCore.IFields, _
                     theFeatureClass As esriCore.IFeatureClass, _
                     theLayer As esriCore.IFeatureLayer)
'
```

```
' * * * * * * * * * * * * * * * * * * * * * * * * * * * * * * * * * *
' *                                                                  *
' *  Name: avGetFeatData                      File Name: avgtfdat.bas *
' *                                                                  *
' *  PURPOSE:  GET THE FEATURE DATA GIVEN A THEME AND AN OBJECT ID    *
' *                                                                  *
' *  GIVEN:     pmxDoc      = the active view                         *
' *             theTheme    = the theme to be processed               *
' *             theObjId    = the object id of the desired feature    *
' *                                                                  *
' *  RETURN:    theFeature = the feature                              *
' *             theShape    = the geometry of a feature               *
' *             shapeType   = the shape type of a feature             *
' *                                                                  *
' *  Dim pmxDoc As IMxDocument                                        *
' *  Dim theTheme As Variant                                          *
' *  Dim theObjId As Long                                             *
' *  Dim theFeature As IFeature                                       *
' *  Dim theShape As IGeometry                                        *
' *  Dim shapeType As esriGeometryType                                *
' *                                                                  *
' * * * * * * * * * * * * * * * * * * * * * * * * * * * * * * * * * *
'
Public Sub avGetFeatData(pmxDoc As esriCore.IMxDocument, _
                         theTheme, theObjId, _
                         theFeature As esriCore.IFeature, _
                         theShape As esriCore.IGeometry, _
                         shapeType As esriCore.esriGeometryType)
'
' * * * * * * * * * * * * * * * * * * * * * * * * * * * * * * * * * *
' *                                                                  *
' *  Name: avGetFeature                       File Name: avgtfeat.bas *
' *                                                                  *
' *  PURPOSE:  GET THE FEATURE GIVEN A THEME AND AN OBJECT ID         *
' *                                                                  *
' *  GIVEN:     pmxDoc      = the active view                         *
' *             theTheme    = the theme to be processed               *
' *             theObjId    = the object id of the desired feature    *
' *                                                                  *
' *  RETURN:    theFeature = the feature                              *
' *                                                                  *
' *  Dim pmxDoc As IMxDocument                                        *
' *  Dim theTheme As Variant                                          *
' *  Dim theObjId As Long                                             *
' *  Dim theFeature As IFeature                                       *
' *                                                                  *
' * * * * * * * * * * * * * * * * * * * * * * * * * * * * * * * * * *
'
Public Sub avGetFeature(pmxDoc As esriCore.IMxDocument, _
                        theTheme, theObjId, _
                        theFeature As esriCore.IFeature)
'
' * * * * * * * * * * * * * * * * * * * * * * * * * * * * * * * * * *
' *                                                                  *
' *  Name: avGetFieldNames                    File Name: avgtfnam.bas *
' *                                                                  *
' *  PURPOSE:  GET THE FIELD NAMES USED TO CLASSIFY A THEME           *
' *                                                                  *
' *  GIVEN:     theLegend = legend to be processed                    *
' *                                                                  *
' *  RETURN:    nameList  = list of the field names that were used in *
' *                         a classification (empty for SingleSymbol  *
' *                         legends or any other type of legend that  *
' *                         does not require a field name)            *
' *                                                                  *
' *  Dim theLegend As IFeatureRenderer                                *
' *  Dim nameList As New Collection                                   *
' *                                                                  *
' * * * * * * * * * * * * * * * * * * * * * * * * * * * * * * * * * *
'
Public Sub avGetFieldNames(thelegend As esriCore.IFeatureRenderer, _
                           nameList)
'
```

```
' * * * * * * * * * * * * * * * * * * * * * * * * * * * * * * * * * * * *
' *                                                                       *
' *  Name: avGetFields                           File Name: avgtflds.bas  *
' *                                                                       *
' *  PURPOSE:  GET A LIST OF FIELD NAMES FOR A LAYER OR TABLE             *
' *                                                                       *
' *  GIVEN:    theVTab = field list for the theme or table                *
' *                                                                       *
' *  RETURN:   theList = list of field names for an attribute table,      *
' *                      these are not the alias names for the fields     *
' *                                                                       *
' *  Dim theVTab As IFields                                               *
' *  Dim theList As New Collection                                        *
' *                                                                       *
' * * * * * * * * * * * * * * * * * * * * * * * * * * * * * * * * * * * *
'
Public Sub avGetFields(theVTab As esriCore.IFields, theList)
'
' * * * * * * * * * * * * * * * * * * * * * * * * * * * * * * * * * * * *
' *                                                                       *
' *  Name: avGetGeometry                         File Name: avgtgeom.bas  *
' *                                                                       *
' *  PURPOSE:  GET THE GEOMETRY OF A FEATURE GIVEN ITS THEME AND          *
' *            OBJECT ID                                                  *
' *                                                                       *
' *  GIVEN:    pmxDoc   = the active view                                 *
' *            theTheme = the theme to be processed                       *
' *            theObjId = the object id of the desired feature            *
' *                                                                       *
' *  RETURN:   theShape = the geometry of a feature                       *
' *                                                                       *
' *  Dim pmxDoc As IMxDocument                                            *
' *  Dim theTheme As Variant                                              *
' *  Dim theObjId As Long                                                 *
' *  Dim theShape As IGeometry                                            *
' *                                                                       *
' * * * * * * * * * * * * * * * * * * * * * * * * * * * * * * * * * * * *
'
Public Sub avGetGeometry(pmxDoc As esriCore.IMxDocument, _
                         theTheme, theObjId, _
                         theShape As esriCore.IGeometry)
'
' * * * * * * * * * * * * * * * * * * * * * * * * * * * * * * * * * * * *
' *                                                                       *
' *  Name: avGetGraphicList                      File Name: avgtgral.bas  *
' *                                                                       *
' *  PURPOSE:  TO GET A LIST OF THE GRAPHICS IN A GRAPHICS LAYER          *
' *                                                                       *
' *  GIVEN:    pCurGraLyr = the graphics layer containing the user        *
' *                         programmed graphics                           *
' *                                                                       *
' *  RETURN:   graList    = list of graphic elements in the graphics      *
' *                         layer                                         *
' *                                                                       *
' *  Dim pCurGraLyr As IGraphicsLayer                                     *
' *  Dim graList As New Collection                                        *
' *                                                                       *
' * * * * * * * * * * * * * * * * * * * * * * * * * * * * * * * * * * * *
'
Public Sub avGetGraphicList(pCurGraLyr As esriCore.IGraphicsLayer, _
                            graList)
'
' * * * * * * * * * * * * * * * * * * * * * * * * * * * * * * * * * * * *
' *                                                                       *
' *  Name: avGetLayerIndx                        File Name: avgtlyri.bas  *
' *                                                                       *
' *  PURPOSE:  TO DETERMINE THE INDEX OF THE LAYER OR TABLE WE ARE        *
' *            DEALING WITH                                               *
' *                                                                       *
' *  GIVEN:    theTheme       = the layer or table to be processed        *
' *                                                                       *
' *  RETURN:   avGetLayerIndx = index of the layer or table               *
' *                                                                       *
```

```
' *  Dim theTheme As Variant                                              *
' *  Dim avGetLayerIndx As Long                                           *
' *                                                                       *
' * * * * * * * * * * * * * * * * * * * * * * * * * * * * * * * * * * * *
'
Public Function avGetLayerIndx(theTheme) As Long
'
' * * * * * * * * * * * * * * * * * * * * * * * * * * * * * * * * * * * *
' *                                                                       *
' *  Name: avGetLayerType                        File Name: avgtlyrt.bas  *
' *                                                                       *
' *  PURPOSE:  TO DETERMINE THE TYPE OF LAYER WE ARE DEALING WITH         *
' *                                                                       *
' *  GIVEN:    pSelected = the data layer object to be processed          *
' *                                                                       *
' *  RETURN:   aName     = name of the data layer object (uppercase)      *
' *            aType     = type of data layer object                      *
' *                          0 = unknown                                  *
' *                          1 = standalone table                         *
' *                          2 = raster layer                             *
' *                          3 = tin layer                                *
' *                          4 = annotation layer                         *
' *                          5 = feature layer                            *
' *                          6 = CAD annotation layer                     *
' *                          7 = CAD layer                                *
' *                                                                       *
' *  Dim pSelected As IUnknown                                            *
' *  Dim aName As Variant                                                 *
' *  Dim aType As Integer                                                 *
' *                                                                       *
' * * * * * * * * * * * * * * * * * * * * * * * * * * * * * * * * * * * *
'
Public Sub avGetLayerType(pSelected As IUnknown, aName, aType)
'
' * * * * * * * * * * * * * * * * * * * * * * * * * * * * * * * * * * * *
' *                                                                       *
' *  Name: avGetLegend                           File Name: avgetleg.bas  *
' *                                                                       *
' *  PURPOSE:  TO GET THE LEGEND THAT IS ASSOCIATED WITH A THEME          *
' *                                                                       *
' *  GIVEN:    pmxDoc   = the active view                                 *
' *            theTheme = theme to be processed                           *
' *                                                                       *
' *  RETURN:   aLegend  = theme legend                                    *
' *                                                                       *
' *  Dim pmxDoc As IMxDocument                                            *
' *  Dim theTheme As Variant                                              *
' *  Dim aLegend As IFeatureRenderer                                      *
' *                                                                       *
' * * * * * * * * * * * * * * * * * * * * * * * * * * * * * * * * * * * *
'
Public Sub avGetLegend(pmxDoc As esriCore.IMxDocument, theTheme, _
                      aLegend As esriCore.IFeatureRenderer)
'
' * * * * * * * * * * * * * * * * * * * * * * * * * * * * * * * * * * * *
' *                                                                       *
' *  Name: avGetLegendType                       File Name: avlgttyp.bas  *
' *                                                                       *
' *  PURPOSE:  DETERMINE THE TYPE OF LEGEND IN USE                        *
' *                                                                       *
' *  GIVEN:    theLegend        = legend to be processed                  *
' *                                                                       *
' *  RETURN:   avGetLegendType = type of legend (renderer) in use         *
' *                              UNIQUE       : Unique Value              *
' *                              SIMPLE       : Simple                    *
' *                              SCALE        : ScaleDependent            *
' *                              PROPORTIONAL : Proportional Symbol       *
' *                              BIVARIATE    : Bivariate                 *
' *                              CHART        : Chart                     *
' *                              CLASS        : Class Breaks              *
' *                              DOT          : Dot Density               *
' *                                                                       *
```

```
' *  Dim theLegend As IFeatureRenderer                                  *
' *  Dim avGetLegendType As String                                      *
' *                                                                     *
' * * * * * * * * * * * * * * * * * * * * * * * * * * * * * * * * * * * *
'
Public Function avGetLegendType(thelegend _
                              As esriCore.IFeatureRenderer) As String
'
' * * * * * * * * * * * * * * * * * * * * * * * * * * * * * * * * * * * *
' *                                                                     *
' *  Name: avGetName                          File Name: avgetnam.bas  *
' *                                                                     *
' *  PURPOSE:  TO GET THE CAPTION OF THE APPLICATION                    *
' *                                                                     *
' *  GIVEN:    nothing                                                  *
' *                                                                     *
' *  RETURN:   aTitle  = name of the application appearing in the      *
' *                      upper left corner of the application window   *
' *                                                                     *
' *  Dim aTitle As String                                               *
' *                                                                     *
' * * * * * * * * * * * * * * * * * * * * * * * * * * * * * * * * * * * *
'
Public Sub avGetName(aTitle)
'
' * * * * * * * * * * * * * * * * * * * * * * * * * * * * * * * * * * * *
' *                                                                     *
' *  Name: avGetNumClasses                    File Name: avgtncls.bas  *
' *                                                                     *
' *  PURPOSE:  DETERMINE THE NUMBER OF CLASSES IN A LEGEND              *
' *                                                                     *
' *  GIVEN:    theLegend       = legend to be processed                 *
' *                                                                     *
' *  RETURN:   avGetNumClasses = number of classes in the legend       *
' *                                                                     *
' *  Dim theLegend As IFeatureRenderer                                  *
' *  Dim avGetNumClasses As Long                                        *
' *                                                                     *
' * * * * * * * * * * * * * * * * * * * * * * * * * * * * * * * * * * * *
'
Public Function avGetNumClasses(thelegend _
                              As esriCore.IFeatureRenderer) As Long
'
' * * * * * * * * * * * * * * * * * * * * * * * * * * * * * * * * * * * *
' *                                                                     *
' *  Name: avGetNumRecords                    File Name: avgtnrec.bas  *
' *                                                                     *
' *  PURPOSE:  GET THE NUMBER OF RECORDS IN A LAYER OR TABLE            *
' *                                                                     *
' *  GIVEN:    pmxDoc          = the active view                        *
' *            theTheme        = the theme or table to be processed     *
' *                                                                     *
' *  RETURN:   avGetNumRecords = number of records in theme or table   *
' *                                                                     *
' *  Dim pmxDoc As IMxDocument                                          *
' *  Dim theTheme As Variant                                            *
' *  Dim avGetNumRecords As Long                                        *
' *                                                                     *
' * * * * * * * * * * * * * * * * * * * * * * * * * * * * * * * * * * * *
'
Public Function avGetNumRecords(pmxDoc As esriCore.IMxDocument, _
                               theTheme) As Long
'
' * * * * * * * * * * * * * * * * * * * * * * * * * * * * * * * * * * * *
' *                                                                     *
' *  Name: avGetPalette                       File Name: avgtpalt.bas  *
' *                                                                     *
' *  PURPOSE:  GET THE VARIOUS PALETTES AVAILABLE IN THE APPLICATION   *
' *                                                                     *
' *  GIVEN:    pmxDoc        = the active view                          *
' *                                                                     *
' *  RETURN:   avGetPalette = the available style galleries            *
' *                           Reference Systems :  12 items            *
```

```
' *                          Shadows          :  12 items          *
' *                          Area Patches     :   8 items          *
' *                          Line Patches     :   9 items          *
' *                          Labels           :  20 items          *
' *                          North Arrows     :  97 items          *
' *                          Scale Bars       :  11 items          *
' *                          Legend Items     :  18 items          *
' *                          Scale Text       :   7 items          *
' *                          Color Ramps      :  78 items          *
' *                          Borders          :  16 items          *
' *                          Backgrounds      :  18 items          *
' *                          Colors           : 120 items          *
' *                          Fill Symbols     :  53 items          *
' *                          Line Symbols     :  86 items          *
' *                          Marker Symbols   : 114 items          *
' *                          Text Symbols     :  35 items          *
' *                                                                *
' *  NOTE:     Depending upon the installation, the total number of *
' *            galleries and items within each gallery may vary    *
' *                                                                *
' *  Dim pmxDoc As IMxDocument                                     *
' *  Dim avGetPalette As IStyleGallery                             *
' *                                                                *
' * * * * * * * * * * * * * * * * * * * * * * * * * * * * * * * * * *
'
Public Function avGetPalette(pmxDoc As esriCore.IMxDocument) _
                                         As esriCore.IStyleGallery
'
' * * * * * * * * * * * * * * * * * * * * * * * * * * * * * * * * * *
' *                                                                *
' *  Name: avGetPrecision                 File Name: avgprecs.bas  *
' *                                                                *
' *  PURPOSE:  GET THE PRECISION OF A FIELD                        *
' *                                                                *
' *  GIVEN:    theFTab        = the FTab or VTab to be processed   *
' *            theField       = index into FTab or VTab representing *
' *                             the field to be processed          *
' *                                                                *
' *  RETURN:   avGetPrecision = number of digits to the right of the *
' *                             decimal point                      *
' *                                                                *
' *  NOTE:     This function always returns 0 for fields contained in *
' *            a personal geodatabase                              *
' *                                                                *
' *  Dim theFTab As IFields                                        *
' *  Dim theField As Long                                          *
' *  Dim avGetPrecision As Long                                    *
' *                                                                *
' * * * * * * * * * * * * * * * * * * * * * * * * * * * * * * * * * *
'
Public Function avGetPrecision(theFTab As esriCore.IFields, _
                               theField) As Long
'
' * * * * * * * * * * * * * * * * * * * * * * * * * * * * * * * * * *
' *                                                                *
' *  Name: avGetPathName                  File Name: avgpthnm.bas  *
' *                                                                *
' *  PURPOSE:  GET THE PATH NAME THAT APPEARS IN A STRING          *
' *                                                                *
' *  GIVEN:    aPath          = the full path name to be processed *
' *                                                                *
' *  RETURN:   avGetPathName = path name appearing in a string minus *
' *                            the last component in the string   *
' *                                given              return      *
' *                            c:\test\vb\aFile.shp   c:\test\vb  *
' *                            c:\test\vb\aFile       c:\test\vb  *
' *                            c:\test\vb\            c:\test     *
' *                            c:\test\vb             c:\test     *
' *                            c:\a                   c:\         *
' *                            c:\                                *
' *                            aFile.shp                          *
' *                            The last two examples will yield empty *
' *                            strings ("")                       *
```

```
' *                                                                      *
' *  Dim aPath As String                                                 *
' *  Dim avGetPathName As String                                         *
' *                                                                      *
' * * * * * * * * * * * * * * * * * * * * * * * * * * * * * * * * * * * *
'
Public Function avGetPathName(aPath) As String
'
' * * * * * * * * * * * * * * * * * * * * * * * * * * * * * * * * * * * *
' *                                                                      *
' *  Name: avGetProjectName                    File Name: avgtpnam.bas   *
' *                                                                      *
' *  PURPOSE:  GET THE NAME OF THE CURRENT DOCUMENT                      *
' *                                                                      *
' *  GIVEN:    nothing                                                   *
' *                                                                      *
' *  RETURN:   aTitle  = the name of the current document, this will    *
' *                      include the .mxd extension (i.e. sample.mxd)    *
' *                                                                      *
' *  Dim aTitle As String                                                *
' *                                                                      *
' * * * * * * * * * * * * * * * * * * * * * * * * * * * * * * * * * * * *
'
Public Sub avGetProjectName(aTitle)
'
' * * * * * * * * * * * * * * * * * * * * * * * * * * * * * * * * * * * *
' *                                                                      *
' *  Name: avGetSelFeatures                    File Name: avgtsfet.bas   *
' *                                                                      *
' *  PURPOSE:  PRESERVE THE THEMES AND RECORD NUMBERS OF THEMES WITH     *
' *            SELECTED FEATURES                                         *
' *                                                                      *
' *  GIVEN:    pmxDoc     = the active view                              *
' *            themeList  = list of themes to be processed               *
' *            selMode    = mode of selection                            *
' *                          0  = all features                           *
' *                          1  = point features                         *
' *                          2  = polyline features                      *
' *                          3  = polygon features                       *
' *                          4  = polyline and polygon features          *
' *                          10 = same as 0 and include themes that      *
' *                               do not have selected features          *
' *                                                                      *
' *  RETURN:   selThmList = list of themes with selected features        *
' *            selRecList = list of selected features record numbers     *
' *                                                                      *
' *  NOTE:     (a) structure of selThmList is:                           *
' *                Item 1: name of theme 1                               *
' *                Item 2: number of selected features in theme 1        *
' *                Item 3: name of theme 2                               *
' *                Item 4: number of selected features in theme 2        *
' *                Repeat Items 1 and 2 for each theme                   *
' *            (b) structure of selRecList is:                           *
' *                Item 1: selected feature 1 OID in theme 1             *
' *                Item 2: selected feature 2 OID in theme 1             *
' *                Repeat Item 1 for each selected feature in theme 1    *
' *                Item 3: selected feature 1 OID in theme 2             *
' *                Item 4: selected feature 2 OID in theme 2             *
' *                Repeat Item 3 for each selected feature in theme 2    *
' *                                                                      *
' *  Dim pmxDoc As IMxDocument                                           *
' *  Dim themeList As New Collection                                     *
' *  Dim selMode As Integer                                              *
' *  Dim selThmList As New Collection                                    *
' *  Dim selRecList As New Collection                                    *
' *                                                                      *
' * * * * * * * * * * * * * * * * * * * * * * * * * * * * * * * * * * * *
'
Public Sub avGetSelFeatures(pmxDoc As esriCore.IMxDocument, _
                            themeList, selMode, _
                            selThmList, selRecList)
'
```

```
' * * * * * * * * * * * * * * * * * * * * * * * * * * * * * * * * * * *
' *                                                                   *
' *  Name: avGetSelectedExtent              File Name: avgtsext.bas   *
' *                                                                   *
' *  PURPOSE:  GET THE ENCLOSING RECTANGLE FOR THE SELECTED SET OF A  *
' *            THEME, OR THE ENCLOSING RECTANGLE FOR THE ENTIRE THEME *
' *            IF THE SELECTED SET IS EMPTY (NO SELECTED FEATURES)    *
' *                                                                   *
' *  GIVEN:    pmxDoc   = the active view                             *
' *            theTheme = the theme to be processed                   *
' *                                                                   *
' *  RETURN:   theRect  = the enclosing rectangle encompassing the    *
' *                       selected features, or all of the features,  *
' *                       if no features are selected                 *
' *                                                                   *
' *  Dim pmxDoc As IMxDocument                                        *
' *  Dim theTheme As Variant                                          *
' *  Dim theRect As IEnvelope                                         *
' *                                                                   *
' * * * * * * * * * * * * * * * * * * * * * * * * * * * * * * * * * * *
'
Public Sub avGetSelectedExtent(pmxDoc As esriCore.IMxDocument, _
                              theTheme, theRect As esriCore.IEnvelope)
'
' * * * * * * * * * * * * * * * * * * * * * * * * * * * * * * * * * * *
' *                                                                   *
' *  Name: avGetSelected                    File Name: avgtmsel.bas   *
' *                                                                   *
' *  PURPOSE:  GET THE SELECTED GRAPHICS IN THE MAP                   *
' *                                                                   *
' *  GIVEN:    pmxDoc       = the active view                         *
' *                                                                   *
' *  RETURN:   selGraphList = list containing the selected graphic    *
' *                           elements                                *
' *                                                                   *
' *  Dim pmxDoc As IMxDocument                                        *
' *  Dim selGraphList As New Collection                               *
' *                                                                   *
' * * * * * * * * * * * * * * * * * * * * * * * * * * * * * * * * * * *
'
Public Sub avGetSelected(pmxDoc As esriCore.IMxDocument, selGraphList)
'
' * * * * * * * * * * * * * * * * * * * * * * * * * * * * * * * * * * *
' *                                                                   *
' *  Name: avGetSelectionClear              File Name: avgtsclr.bas   *
' *                                                                   *
' *  PURPOSE:  REMOVE A RECORD FROM THE SELECTED SET FOR A LAYER OR   *
' *            TABLE                                                  *
' *                                                                   *
' *  GIVEN:    pmxDoc   = the active view                             *
' *            theTheme = the theme or table to be processed          *
' *            theRcrd  = the record to be removed from the selection *
' *                                                                   *
' *  RETURN:   nothing                                                *
' *                                                                   *
' *  Dim pmxDoc As IMxDocument                                        *
' *  Dim theTheme As Variant                                          *
' *  Dim theRcrd As Long                                              *
' *                                                                   *
' * * * * * * * * * * * * * * * * * * * * * * * * * * * * * * * * * * *
'
Public Sub avGetSelectionClear(pmxDoc As esriCore.IMxDocument, _
                               theTheme, theRcrd)
'
' * * * * * * * * * * * * * * * * * * * * * * * * * * * * * * * * * * *
' *                                                                   *
' *  Name: avGetSelectionIDs                File Name: avgtsids.bas   *
' *                                                                   *
' *  PURPOSE:  BUILD A COLLECTION OF OIDS FROM A SELECTION SET        *
' *                                                                   *
' *  GIVEN:    psTableSel  = the selection set for a theme or table   *
' *                                                                   *
' *  RETURN:   selRecsList = the list of OIDs for the selection set   *
```

```
' *                                                                    *
' *  Dim psTableSel As ISelectionSet                                   *
' *  Dim selRecsList as New Collection                                 *
' *                                                                    *
' * * * * * * * * * * * * * * * * * * * * * * * * * * * * * * * * * * *
'
Public Sub avGetSelectionIDs(psTableSel As esriCore.ISelectionSet, _
'
' * * * * * * * * * * * * * * * * * * * * * * * * * * * * * * * * * * *
' *                                                                    *
' *  Name: avGetSelection                  File Name: avgtselt.bas     *
' *                                                                    *
' *  PURPOSE:  GET THE SELECTED SET FOR A FOR A LAYER OR TABLE         *
' *                                                                    *
' *  GIVEN:    pmxDoc     = the active view                            *
' *            theTheme   = the theme or table to be processed         *
' *                                                                    *
' *  RETURN:   psTableSel = the selection set for the theme            *
' *                                                                    *
' *  Dim pmxDoc As IMxDocument                                         *
' *  Dim theTheme As Variant                                           *
' *  Dim psTableSel As ISelectionSet                                   *
' *                                                                    *
' * * * * * * * * * * * * * * * * * * * * * * * * * * * * * * * * * * *
'
Public Sub avGetSelection(pmxDoc As esriCore.IMxDocument, theTheme, _
                      psTableSel As esriCore.ISelectionSet)
'
' * * * * * * * * * * * * * * * * * * * * * * * * * * * * * * * * * * *
' *                                                                    *
' *  Name: avGetShapeType                  File Name: avgtshtp.bas     *
' *                                                                    *
' *  PURPOSE:  GET THE DEFAULT SHAPE TYPE FOR A THEME                  *
' *                                                                    *
' *  GIVEN:    pmxDoc          = the active view                       *
' *            theTheme        = the theme to be processed             *
' *                                                                    *
' *  RETURN:   avGetShapeType = the default shape type                 *
' *                                                                    *
' *  Dim pmxDoc As IMxDocument                                         *
' *  Dim theTheme As Variant                                           *
' *  Dim avGetShapeType As esriGeometryType                            *
' *                                                                    *
' * * * * * * * * * * * * * * * * * * * * * * * * * * * * * * * * * * *
'
Public Function avGetShapeType(pmxDoc As esriCore.IMxDocument, _
                          theTheme) As esriCore.esriGeometryType
'
' * * * * * * * * * * * * * * * * * * * * * * * * * * * * * * * * * * *
' *                                                                    *
' *  Name: avGetThemeExtent                File Name: avthmext.bas     *
' *                                                                    *
' *  PURPOSE:  GET THE ENCLOSING RECTANGLE FOR A THEME                 *
' *                                                                    *
' *  GIVEN:    pmxDoc   = the active view                              *
' *            theTheme = the theme to be processed                    *
' *                                                                    *
' *  RETURN:   theRect  = the enclosing rectangle encompassing all     *
' *                       of the features in the theme                 *
' *                                                                    *
' *  Dim pmxDoc As IMxDocument                                         *
' *  Dim theTheme As Variant                                           *
' *  Dim theRect As IEnvelope                                          *
' *                                                                    *
' * * * * * * * * * * * * * * * * * * * * * * * * * * * * * * * * * * *
'
Public Sub avGetThemeExtent(pmxDoc As esriCore.IMxDocument, theTheme, _
                        theRect As esriCore.IEnvelope)
'
```

```
' * * * * * * * * * * * * * * * * * * * * * * * * * * * * * * * * * * *
' *                                                                   *
' *  Name: avGetThemes                        File Name: avgtthem.bas  *
' *                                                                   *
' *  PURPOSE:  GET A LIST OF THEMES OR TABLES                         *
' *                                                                   *
' *  GIVEN:    pmxDoc      = the active view                          *
' *            opmode      = mode of operation                        *
' *                            0 = find all layers                    *
' *                            1 = find only feature layers           *
' *                            2 = find all tables                    *
' *                                                                   *
' *  RETURN:   ThemesList = list of themes                            *
' *                                                                   *
' *  Dim pmxDoc As IMxDocument                                        *
' *  Dim opmode As Integer                                            *
' *  Dim ThemesList As New Collection                                 *
' *                                                                   *
' * * * * * * * * * * * * * * * * * * * * * * * * * * * * * * * * * * *
'
Public Sub avGetThemes(pmxDoc As esriCore.IMxDocument, opmode, _
                       ThemesList)
'
' * * * * * * * * * * * * * * * * * * * * * * * * * * * * * * * * * * *
' *                                                                   *
' *  Name: avGetVisibleCADLayers              File Name: avgtvcly.bas  *
' *                                                                   *
' *  PURPOSE:  GET A LIST OF THE VISIBLE CAD ANNOTATION LAYERS AND    *
' *            VISIBLE CAD LAYERS                                     *
' *                                                                   *
' *  GIVEN:    pmxDoc      = the active view                          *
' *                                                                   *
' *  RETURN:   vThemesList = list of visible CAD layers               *
' *                                                                   *
' *  Dim pmxDoc As IMxDocument                                        *
' *  Dim vThemesList As New Collection                                *
' *                                                                   *
' * * * * * * * * * * * * * * * * * * * * * * * * * * * * * * * * * * *
'
Public Sub avGetVisibleCADLayers(pmxDoc As esriCore.IMxDocument, _
                                 vThemesList)
'
' * * * * * * * * * * * * * * * * * * * * * * * * * * * * * * * * * * *
' *                                                                   *
' *  Name: avGetVisibleThemes                 File Name: avgtvthm.bas  *
' *                                                                   *
' *  PURPOSE:  GET A LIST OF THE VISIBLE THEMES                       *
' *                                                                   *
' *  GIVEN:    pmxDoc      = the active view                          *
' *                                                                   *
' *  RETURN:   vThemesList = list of visible themes                   *
' *                                                                   *
' *  NOTE:     Only annotation and feature layers are processed       *
' *                                                                   *
' *  Dim pmxDoc As IMxDocument                                        *
' *  Dim vThemesList As New Collection                                *
' *                                                                   *
' * * * * * * * * * * * * * * * * * * * * * * * * * * * * * * * * * * *
'
Public Sub avGetVisibleThemes(pmxDoc As esriCore.IMxDocument, _
                              vThemesList)
'
' * * * * * * * * * * * * * * * * * * * * * * * * * * * * * * * * * * *
' *                                                                   *
' *  Name: avGetVTabIDs                       File Name: avgtvtid.bas  *
' *                                                                   *
' *  PURPOSE:  GET A LIST OF OBJECT IDS FOR A TABLE                   *
' *                                                                   *
' *  GIVEN:    pmxDoc      = the active view                          *
' *            theTable    = the table to be processed                *
' *                                                                   *
' *  RETURN:   theRecsList = the list of OIDs for the table           *
' *                                                                   *
```

```
' *  Dim pmxDoc As IMxDocument                                          *
' *  Dim theTable As Variant                                            *
' *  Dim theRecsList as New Collection                                  *
' *                                                                     *
' * * * * * * * * * * * * * * * * * * * * * * * * * * * * * * * * * * * *
'
Public Sub avGetVTabIDs(pmxDoc As esriCore.IMxDocument, theTable, _
                    theRecsList)
'
' * * * * * * * * * * * * * * * * * * * * * * * * * * * * * * * * * * * *
' *                                                                     *
' *  Name: avGetVTab                          File Name: avgtvtab.bas   *
' *                                                                     *
' *  PURPOSE:  GET THE ATTRIBUTE TABLE FOR A LAYER OR TABLE             *
' *                                                                     *
' *  GIVEN:    pmxDoc   = the active view                               *
' *            theTheme = the theme or table to be processed            *
' *                                                                     *
' *  RETURN:   theVTab  = the attribute table for the theme or table    *
' *                                                                     *
' *  Dim pmxDoc As IMxDocument                                          *
' *  Dim theTheme As Variant                                            *
' *  Dim theVTab As IFields                                             *
' *                                                                     *
' * * * * * * * * * * * * * * * * * * * * * * * * * * * * * * * * * * * *
'
Public Sub avGetVTab(pmxDoc As esriCore.IMxDocument, theTheme, _
                 theVTab As esriCore.IFields)
'
' * * * * * * * * * * * * * * * * * * * * * * * * * * * * * * * * * * * *
' *                                                                     *
' *  Name: avGetWinFonts                      File Name: avgtwfnt.bas   *
' *                                                                     *
' *  PURPOSE:  GET A LIST OF THE FONTS INSTALLED ON THE PC              *
' *                                                                     *
' *  GIVEN:    nothing                                                  *
' *                                                                     *
' *  RETURN:   aColl    = an alphabetically sorted list of the fonts    *
' *                       that are installed on the computer            *
' *                                                                     *
' *  Dim aColl As New Collection                                        *
' *                                                                     *
' * * * * * * * * * * * * * * * * * * * * * * * * * * * * * * * * * * * *
'
Public Sub avGetWinFonts(aColl)
'
' * * * * * * * * * * * * * * * * * * * * * * * * * * * * * * * * * * * *
' *                                                                     *
' *  Name: avGetWinFonts2                     File Name: avgtwft2.bas   *
' *                                                                     *
' *  PURPOSE:  USE THE WINDOWS API TO GET A LIST OF THE FONTS THAT      *
' *            ARE INSTALLED ON THE PC                                  *
' *                                                                     *
' *  GIVEN:    hWnd  = Windows handle that is associated with the       *
' *                    calling procedure                                *
' *                                                                     *
' *  RETURN:   aColl = an alphabetically sorted list of the fonts       *
' *                    that are installed on the computer               *
' *                                                                     *
' *  NOTE:     The third argument in the definition of the              *
' *            EnumFontFamilies function is a Long that represents a    *
' *            procedure. The argument must contain the address of      *
' *            the procedure, rather than the value that the            *
' *            procedure returns. In the call to EnumFontFamilies,      *
' *            the third argument requires the AddressOf operator to    *
' *            return the address of the EnumFontFamProc procedure,     *
' *            which is the name of the callback procedure you supply   *
' *            when using the Windows API function, EnumFontFamilies.   *
' *            Windows calls EnumFontFamProc once for each of the       *
' *            font families on the system when you pass AddressOf      *
' *            EnumFontFamProc to EnumFontFamilies. The last argument   *
' *            passed to EnumFontFamilies specifies the collection in   *
' *            which the information is to be added to.                 *
```

```
' *                                                         *
' * * * * * * * * * * * * * * * * * * * * * * * * * * * * * *
'


'
' * * * * * * * * * * * * * * * * * * * * * * * * * * * * * *
' *                                                         *
' *  Name: avGetWorkDir                 File Name: avgtwdir.bas *
' *                                                         *
' *  PURPOSE:  GET THE CURRENT WORKING DIRECTORY            *
' *                                                         *
' *  GIVEN:    nothing                                      *
' *                                                         *
' *  RETURN:   theWorkDir = current working directory       *
' *                                                         *
' *  Dim theWorkDir As String                               *
' *                                                         *
' * * * * * * * * * * * * * * * * * * * * * * * * * * * * * *
'
Public Sub avGetWorkDir(theWorkDir)
'
' * * * * * * * * * * * * * * * * * * * * * * * * * * * * * *
' *                                                         *
' *  Name: avGraphicGetShape            File Name: avggtshp.bas *
' *                                                         *
' *  PURPOSE:  TO GET THE GEOMETRY THAT IS ASSOCIATED WITH A GRAPHIC *
' *                                                         *
' *  GIVEN:    pElement          = the graphic to be processed *
' *                                                         *
' *  RETURN:   avGraphicGetShape = geometry describing the graphic *
' *                                                         *
' *  NOTE:     This routine will process PEN, MARKER, FILL and TEXT *
' *            symbols since they all share the IElement interface *
' *                                                         *
' *  Dim pElement As IElement                               *
' *  Dim avGraphicGetShape As IGeometry                     *
' *                                                         *
' * * * * * * * * * * * * * * * * * * * * * * * * * * * * * *
'
Public Function avGraphicGetShape(pElement As esriCore.IElement) _
                                          As esriCore.IGeometry
'
' * * * * * * * * * * * * * * * * * * * * * * * * * * * * * *
' *                                                         *
' *  Name: avGraphicGetSymbol           File Name: avggtsym.bas *
' *                                                         *
' *  PURPOSE:  TO GET THE SYMBOL THAT IS ASSOCIATED WITH A GRAPHIC *
' *                                                         *
' *  GIVEN:    aSymTyp           = type of graphic to be assigned *
' *                                PEN    : line symbol      *
' *                                MARKER : point symbol     *
' *                                FILL   : polygon symbol   *
' *            pElement          = the graphic to be processed *
' *                                                         *
' *  RETURN:   avGraphicGetSymbol = symbol describing the graphic *
' *                                element, its properties such as *
' *                                its color                 *
' *                                                         *
' *  Dim aSymTyp As String                                  *
' *  Dim pElement As IElement                               *
' *  Dim avGraphicGetSymbol As ISymbol                      *
' *                                                         *
' * * * * * * * * * * * * * * * * * * * * * * * * * * * * * *
'
Public Function avGraphicGetSymbol(aSymTyp, _
                              pElement As esriCore.IElement) _
                                          As esriCore.ISymbol
'
```

```
' * * * * * * * * * * * * * * * * * * * * * * * * * * * * * * * * * * *
' *                                                                     *
' *  Name: avGraphicInvalidate               File Name: avginval.bas   *
' *                                                                     *
' *  PURPOSE:  TO REDRAW OR UPDATE A GRAPHIC ELEMENT                    *
' *                                                                     *
' *  GIVEN:    pElement = graphic to be redrawn due to a change that   *
' *                       has been made to it                           *
' *                                                                     *
' *  RETURN:   nothing                                                  *
' *                                                                     *
' *  NOTE:     Use this routine to redraw a graphic that has already   *
' *            been added to a graphics layer and subsequently          *
' *            modified                                                 *
' *                                                                     *
' *  Dim pElement As IElement                                           *
' *                                                                     *
' * * * * * * * * * * * * * * * * * * * * * * * * * * * * * * * * * * *
'
Public Sub avGraphicInvalidate(pElement As esriCore.IElement)
'
' * * * * * * * * * * * * * * * * * * * * * * * * * * * * * * * * * * *
' *                                                                     *
' *  Name: avGraphicListDelete               File Name: avgldele.bas   *
' *                                                                     *
' *  PURPOSE:  TO DELETE A GRAPHICS LAYER                               *
' *                                                                     *
' *  GIVEN:    pCurGraLyr = the graphics layer containing the user     *
' *                         programmed graphics                         *
' *                                                                     *
' *  RETURN:   nothing                                                  *
' *                                                                     *
' *  NOTE:     The basic graphics layer can not be deleted, only user  *
' *            created graphics layers can be deleted with this macro  *
' *                                                                     *
' *  Dim pCurGraLyr As IGraphicsLayer                                   *
' *                                                                     *
' * * * * * * * * * * * * * * * * * * * * * * * * * * * * * * * * * * *
'
Public Sub avGraphicListDelete(pCurGraLyr As esriCore.IGraphicsLayer)
'
' * * * * * * * * * * * * * * * * * * * * * * * * * * * * * * * * * * *
' *                                                                     *
' *  Name: avGraphicListEmpty                File Name: avglempt.bas   *
' *                                                                     *
' *  PURPOSE:  TO DELETE THE GRAPHICS FROM A GRAPHICS LAYER             *
' *                                                                     *
' *  GIVEN:    pCurGraLyr = the graphics layer containing the user     *
' *                         programmed graphics                         *
' *                                                                     *
' *  RETURN:   nothing                                                  *
' *                                                                     *
' *  NOTE:     The graphics within the graphics layer is deleted, the  *
' *            graphics layer itself is not deleted so that graphics   *
' *            can be added to the graphics layer at another time      *
' *                                                                     *
' *  Dim pCurGraLyr As IGraphicsLayer                                   *
' *                                                                     *
' * * * * * * * * * * * * * * * * * * * * * * * * * * * * * * * * * * *
'
Public Sub avGraphicListEmpty(pCurGraLyr As esriCore.IGraphicsLayer)
'
' * * * * * * * * * * * * * * * * * * * * * * * * * * * * * * * * * * *
' *                                                                     *
' *  Name: avGraphicSetShape                 File Name: avgstshp.bas   *
' *                                                                     *
' *  PURPOSE:  TO SET THE GEOMETRY THAT IS ASSOCIATED WITH A GRAPHIC   *
' *                                                                     *
' *  GIVEN:    pElement = the graphic to be modified                   *
' *            theGeom  = the new geometry that describes the graphic  *
' *                                                                     *
' *  RETURN:   nothing                                                  *
' *                                                                     *
```

```
' *  NOTE:     The geometry of the given element is modified by this  *
' *            routine, so that pElement is different after calling  *
' *             this routine                                          *
' *                                                                   *
' *  Dim pElement As IElement                                         *
' *  Dim theGeom As IGeometry                                         *
' *                                                                   *
' * * * * * * * * * * * * * * * * * * * * * * * * * * * * * * * * * * *
'
Public Sub avGraphicSetShape(pElement As esriCore.IElement, _
                             theGeom As esriCore.IGeometry)
'
' * * * * * * * * * * * * * * * * * * * * * * * * * * * * * * * * * * *
' *                                                                   *
' *  Name: avGraphicSetSymbol                  File Name: avgstsym.bas  *
' *                                                                   *
' *  PURPOSE:  TO ASSIGN A SYMBOL TO A GRAPHIC                         *
' *                                                                   *
' *  GIVEN:    aSymTyp  = type of graphic to be assigned               *
' *                       PEN    : line symbol                         *
' *                       MARKER : point symbol                        *
' *                       FILL   : polygon symbol                      *
' *            pElement = graphic for which the given symbol is to be *
' *                       assigned to                                  *
' *            pSymbol  = symbol to be assigned to the graphic         *
' *                                                                   *
' *  RETURN:   nothing                                                 *
' *                                                                   *
' *  Dim aSymTyp As String                                             *
' *  Dim pElement As IElement                                          *
' *  Dim pSymbol As ISymbol                                            *
' *                                                                   *
' * * * * * * * * * * * * * * * * * * * * * * * * * * * * * * * * * * *
'
Public Sub avGraphicSetSymbol(aSymTyp, _
                              pElement As esriCore.IElement, _
                              pSymbol As esriCore.ISymbol)
'
' * * * * * * * * * * * * * * * * * * * * * * * * * * * * * * * * * * *
' *                                                                   *
' *  Name: avGraphicShapeMake                  File Name: avgsmake.bas  *
' *                                                                   *
' *  PURPOSE:  TO CREATE A GRAPHIC SHAPE THAT CAN BE ADDED TO THE      *
' *            GRAPHICS LIST                                           *
' *                                                                   *
' *  GIVEN:    aSymTyp             = type of graphic to be created     *
' *                                  PEN    : line symbol              *
' *                                  MARKER : point symbol             *
' *                                  FILL   : polygon symbol           *
' *            theGeom             = geometry describing the graphic   *
' *                                                                   *
' *  RETURN:   avGraphicShapeMake = the graphic that can be added to  *
' *                                  the graphics layer                *
' *                                                                   *
' *  Dim aSymTyp As String                                             *
' *  Dim theGeom As IGeometry                                          *
' *  Dim avGraphicShapeMake As IElement                                *
' *                                                                   *
' * * * * * * * * * * * * * * * * * * * * * * * * * * * * * * * * * * *
'
Public Function avGraphicShapeMake(aSymTyp, _
                                   theGeom As esriCore.IGeometry) _
                                   As esriCore.IElement
'
' * * * * * * * * * * * * * * * * * * * * * * * * * * * * * * * * * * *
' *                                                                   *
' *  Name: avGraphicTextGetAngle               File Name: avgtgang.bas  *
' *                                                                   *
' *  PURPOSE:  TO GET THE ANGLE ASSOCIATED WITH A GRAPHIC TEXT         *
' *                                                                   *
' *  GIVEN:    aGraphicText = graphic text to be processed             *
' *                                                                   *
' *  RETURN:   aAngle       = graphic text angle (degrees)             *
```

```
' *                                                                      *
' *  Dim aGraphicText As IElement                                        *
' *  Dim avGraphicTextGetAngle As Double                                 *
' *                                                                      *
' * * * * * * * * * * * * * * * * * * * * * * * * * * * * * * * * * * * *
'
Public Function avGraphicTextGetAngle(aGraphicText _
                                  As esriCore.IElement) As Double
'
' * * * * * * * * * * * * * * * * * * * * * * * * * * * * * * * * * * * *
' *                                                                      *
' *  Name: avGraphicTextGetStyle             File Name: avgtgsty.bas    *
' *                                                                      *
' *  PURPOSE:  TO GET THE TEXT STYLE ATTRIBUTES THAT ARE ASSOCIATED     *
' *            WITH A GRAPHIC TEXT, THIS INCLUDES THE ATTRIBUTES FOR    *
' *            FONT, SIZE, BOLD, ITALIC AND COLOR                       *
' *                                                                      *
' *  GIVEN:    aGraphicText = graphic text to be processed              *
' *                                                                      *
' *  RETURN:   aFont        = font name                                 *
' *            aSize        = font size                                 *
' *            aBold        = font style (1 = normal, 2 = bold)         *
' *            aItalic      = font style (1 = normal, 2 = italic)       *
' *            aColor       = font color (color object)                 *
' *                                                                      *
' *  Dim aGraphicText As IElement                                        *
' *  Dim aFont As String                                                 *
' *  Dim aSize As Double                                                 *
' *  Dim aBold, aItalic As Integer                                       *
' *  Dim aColor As IColor                                                *
' *                                                                      *
' * * * * * * * * * * * * * * * * * * * * * * * * * * * * * * * * * * * *
'
Public Sub avGraphicTextGetStyle(aGraphicText As esriCore.IElement, _
                              aFont, aSize, aBold, aItalic, _
                              aColor As esriCore.iColor)
'
' * * * * * * * * * * * * * * * * * * * * * * * * * * * * * * * * * * * *
' *                                                                      *
' *  Name: avGraphicTextGetSymbol            File Name: avgtgsym.bas    *
' *                                                                      *
' *  PURPOSE:  TO GET THE SYMBOL ASSOCIATED WITH A GRAPHIC TEXT         *
' *                                                                      *
' *  GIVEN:    aGraphicText           = graphic text to be processed    *
' *                                                                      *
' *  RETURN:   avGraphicTextGetSymbol = symbol describing the graphic   *
' *                                     text, its properties such as    *
' *                                     font, size, bold, italic and    *
' *                                     color                           *
' *                                                                      *
' *  Dim aGraphicText As IElement                                        *
' *  Dim avGraphicTextGetSymbol As ISymbol                               *
' *                                                                      *
' * * * * * * * * * * * * * * * * * * * * * * * * * * * * * * * * * * * *
'
Public Function avGraphicTextGetSymbol(aGraphicText _
                          As esriCore.IElement) As esriCore.ISymbol
'
' * * * * * * * * * * * * * * * * * * * * * * * * * * * * * * * * * * * *
' *                                                                      *
' *  Name: avGraphicTextGetText              File Name: avgtgtxt.bas    *
' *                                                                      *
' *  PURPOSE:  TO GET THE TEXT STRING ASSOCIATED WITH A GRAPHIC TEXT    *
' *                                                                      *
' *  GIVEN:    aGraphicText          = graphic text to be processed     *
' *                                                                      *
' *  RETURN:   avGraphicTextGetAngle = the text string                  *
' *                                                                      *
' *  Dim aGraphicText As IElement                                        *
' *  Dim avGraphicTextGetText As String                                  *
' *                                                                      *
' * * * * * * * * * * * * * * * * * * * * * * * * * * * * * * * * * * * *
'
```

```
Public Function avGraphicTextGetText(aGraphicText As esriCore.IElement)
'
' * * * * * * * * * * * * * * * * * * * * * * * * * * * * * * * * * * *
' *                                                                   *
' *  Name: avGraphicTextMake                  File Name: avgtmake.bas *
' *                                                                   *
' *  PURPOSE:  TO CREATE A GRAPHIC TEXT THAT CAN BE ADDED TO THE      *
' *            GRAPHICS LAYER                                         *
' *                                                                   *
' *  GIVEN:    aString           = text string that is associated     *
' *                                with the graphic text              *
' *            theGeom           = geometry describing the graphic    *
' *                                                                   *
' *  RETURN:   avGraphicTextMake = the graphic that can be added to   *
' *                                the graphics layer                 *
' *                                                                   *
' *  Dim aString As String                                            *
' *  Dim theGeom As IGeometry                                         *
' *  Dim avGraphicTextMake As IElement                                *
' *                                                                   *
' * * * * * * * * * * * * * * * * * * * * * * * * * * * * * * * * * * *
'
Public Function avGraphicTextMake(aString, _
                                  theGeom As esriCore.IGeometry) _
                                  As esriCore.IElement
'
' * * * * * * * * * * * * * * * * * * * * * * * * * * * * * * * * * * *
' *                                                                   *
' *  Name: avGraphicTextSetAngle              File Name: avgtsang.bas *
' *                                                                   *
' *  PURPOSE:  TO SET THE ANGLE ASSOCIATED WITH A GRAPHIC TEXT        *
' *                                                                   *
' *  GIVEN:    aGraphicText = graphic text to be processed            *
' *            aAngle       = angle to be assigned (degrees)          *
' *                                                                   *
' *  RETURN:   nothing                                                *
' *                                                                   *
' *  Dim aGraphicText As IElement                                     *
' *  Dim aAngle As Double                                             *
' *                                                                   *
' * * * * * * * * * * * * * * * * * * * * * * * * * * * * * * * * * * *
'
Public Sub avGraphicTextSetAngle(aGraphicText As esriCore.IElement, _
                                 aAngle)
'
' * * * * * * * * * * * * * * * * * * * * * * * * * * * * * * * * * * *
' *                                                                   *
' *  Name: avGraphicTextSetStyle              File Name: avgtssty.bas *
' *                                                                   *
' *  PURPOSE:  TO SET THE TEXT STYLE ATTRIBUTES THAT ARE ASSOCIATED   *
' *            WITH A GRAPHIC TEXT, THIS INCLUDES THE ATTRIBUTES FOR  *
' *            FONT, SIZE, BOLD, ITALIC AND COLOR                     *
' *                                                                   *
' *  GIVEN:    aGraphicText = graphic text to be processed            *
' *            aFont        = font name                               *
' *            aSize        = font size                               *
' *            aBold        = font style (1 = normal, 2 = bold)       *
' *            aItalic      = font style (1 = normal, 2 = italic)     *
' *            aColor       = font color (color object)               *
' *                                                                   *
' *  RETURN:   nothing                                                *
' *                                                                   *
' *  Dim aGraphicText As IElement                                     *
' *  Dim aFont As String                                              *
' *  Dim aSize As Double                                              *
' *  Dim aBold, aItalic As Integer                                    *
' *  Dim aColor As IColor                                             *
' *                                                                   *
' * * * * * * * * * * * * * * * * * * * * * * * * * * * * * * * * * * *
'
Public Sub avGraphicTextSetStyle(aGraphicText As esriCore.IElement, _
                                 aFont, aSize, aBold, aItalic, _
                                 aColor As esriCore.iColor)
```

```
'
' * * * * * * * * * * * * * * * * * * * * * * * * * * * * * * * * * * *
' *                                                                     *
' *  Name: avGraphicTextSetSymbol          File Name: avstgsym.bas     *
' *                                                                     *
' *  PURPOSE:  TO SET THE SYMBOL ASSOCIATED WITH A GRAPHIC TEXT         *
' *                                                                     *
' *  GIVEN:    aGraphicText = graphic text to be processed             *
' *           pTextSymbol  = symbol describing the graphic text, its  *
' *                          properties such as font, size, bold,     *
' *                           italic and color                         *
' *                                                                     *
' *  RETURN:   nothing                                                 *
' *                                                                     *
' *  Dim aGraphicText As IElement                                      *
' *  Dim pTextSymbol As ISymbol                                        *
' *                                                                     *
' * * * * * * * * * * * * * * * * * * * * * * * * * * * * * * * * * * *
'
Public Sub avGraphicTextSetSymbol(aGraphicText As esriCore.IElement, _
                                  pTextSymbol As esriCore.ISymbol)
'
' * * * * * * * * * * * * * * * * * * * * * * * * * * * * * * * * * * *
' *                                                                     *
' *  Name: avGraphicTextSetText            File Name: avstgtxt.bas     *
' *                                                                     *
' *  PURPOSE:  TO SET THE TEXT STRING ASSOCIATED WITH A GRAPHIC TEXT   *
' *                                                                     *
' *  GIVEN:    aGraphicText = graphic text to be processed             *
' *            aString      = the text string to be assigned           *
' *                                                                     *
' *  RETURN:   nothing                                                 *
' *                                                                     *
' *  Dim aGraphicText As IElement                                      *
' *  Dim aString As String                                             *
' *                                                                     *
' * * * * * * * * * * * * * * * * * * * * * * * * * * * * * * * * * * *
'
Public Sub avGraphicTextSetText(aGraphicText As esriCore.IElement, _
                                aString)
'
' * * * * * * * * * * * * * * * * * * * * * * * * * * * * * * * * * * *
' *                                                                     *
' *  Name: avInit                          File Name: avinitgv.bas     *
' *                                                                     *
' *  PURPOSE:  TO INITIALIZE THE GLOBAL VARIABLES THAT ARE USED BY     *
' *            THE AVENUE WRAPS                                        *
' *                                                                     *
' *  GIVEN:    nothing                                                 *
' *                                                                     *
' *  RETURN:   nothing                                                 *
' *                                                                     *
' *  NOTE:     This routine needs to be called only once, typically   *
' *            when the project (*.mxd) is opened or when the         *
' *            extension is initially loaded.  When the Avenue Wraps   *
' *            are used in a project file (*.mxd) a call to avInit     *
' *            can be made in the OpenDocument event for the           *
' *            procedure MxDocument. In so doing when the project is   *
' *            opened, avInit will be called to initialize the global *
' *            variables referenced in avInit                          *
' *                                                                     *
' * * * * * * * * * * * * * * * * * * * * * * * * * * * * * * * * * * *
'
Public Sub avInit()
'
' * * * * * * * * * * * * * * * * * * * * * * * * * * * * * * * * * * *
' *                                                                     *
' *  Name: avIntersects                    File Name: avintsec.bas     *
' *                                                                     *
' *  PURPOSE:  TO CHECK IF TWO SHAPES INTERSECT EACH OTHER              *
' *                                                                     *
' *  GIVEN:    aShape1      = base shape                               *
```

```
' *           aShape2      = second shape to be intersected with the *
' *                          base shape                               *
' *                                                                   *
' *  RETURN:    avIntersects = intersection state of the input objects *
' *                          true = intersect, false = do not         *
' *                                                                   *
' *  Dim aShape1 As IGeometry                                         *
' *  Dim aShape2 As IGeometry                                         *
' *  Dim avIntersects As Boolean                                      *
' *                                                                   *
' * * * * * * * * * * * * * * * * * * * * * * * * * * * * * * * * * * *
'
Public Function avIntersects(aShape1 As esriCore.IGeometry, _
                             aShape2 As esriCore.IGeometry) As Boolean
'
' * * * * * * * * * * * * * * * * * * * * * * * * * * * * * * * * * * *
' *                                                                   *
' *  Name: avInterval                        File Name: avintrvl.bas  *
' *                                                                   *
' *  PURPOSE:  TO SET THE LEGEND THAT IS ASSOCIATED WITH A THEME      *
' *            TO BE OF QUANTILE TYPE WITH THE CLASSES DETERMINED BY  *
' *            USING AN EQUAL INTERVAL METHOD                         *
' *                                                                   *
' *  GIVEN:    pmxDoc    = the active view                            *
' *            theTheme  = theme to be processed                      *
' *            aField    = field name that theme is to be classified  *
' *                        upon                                       *
' *            numClass  = number of classes to be generated          *
' *                                                                   *
' *  RETURN:   nothing                                                *
' *                                                                   *
' *  NOTE:     Divides the features in the theme into numClass        *
' *            classifications of equal size using the values in      *
' *            aField. This is only supported for numeric fields      *
' *                                                                   *
' *  Dim pmxDoc As IMxDocument                                        *
' *  Dim theTheme As Variant                                          *
' *  Dim aField As String                                             *
' *  Dim numClass As Long                                             *
' *                                                                   *
' * * * * * * * * * * * * * * * * * * * * * * * * * * * * * * * * * * *
'
Public Sub avInterval(pmxDoc As esriCore.IMxDocument, _
                     theTheme, aField, numClass)
'
' * * * * * * * * * * * * * * * * * * * * * * * * * * * * * * * * * * *
' *                                                                   *
' *  Name: avInvalidateTOC                   File Name: avinvtoc.bas  *
' *                                                                   *
' *  PURPOSE:  REFRESH THE TABLE OF CONTENTS                          *
' *                                                                   *
' *  GIVEN:    name    = name of theme or table in the Table of       *
' *                      Contents to be refreshed, if NULL the entire *
' *                      Table of Contents will be refreshed          *
' *                                                                   *
' *  RETURN:   nothing                                                *
' *                                                                   *
' *  Dim name As Variant                                              *
' *                                                                   *
' * * * * * * * * * * * * * * * * * * * * * * * * * * * * * * * * * * *
'
Public Sub avInvalidateTOC(name)
'
' * * * * * * * * * * * * * * * * * * * * * * * * * * * * * * * * * * *
' *                                                                   *
' *  Name: avInvSelFeatures                  File Name: avinsfet.bas  *
' *                                                                   *
' *  PURPOSE:  TO REDRAW THE SELECTED FEATURES FOR A SET OF THEMES    *
' *                                                                   *
' *  GIVEN:    pmxDoc      = the active view                          *
' *            selThmList = list of themes with selected features     *
' *                                                                   *
' *  RETURN:   nothing                                                *
```

```
' *                                                                    *
' *  NOTE:     Structure of selThmList is a sequential list with two  *
' *            items per theme, name of the theme and the number of   *
' *            selected features in the theme, so that:              *
' *             Item 1: name of theme 1                               *
' *            Item 2: number of selected features in theme 1        *
' *             Item 3: name of theme 2                               *
' *            Item 4: number of selected features in theme 2        *
' *            Repeat Items 1 and 2 for each theme                   *
' *                                                                    *
' *  Dim pmxDoc As IMxDocument                                         *
' *  Dim selThmList As New Collection                                  *
' *                                                                    *
' * * * * * * * * * * * * * * * * * * * * * * * * * * * * * * * * * * *
'
Public Sub avInvSelFeatures(pmxDoc As esriCore.IMxDocument, selThmList)
'
' * * * * * * * * * * * * * * * * * * * * * * * * * * * * * * * * * * *
' *                                                                    *
' *  Name: avIsEditable                      File Name: avisedbl.bas  *
' *                                                                    *
' *  PURPOSE:  DETERMINE IF A LAYER OR TABLE IS EDITABLE OR NOT        *
' *                                                                    *
' *  GIVEN:    name          = name of theme or table for which its   *
' *                            editability status is to be checked    *
' *                                                                    *
' *  RETURN:   avIsEditable = editability state of the layer or table *
' *                            true = editable, false = not editable  *
' *                                                                    *
' *  Dim name As Variant                                               *
' *  Dim avIsEditable As Boolean                                       *
' *                                                                    *
' * * * * * * * * * * * * * * * * * * * * * * * * * * * * * * * * * * *
'
Public Function avIsEditable(name) As Boolean
'
' * * * * * * * * * * * * * * * * * * * * * * * * * * * * * * * * * * *
' *                                                                    *
' *  Name: avIsFTheme                        File Name: avisfthm.bas  *
' *                                                                    *
' *  PURPOSE:  DETERMINE IF A LAYER IS OF FEATURE LAYER TYPE OR NOT    *
' *                                                                    *
' *  GIVEN:    name        = name of input object for which its       *
' *                          feature layer type is to be checked      *
' *                                                                    *
' *  RETURN:   avIsFTheme = flag denoting whether the input object    *
' *                          is a feature layer or not                *
' *                          true = is, false = not                   *
' *                                                                    *
' *  Dim name As Variant                                               *
' *  Dim avIsFTheme As Boolean                                         *
' *                                                                    *
' * * * * * * * * * * * * * * * * * * * * * * * * * * * * * * * * * * *
'
Public Function avIsFTheme(name)
'
' * * * * * * * * * * * * * * * * * * * * * * * * * * * * * * * * * * *
' *                                                                    *
' *  Name: avIsVisible                       File Name: avisvibl.bas  *
' *                                                                    *
' *  PURPOSE:  DETERMINE IF AN OBJECT IS VISIBLE OR NOT                *
' *                                                                    *
' *  GIVEN:    name         = name of input object for which its      *
' *                           visibility status is to be checked      *
' *                                                                    *
' *  RETURN:   avIsVisible = visible state of the input object        *
' *                           true = visible, false = not visible     *
' *                                                                    *
' *  Dim name As Variant                                               *
' *  Dim avIsVisible As Boolean                                        *
' *                                                                    *
' * * * * * * * * * * * * * * * * * * * * * * * * * * * * * * * * * * *
'
```

```
Public Function avIsVisible(name) As Boolean
'
' * * * * * * * * * * * * * * * * * * * * * * * * * * * * * * * * * * *
' *                                                                     *
' *  Name: avLegendGetSymbols                   File Name: avlgtsym.bas *
' *                                                                     *
' *  PURPOSE:  GET A LIST OF SYMBOLS APPEARING IN A LEGEND              *
' *                                                                     *
' *  GIVEN:    theLegend = legend to be processed                       *
' *                                                                     *
' *  RETURN:   symbList  = list of symbols used in the legend           *
' *                                                                     *
' *  Dim theLegend As IFeatureRenderer                                  *
' *  Dim symbList As New Collection                                     *
' *                                                                     *
' * * * * * * * * * * * * * * * * * * * * * * * * * * * * * * * * * * *
'
Public Sub avLegendGetSymbols(thelegend As esriCore.IFeatureRenderer, _
                              symbList)
'
' * * * * * * * * * * * * * * * * * * * * * * * * * * * * * * * * * * *
' *                                                                     *
' *  Name: avLegendSetSymbols                   File Name: avlstsym.bas *
' *                                                                     *
' *  PURPOSE:  TO SET THE SYMBOLS USED IN THE CLASSIFICATIONS WHICH     *
' *            APPEAR IN A LEGEND                                       *
' *                                                                     *
' *  GIVEN:    theLegend = legend to be processed                       *
' *            symbList  = list of symbols that are to be assigned      *
' *                        to the classifications in a legend           *
' *                                                                     *
' *  RETURN:   nothing                                                  *
' *                                                                     *
' *  Dim theLegend As IFeatureRenderer                                  *
' *  Dim symbList As New Collection                                     *
' *                                                                     *
' * * * * * * * * * * * * * * * * * * * * * * * * * * * * * * * * * * *
'
Public Sub avLegendSetSymbols(thelegend As esriCore.IFeatureRenderer, _
                              symbList)
'
' * * * * * * * * * * * * * * * * * * * * * * * * * * * * * * * * * * *
' *                                                                     *
' *  Name: avLineFileClose                      File Name: avlfclos.bas *
' *                                                                     *
' *  PURPOSE:  CLOSE A FILE CONNECTION                                  *
' *                                                                     *
' *  GIVEN:    aFile   = textstream object to be processed              *
' *                                                                     *
' *  RETURN:   nothing                                                  *
' *                                                                     *
' *  Dim aFile                                                          *
' *                                                                     *
' * * * * * * * * * * * * * * * * * * * * * * * * * * * * * * * * * * *
'
Public Sub avLineFileClose(aFile)
'
' * * * * * * * * * * * * * * * * * * * * * * * * * * * * * * * * * * *
' *                                                                     *
' *  Name: avLineFileMake                       File Name: avlfmake.bas *
' *                                                                     *
' *  PURPOSE:  OPEN A FILE CONNECTION FOR READING AND/OR WRITING        *
' *                                                                     *
' *  GIVEN:    aFileName      = the name of the file to be processed    *
' *            aFilePerm      = the type of file connection desired     *
' *                              READ   : open file for reading         *
' *                              WRITE  : open file for writing         *
' *                              APPEND : open file for appending       *
' *                                                                     *
' *  RETURN:   avLineFileMake = textstream object that can be used      *
' *                             for reading and/or writing operations   *
' *                                                                     *
```

```
' *  Dim aFileName As String                                           *
' *  Dim aFilePerm As String                                           *
' *  Dim avLineFileMake                                                *
' *                                                                    *
' * * * * * * * * * * * * * * * * * * * * * * * * * * * * * * * * * * *
'
Public Function avLineFileMake(aFileName, aFilePerm)
'
' * * * * * * * * * * * * * * * * * * * * * * * * * * * * * * * * * * *
' *                                                                    *
' *  Name: avListFiles                     File Name: avlstfil.bas     *
' *                                                                    *
' *  PURPOSE:  GET A LIST OF FILES IN A DIRECTORY                      *
' *                                                                    *
' *  GIVEN:    aDir    = the directory to be scanned                   *
' *            filType = the type of files to be searched for          *
' *                   vbNormal     0 Specifies files with no           *
' *                                  attributes                        *
' *                   vbReadOnly   1 Specifies read-only files in      *
' *                                  addition to files with no         *
' *                                  attributes                        *
' *                   vbHidden     2 Specifies hidden files in         *
' *                                  addition to files with no         *
' *                                  attributes                        *
' *                   VbSystem     4 Specifies system files in         *
' *                                  addition to files with no         *
' *                                  attributes                        *
' *                   vbVolume     8 Specifies volume label; if        *
' *                                  any other attribute is given      *
' *                                  vbVolume is ignored               *
' *                   vbDirectory 16 Specifies directories or          *
' *                                  folders in addition to files      *
' *                                  with no attributes                *
' *                                                                    *
' *  RETURN:   filList = list of files that were found as specified    *
' *                      by the filType argument                       *
' *                                                                    *
' *  NOTE:     The filType argument can be specified either as the     *
' *            numeric value, shown above, or as the VB keyword that   *
' *            is shown                                                *
' *                                                                    *
' *  Dim aDir As String                                                *
' *  Dim filType As Integer                                            *
' *  Dim filList As New Collection                                     *
' *                                                                    *
' * * * * * * * * * * * * * * * * * * * * * * * * * * * * * * * * * * *
'
Public Sub avListFiles(aDIR, filType, filList)
'
' * * * * * * * * * * * * * * * * * * * * * * * * * * * * * * * * * * *
' *                                                                    *
' *  Name: avMoveTo                        File Name: avmoveto.bas     *
' *                                                                    *
' *  PURPOSE:  TO REPOSITION A WINDOW OBJECT                           *
' *                                                                    *
' *  GIVEN:    aDoc    = the window object                             *
' *            aLeft   = distance from the left side of the screen     *
' *            aTop    = distance from the top of the screen           *
' *                                                                    *
' *  RETURN:   nothing                                                 *
' *                                                                    *
' *  Dim aDoc As IUnknown                                              *
' *  Dim aLeft, aTop As Long                                           *
' *                                                                    *
' * * * * * * * * * * * * * * * * * * * * * * * * * * * * * * * * * * *
'
Public Sub avMoveTo(aDoc As IUnknown, aLeft, aTop)
'
```

```
'************************************************************
'*                                                          *
'*  Name: avMsgBoxChoice               File Name: avmbchoi.bas *
'*                                                          *
'*  PURPOSE:  DISPLAY A CHOICE MESSAGE BOX WHICH CONTAINS A LIST OF *
'*            STRING ITEMS                                  *
'*                                                          *
'*  GIVEN:    aList   = the list of items to be displayed   *
'*            aMsg    = the message to be displayed         *
'*            Heading = message box caption                 *
'*                                                          *
'*  RETURN:   ians    = the item selected by the user, if the user *
'*                      Cancels the command: ians will be equal to *
'*                      vbCancel                            *
'*                                                          *
'*  Dim aList As New Collection                             *
'*  Dim aMsg, Heading As Variant                            *
'*  Dim ians As Variant                                     *
'*                                                          *
'************************************************************
'
Public Sub avMsgBoxChoice(aList, aMsg, Heading, ians)
'
'************************************************************
'*                                                          *
'*  Name: avMsgBoxInfo                 File Name: avmbinfo.bas *
'*                                                          *
'*  PURPOSE:  DISPLAY AN INFORMATION MESSAGE BOX            *
'*                                                          *
'*  GIVEN:    aMessage = the message to be displayed        *
'*            Heading  = message box caption                *
'*                                                          *
'*  RETURN:   nothing                                       *
'*                                                          *
'*  Dim aMessage, Heading As Variant                        *
'*                                                          *
'************************************************************
'
Public Sub avMsgBoxInfo(aMessage, Heading)
'
'************************************************************
'*                                                          *
'*  Name: avMsgBoxInput                File Name: avmbinpu.bas *
'*                                                          *
'*  PURPOSE:  DISPLAY A SINGLE LINE INPUT MESSAGE BOX       *
'*                                                          *
'*  GIVEN:    aMsg     = the message to be displayed        *
'*            Heading  = message box caption                *
'*            aDefault = the default button value           *
'*                                                          *
'*  RETURN:   ians     = the response from the user, if the user *
'*                       Cancels the command: ians will be equal to *
'*                       NULL                               *
'*                                                          *
'*  Dim aMsg, Heading, aDefault As Variant                  *
'*  Dim ians As Variant                                     *
'*                                                          *
'************************************************************
'
Public Sub avMsgBoxInput(aMsg, Heading, aDefault, ians)
'
'************************************************************
'*                                                          *
'*  Name: avMsgBoxList                 File Name: avmblist.bas *
'*                                                          *
'*  PURPOSE:  DISPLAY A LIST OF STRINGS WITHIN A MESSAGE BOX *
'*                                                          *
'*  GIVEN:    aList   = the list of strings to be displayed *
'*            aMsg    = the message to be displayed         *
'*            Heading = message box caption                 *
'*                                                          *
'*  RETURN:   ians    = the button that was selected, ians will be *
'*                      equal to vbOK or vbCancel           *
```

```
' *                                                                   *
' *  Dim aList As New Collection                                      *
' *  Dim aMsg, Heading As Variant                                     *
' *  Dim ians As Integer                                              *
' *                                                                   *
' * * * * * * * * * * * * * * * * * * * * * * * * * * * * * * * * * * *
'
Public Sub avMsgBoxList(aList, aMsg, Heading, ians)
'
' * * * * * * * * * * * * * * * * * * * * * * * * * * * * * * * * * * *
' *                                                                   *
' *  Name: avMsgBoxMultiInput                File Name: avmbminp.bas  *
' *                                                                   *
' *  PURPOSE:  DISPLAY A MULTI-INPUT LINE MESSAGE BOX                 *
' *                                                                   *
' *  GIVEN:    aMsg     = the message to be displayed                 *
' *            Heading  = message box caption                         *
' *            labels   = list of labels for each of the items that   *
' *                       are prompted for                            *
' *            defaults = list of default values for each of the      *
' *                       items that are prompted for                 *
' *                                                                   *
' *  RETURN:   aList    = list of responses for each of the items     *
' *                       that were displayed, if the user Cancels    *
' *                       the command: aList will be an empty list,   *
' *                       that is, aList.Count will equal 0           *
' *                                                                   *
' *  NOTE:     There is no limit to the number of items that can be   *
' *            prompted for, the difficulty however will be that the  *
' *            dialog box may exceed the visible area of the screen.  *
' *            Recommend using avMsgBoxMultiInput2 when more than 12  *
' *            to 15 items are to be displayed.                       *
' *                                                                   *
' *  Dim aMsg, Heading As Variant                                     *
' *  Dim labels As New Collection                                     *
' *  Dim defaults As New Collection                                   *
' *  Dim aList As New Collection                                      *
' *                                                                   *
' * * * * * * * * * * * * * * * * * * * * * * * * * * * * * * * * * * *
'
Public Sub avMsgBoxMultiInput(aMsg, Heading, Labels, Defaults, aList)
'
' * * * * * * * * * * * * * * * * * * * * * * * * * * * * * * * * * * *
' *                                                                   *
' *  Name: avMsgBoxMultiList                 File Name: avmbmlst.bas  *
' *                                                                   *
' *  PURPOSE:  DISPLAY A LIST OF STRINGS WITHIN A MESSAGE BOX WITH    *
' *            THE ABILITY TO SELECT MULTIPLE ITEMS                   *
' *                                                                   *
' *  GIVEN:    aList    = the list of strings to be displayed         *
' *            aMsg     = the message to be displayed                 *
' *            Heading  = message box caption                         *
' *                                                                   *
' *  RETURN:   itemList = the list of items selected by the user,     *
' *                       will be an empty list if the user cancels   *
' *                       the operation                               *
' *                                                                   *
' *  Dim aList As New Collection                                      *
' *  Dim aMsg, Heading As Variant                                     *
' *  Dim itemList As New Collection                                   *
' *                                                                   *
' * * * * * * * * * * * * * * * * * * * * * * * * * * * * * * * * * * *
'
Public Sub avMsgBoxMultiList(aList, aMsg, Heading, itemList)
'
' * * * * * * * * * * * * * * * * * * * * * * * * * * * * * * * * * * *
' *                                                                   *
' *  Name: avMsgBoxMultiInput2               File Name: avmbmin2.bas  *
' *                                                                   *
' *  PURPOSE:  DISPLAY A MULTI-INPUT LINE MESSAGE BOX WITH A BACK     *
' *            BUTTON WHEN MORE THAN 10 ITEMS ARE TO BE DISPLAYED     *
' *                                                                   *
```

```
' *  GIVEN:     aMsg      = the message to be displayed              *
' *             Heading   = message box caption                      *
' *             labels    = list of labels for each of the items that *
' *                         are prompted for                         *
' *             defaults  = list of default values for each of the   *
' *                         items that are prompted for              *
' *                                                                  *
' *  RETURN:    aList     = list of responses for each of the items  *
' *                         that were displayed, if the user Cancels *
' *                         the command: aList will be an empty list, *
' *                         that is, aList.Count will equal 0        *
' *                                                                  *
' *  Dim aMsg, Heading As Variant                                    *
' *  Dim labels As New Collection                                    *
' *  Dim defaults As New Collection                                  *
' *  Dim aList As New Collection                                     *
' *                                                                  *
' * * * * * * * * * * * * * * * * * * * * * * * * * * * * * * * * * *
'
Public Sub avMsgBoxMultiInput2(aMsg, Heading, Labels, Defaults, aList)
'
' * * * * * * * * * * * * * * * * * * * * * * * * * * * * * * * * * *
' *                                                                  *
' *  Name: avMsgBoxWarning                File Name: avmbwarn.bas    *
' *                                                                  *
' *  PURPOSE:  DISPLAY A WARNING MESSAGE BOX                         *
' *                                                                  *
' *  GIVEN:     aMessage = the message to be displayed               *
' *             Heading  = message box caption                       *
' *                                                                  *
' *  RETURN:    nothing                                              *
' *                                                                  *
' *  Dim aMessage, Heading As Variant                                *
' *                                                                  *
' * * * * * * * * * * * * * * * * * * * * * * * * * * * * * * * * * *
'
Public Sub avMsgBoxWarning(aMessage, Heading)
'
' * * * * * * * * * * * * * * * * * * * * * * * * * * * * * * * * * *
' *                                                                  *
' *  Name: avMsgBoxYesNoCancel            File Name: avmbysnc.bas    *
' *                                                                  *
' *  PURPOSE:  DISPLAY A YES, NO, CANCEL MESSAGE BOX                 *
' *                                                                  *
' *  GIVEN:     aMessage = the message to be displayed               *
' *             Heading  = message box caption                       *
' *             aDefault = the default button setting                *
' *                        true = yes, false = no                    *
' *                                                                  *
' *  RETURN:    ians     = the button that was selected, ians will be *
' *                        equal to vbYes, vbNo or vbCancel          *
' *                                                                  *
' *  Dim aMessage, Heading As Variant                                *
' *  Dim aDefault As Boolean                                         *
' *  Dim ians As Integer                                             *
' *                                                                  *
' * * * * * * * * * * * * * * * * * * * * * * * * * * * * * * * * * *
'
Public Sub avMsgBoxYesNoCancel(aMessage, Heading, aDefault, ians)
'
' * * * * * * * * * * * * * * * * * * * * * * * * * * * * * * * * * *
' *                                                                  *
' *  Name: avMsgBoxYesNo                  File Name: avmbysno.bas    *
' *                                                                  *
' *  PURPOSE:  DISPLAY A YES, NO MESSAGE BOX                         *
' *                                                                  *
' *  GIVEN:     aMessage = the message to be displayed               *
' *             Heading  = message box caption                       *
' *             aDefault = the default button setting                *
' *                        true = yes, false = no                    *
' *                                                                  *
' *  RETURN:    ians     = the button that was selected, ians will be *
' *                        equal to vbYes or vbNo                    *
```

```
' *                                                                   *
' *  Dim aMessage, Heading As Variant                                 *
' *  Dim aDefault As Boolean                                          *
' *  Dim ians As Integer                                              *
' *                                                                   *
' * * * * * * * * * * * * * * * * * * * * * * * * * * * * * * * * * * *
'
Public Sub avMsgBoxYesNo(aMessage, Heading, aDefault, ians)
'
' * * * * * * * * * * * * * * * * * * * * * * * * * * * * * * * * * * *
' *                                                                   *
' *  Name: avNatural                         File Name: avnatrul.bas  *
' *                                                                   *
' *  PURPOSE:  TO SET THE LEGEND THAT IS ASSOCIATED WITH A THEME      *
' *            TO BE OF QUANTILE TYPE WITH THE CLASSES DETERMINED BY  *
' *            USING THE NATURAL BREAK METHOD                         *
' *                                                                   *
' *  GIVEN:    pmxDoc     = the active view                           *
' *            theTheme   = theme to be processed                     *
' *            aField     = field name that theme is to be classified *
' *                         upon                                      *
' *            numClass   = number of classes to be generated         *
' *                                                                   *
' *  RETURN:   nothing                                                *
' *                                                                   *
' *  NOTE:     Divides the features in the theme into numClass        *
' *            classifications of equal size using the values in      *
' *            aField. This is only supported for numeric fields      *
' *                                                                   *
' *  Dim pmxDoc As IMxDocument                                        *
' *  Dim theTheme As Variant                                          *
' *  Dim aField As String                                             *
' *  Dim numClass As Long                                             *
' *                                                                   *
' * * * * * * * * * * * * * * * * * * * * * * * * * * * * * * * * * * *
'
Public Sub avNatural(pmxDoc As esriCore.IMxDocument, _
                     theTheme, aField, numClass)
'
' * * * * * * * * * * * * * * * * * * * * * * * * * * * * * * * * * * *
' *                                                                   *
' *  Name: avObjGetName                      File Name: avobjgnm.bas  *
' *                                                                   *
' *  PURPOSE:  TO GET THE NAME OF AN OBJECT                           *
' *                                                                   *
' *  GIVEN:    theObject = object to be processed                     *
' *                                                                   *
' *  RETURN:   aName     = name assigned to the object                *
' *                                                                   *
' *  NOTE:     This subroutine checks if the object matches a known   *
' *            type, if so, the appropriate interface is selected     *
' *            and the given name assigned to the object is extracted *
' *            If a match is not made, the name will be NULL          *
' *                                                                   *
' *  Dim theObject As IUnknown                                        *
' *  Dim aName As Variant                                             *
' *                                                                   *
' * * * * * * * * * * * * * * * * * * * * * * * * * * * * * * * * * * *
'
Public Sub avObjGetName(theObject As IUnknown, aName)
'
' * * * * * * * * * * * * * * * * * * * * * * * * * * * * * * * * * * *
' *                                                                   *
' *  Name: avObjSetName                      File Name: avobjsnm.bas  *
' *                                                                   *
' *  PURPOSE:  TO SET THE NAME FOR AN OBJECT                          *
' *                                                                   *
' *  GIVEN:    theObject = object which is to be named                *
' *            aName     = name to be assigned to the object          *
' *                                                                   *
' *  RETURN:   nothing                                                *
' *                                                                   *
```

```
' *  NOTE:     This subroutine checks if the object matches a known  *
' *            type, if so, the appropriate interface is selected    *
' *            and the given name assigned to the object. If a match *
' *            is not made, the object is left unaltered             *
' *                                                                  *
' *  Dim theObject As IUnknown                                       *
' *  Dim aName As String                                             *
' *                                                                  *
' * * * * * * * * * * * * * * * * * * * * * * * * * * * * * * * * * *
'
Public Sub avObjSetName(theObject As IUnknown, aName)
'
' * * * * * * * * * * * * * * * * * * * * * * * * * * * * * * * * * *
' *                                                                  *
' *  Name: avOpenFeatClass                  File Name: avopenfc.bas  *
' *                                                                  *
' *  PURPOSE:  OPEN A DATASET OR A FEATURECLASS IN A DATASET FOR     *
' *            PROCESSING                                            *
' *                                                                  *
' *  GIVEN:    opmode           = type of dataset to be opened       *
' *                               1 : shapefile                      *
' *                               2 : raster                         *
' *                               3 : tin                            *
' *                               4 : coverage                       *
' *                               5 : access database                *
' *            sDir             = directory location of dataset      *
' *            sName            = name of dataset                    *
' *            aFCtype          = feature class type (only used for  *
' *                               coverages and access databases)    *
' *                               if not to be used specify as NULL  *
' *                                                                  *
' *  RETURN:   avOpenFeatClass = dataset that is opened, if none, the *
' *                               value will be set to NOTHING       *
' *                                                                  *
' *  Dim opmode As Integer                                           *
' *  Dim sDir As String                                              *
' *  Dim sName As String                                             *
' *  Dim aFCtype As String                                           *
' *  Dim avOpenFeatClass As IUnknown                                 *
' *                                                                  *
' * * * * * * * * * * * * * * * * * * * * * * * * * * * * * * * * * *
'
Public Function avOpenFeatClass(opmode, sDir, _
                                sName, aFCtype) As IUnknown
'
' * * * * * * * * * * * * * * * * * * * * * * * * * * * * * * * * * *
' *                                                                  *
' *  Name: avOpenWorkspace                  File Name: avopenwk.bas  *
' *                                                                  *
' *  PURPOSE:  OPEN A WORKSPACE FOR PROCESSING                       *
' *                                                                  *
' *  GIVEN:    opmode           = type of workspace to be opened     *
' *                               1 : shapefile                      *
' *                               2 : raster                         *
' *                               3 : tin                            *
' *                               4 : coverage                       *
' *                               5 : access database                *
' *            sDir             = directory location of workspace    *
' *            sName            = name of workspace                  *
' *                                                                  *
' *  RETURN:   avOpenWorkspace = workspace that is opened, if none,  *
' *                               the value will be set to NOTHING   *
' *                                                                  *
' *  Dim opmode As Integer                                           *
' *  Dim sDir As String                                              *
' *  Dim sName As String                                             *
' *  Dim avOpenWorkspace As IWorkspace                               *
' *                                                                  *
' * * * * * * * * * * * * * * * * * * * * * * * * * * * * * * * * * *
'
Public Function avOpenWorkspace(opmode, _
                                sDir, sName) As esriCore.IWorkspace
'
```

```
' * * * * * * * * * * * * * * * * * * * * * * * * * * * * * * * * * * *
' *                                                                     *
' *  Name: avPaletteGetList                   File Name: avpaltls.bas  *
' *                                                                     *
' *  PURPOSE:  GET A LIST OF THE ITEMS WITHIN A SPECIFIC GALLERY        *
' *                                                                     *
' *  GIVEN:    aGalleryType = type of gallery to be processed           *
' *                          keyword       name of gallery              *
' *                          PEN         : Line Symbols                 *
' *                          MARKER      : Marker Symbols               *
' *                          FILL        : Fill Symbols                 *
' *                          TEXT        : Text Symbols                 *
' *                          COLOR       : Colors                       *
' *                          REFERENCE   : Reference Systems            *
' *                          SHADOWS     : Shadows                      *
' *                          AREAPATCHES : Area Patches                 *
' *                          LINEPATCHES : Line Patches                 *
' *                          LABELS      : Labels                       *
' *                          NORTHARROWS : North Arrows                 *
' *                          SCALEBARS   : Scale Bars                   *
' *                          LEGENDITEMS : Legend Items                 *
' *                          SCALETEXT   : Scale Texts                  *
' *                          COLORRAMPS  : Color Ramps                  *
' *                          BORDERS     : Borders                      *
' *                          BACKGROUNDS : Backgrounds                  *
' *            thePalette   = style gallery, contains all galleries     *
' *                                                                     *
' *  RETURN:   aGalleryList = list of items in the desired gallery      *
' *                                                                     *
' *  Dim aGalleryType As String                                         *
' *  Dim thePalette As IStyleGallery                                    *
' *  Dim aGalleryList As New Collection                                 *
' *                                                                     *
' * * * * * * * * * * * * * * * * * * * * * * * * * * * * * * * * * * *
'
Public Sub avPaletteGetList(aGalleryType, _
                            thePalette As esriCore.IStyleGallery, _
                            aGalleryList)
'
' * * * * * * * * * * * * * * * * * * * * * * * * * * * * * * * * * * *
' *                                                                     *
' *  Name: avPanTo                              File Name: avpanto.bas  *
' *                                                                     *
' *  PURPOSE:  SCRIPT TO CENTER THE DISPLAY ABOUT A POINT               *
' *                                                                     *
' *  GIVEN:    pmxDoc   = the active view                               *
' *            thePoint = point that the display is to be centered      *
' *                       about                                         *
' *                                                                     *
' *  RETURN:   nothing                                                  *
' *                                                                     *
' *  Dim pmxDoc As IMxDocument                                          *
' *  Dim thePoint As IPoint                                             *
' *                                                                     *
' * * * * * * * * * * * * * * * * * * * * * * * * * * * * * * * * * * *
'
Public Sub avPanTo(pmxDoc As esriCore.IMxDocument, _
                  thePoint As esriCore.IPoint)
'
' * * * * * * * * * * * * * * * * * * * * * * * * * * * * * * * * * * *
' *                                                                     *
' *  Name: avPlAsList                          File Name: avpl2lst.bas  *
' *                                                                     *
' *  PURPOSE:  TO CREATE A POINT LIST FROM A GEOMETRY OBJECT            *
' *                                                                     *
' *  GIVEN:    pFeatureGeom = geometry object to be processed           *
' *                                                                     *
' *  RETURN:   shapeList    = list of points comprising the polyline    *
' *                           structure of shapeList is:                *
' *                           Item 1: collection of points in part 1    *
' *                           Repeat Item 1 for each part               *
' *                           So that, shapeList is a list of           *
```

```
' *                           collections with each collection       *
' *                           containing points                      *
' *                                                                  *
' *  NOTE:      (a) This routine can be used for Polyline, Polygon and *
' *                 Multi-point features, it can not be used for Point *
' *                 features                                         *
' *             (b) Use subroutine avAsList when an IFeature object is *
' *                 known and not an IGeometry object                *
' *                                                                  *
' *  Dim pFeatureGeom As IGeometry                                   *
' *  Dim shapeList As New Collection                                 *
' *                                                                  *
' * * * * * * * * * * * * * * * * * * * * * * * * * * * * * * * * * * *
'
Public Sub avPlAsList(pFeatureGeom As esriCore.IGeometry, shapeList)
'
' * * * * * * * * * * * * * * * * * * * * * * * * * * * * * * * * * * *
' *                                                                  *
' *  Name: avPlAsList2                       File Name: avpl2ls2.bas *
' *                                                                  *
' *  PURPOSE:  TO CREATE A POINT LIST FROM A GEOMETRY OBJECT         *
' *                                                                  *
' *  GIVEN:    pFeatureGeom = geometry object to be processed        *
' *                                                                  *
' *  RETURN:   shapeList    = list of points comprising the polyline *
' *                           structure of shapeList is:             *
' *                           Item 1: number of parts                *
' *                           Item 2: number of points in part 1     *
' *                           Item 3: x value of point 1 in part 1   *
' *                           Item 4: y value of point 1 in part 1   *
' *                           Item 5: z value of point 1 in part 1   *
' *                           Item 6: m value of point 1 in part 1   *
' *                           Item 7: id value of point 1 in part 1  *
' *                           Item 8: Repeat Items 3 - 7 for each    *
' *                                   point                          *
' *                           Repeat Items 2 - 8 for each part       *
' *                                                                  *
' *  NOTE:     This routine can be used for Polyline, Polygon and    *
' *            Multi-point features, it can not be used for Point    *
' *            features                                              *
' *                                                                  *
' *  Dim pFeatureGeom As IGeometry                                   *
' *  Dim shapeList As New Collection                                 *
' *                                                                  *
' * * * * * * * * * * * * * * * * * * * * * * * * * * * * * * * * * * *
'
Public Sub avPlAsList2(pFeatureGeom As esriCore.IGeometry, shapeList)
'
' * * * * * * * * * * * * * * * * * * * * * * * * * * * * * * * * * * *
' *                                                                  *
' *  Name: avPlFindVertex                    File Name: avfindvx.bas *
' *                                                                  *
' *  PURPOSE:  TO FIND THE VERTEX CLOSEST TO A LOCATION              *
' *                                                                  *
' *  GIVEN:    ipmode     = the mode of operation                    *
' *                         0 : find the vertex close to a location  *
' *            elmntList = list of points comprising the feature     *
' *                        structure of elmntList is:                *
' *                        Item 1: number of parts                   *
' *                        Item 2: number of points in part 1        *
' *                        Item 3: x value of point 1 in part 1      *
' *                        Item 4: y value of point 1 in part 1      *
' *                        Item 5: z value of point 1 in part 1      *
' *                        Item 6: m value of point 1 in part 1      *
' *                        Item 7: id value of point 1 in part 1     *
' *                        Item 8: Repeat Items 3 - 7 for each       *
' *                                point                             *
' *                        Repeat Items 2 - 8 for each part          *
' *            X,Y        = coordinates of new point                 *
' *                                                                  *
' *  RETURN:   thePart    = the part of the polyline (0 - i)         *
' *                         part numbers begin at zero, not one      *
```

```
' *           thePt     = point number in part closest to location    *
' *                       point numbers begin at one, not zero        *
' *                                                                   *
' *  Dim ipmode As Integer                                            *
' *  Dim elmntList As New Collection                                  *
' *  Dim X, Y As Double                                               *
' *  Dim thePart, thePt As Long                                       *
' *                                                                   *
' * * * * * * * * * * * * * * * * * * * * * * * * * * * * * * * * * * *
'
Public Sub avPlFindVertex(ipmode, elmntList, x, y, thePart, thePt)
'
' * * * * * * * * * * * * * * * * * * * * * * * * * * * * * * * * * * *
' *                                                                   *
' *  Name: avPlGet3Pt                       File Name: avplget3.bas   *
' *                                                                   *
' *  PURPOSE:  TO GET 3 POINTS FROM A FEATURE POINT LIST FOR A        *
' *            SPECIFIC PART IN THE FEATURE                           *
' *                                                                   *
' *  GIVEN:    shapeList = list of points comprising the feature      *
' *                        structure of shapeList is:                 *
' *                        Item 1: number of parts                    *
' *                        Item 2: number of points in part 1         *
' *                        Item 3: x value of point 1 in part 1       *
' *                        Item 4: y value of point 1 in part 1       *
' *                        Item 5: z value of point 1 in part 1       *
' *                        Item 6: m value of point 1 in part 1       *
' *                        Item 7: id value of point 1 in part 1      *
' *                        Item 8: Repeat Items 3 - 7 for each        *
' *                                point                              *
' *                        Repeat Items 2 - 8 for each part           *
' *            thePart   = the part of the polyline (0 - i)           *
' *                        part numbers begin at zero, not one        *
' *                                                                   *
' *  RETURN:   X1,Y1     = start point coordinates of part            *
' *            XM,YM     = mid point coordinates of part              *
' *            X2,Y2     = end point coordinates of part              *
' *                                                                   *
' *  Dim shapeList As New Collection                                  *
' *  Dim thePart As Long                                              *
' *  Dim X1, Y1, XM, YM, X2, Y2 As Double                             *
' *                                                                   *
' * * * * * * * * * * * * * * * * * * * * * * * * * * * * * * * * * * *
'
Public Sub avPlGet3Pt(shapeList, thePart, X1, Y1, Xm, Ym, X2, Y2)
'
' * * * * * * * * * * * * * * * * * * * * * * * * * * * * * * * * * * *
' *                                                                   *
' *  Name: avPlModify                       File Name: avplmdfy.bas   *
' *                                                                   *
' *  PURPOSE:  TO MODIFY A SPECIFIC PART IN A FEATURE POINT LIST      *
' *                                                                   *
' *  GIVEN:    ipmode    = the mode of operation                      *
' *                        0 = change coordinates of a point          *
' *                        1 = insert new point                       *
' *                        2 = delete point                           *
' *            elmntList = list of points comprising the feature      *
' *                        structure of elmntList is:                 *
' *                        Item 1: number of parts                    *
' *                        Item 2: number of points in part 1         *
' *                        Item 3: x value of point 1 in part 1       *
' *                        Item 4: y value of point 1 in part 1       *
' *                        Item 5: z value of point 1 in part 1       *
' *                        Item 6: m value of point 1 in part 1       *
' *                        Item 7: id value of point 1 in part 1      *
' *                        Item 8: Repeat Items 3 - 7 for each        *
' *                                point                              *
' *                        Repeat Items 2 - 8 for each part           *
' *            thePart   = the part of the polyline (0 - i)           *
' *                        part numbers begin at zero, not one        *
' *            iPt       = index in part to be processed              *
' *                        index numbers begin at one, not zero       *
```

```
' *                          if zero is specified the last point in the *
' *                          part will be processed                     *
' *            X,Y,Z     = coordinates of new point                     *
' *                                                                     *
' *  RETURN:   newList   = new list of points comprising the feature    *
' *                                                                     *
' *  Dim ipmode As Integer                                              *
' *  Dim elmntList As New Collection                                    *
' *  Dim thePart, iPt As Long                                           *
' *  Dim X, Y, Z As Double                                              *
' *  Dim newList As New Collection                                      *
' *                                                                     *
' * * * * * * * * * * * * * * * * * * * * * * * * * * * * * * * * * * * *
'
Public Sub avPlModify(ipmode, elmntList, thePart, iPt, x, y, Z, newList)
'
' * * * * * * * * * * * * * * * * * * * * * * * * * * * * * * * * * * * *
' *                                                                     *
' *  Name: avPointMake                        File Name: avpontmk.bas   *
' *                                                                     *
' *  PURPOSE:  TO CREATE A POINT OBJECT FROM COORDINATES                *
' *                                                                     *
' *  GIVEN:    xPt          = x coordinate of point                     *
' *            yPt          = y coordinate of point                     *
' *                                                                     *
' *  RETURN:   avPointMake = the point feature                          *
' *                                                                     *
' *  Dim xPt, yPt As Double                                             *
' *  Dim avPointMake As IPoint                                          *
' *                                                                     *
' * * * * * * * * * * * * * * * * * * * * * * * * * * * * * * * * * * * *
'
Public Function avPointMake(xPt, yPt) As esriCore.IPoint
'
' * * * * * * * * * * * * * * * * * * * * * * * * * * * * * * * * * * * *
' *                                                                     *
' *  Name: avPolygonMake                      File Name: avpolymk.bas   *
' *                                                                     *
' *  PURPOSE:  TO CREATE A POLYGON OBJECT FROM A POINT LIST             *
' *                                                                     *
' *  GIVEN:    shapeList     = the list of points comprising the        *
' *                            polygon                                  *
' *                            structure of shapeList is:               *
' *                            Item 1: collection of points in part 1   *
' *                            Repeat Item 1 for each part              *
' *                            So that, shapeList is a list of          *
' *                            collections with each collection         *
' *                            containing points                        *
' *                                                                     *
' *  RETURN:   avPolygonMake = the polygon feature                      *
' *                                                                     *
' *  NOTE:     If the last point in a part is not the same as the       *
' *            first point in the part, a point will be added to make   *
' *            sure that the part forms a closed polygon                *
' *                                                                     *
' *  Dim shapeList As New Collection                                    *
' *  Dim avPolygonMake As IPolygon                                      *
' *                                                                     *
' * * * * * * * * * * * * * * * * * * * * * * * * * * * * * * * * * * * *
'
Public Function avPolygonMake(shapeList) As esriCore.IPolygon
'
' * * * * * * * * * * * * * * * * * * * * * * * * * * * * * * * * * * * *
' *                                                                     *
' *  Name: avPolygonMake2                     File Name: avpolym2.bas   *
' *                                                                     *
' *  PURPOSE:  TO CREATE A POLYGON OBJECT FROM A POINT LIST             *
' *                                                                     *
' *  GIVEN:    shapeList      = the list of points comprising the       *
' *                             polygon                                 *
' *                             structure of shapeList is:              *
' *                             Item 1: number of parts                 *
' *                             Item 2: number of points in part 1      *
```

```
' *                           Item 3: x value of point 1 in part 1  *
' *                           Item 4: y value of point 1 in part 1  *
' *                           Item 5: z value of point 1 in part 1  *
' *                           Item 6: m value of point 1 in part 1  *
' *                           Item 7: id value of point 1 in part 1 *
' *                           Item 8: Repeat Items 3 - 7 for each   *
' *                                   point                         *
' *                           Repeat Items 2 - 8 for each part      *
' *                                                                 *
' *  RETURN:   avPolygonMake2 = the polygon feature                 *
' *                                                                 *
' *  NOTE:     If the last point in a part is not the same as the   *
' *            first point in the part, a point will be added to make *
' *            sure that the part forms a closed polygon            *
' *                                                                 *
' *  Dim shapeList As New Collection                                *
' *  Dim avPolygonMake2 As IPolygon                                 *
' *                                                                 *
' * * * * * * * * * * * * * * * * * * * * * * * * * * * * * * * * * *
'
Public Function avPolygonMake2(shapeList) As esriCore.IPolygon
'
' * * * * * * * * * * * * * * * * * * * * * * * * * * * * * * * * * *
' *                                                                 *
' *  Name: avPolyline2Pt                     File Name: avp2ptmk.bas *
' *                                                                 *
' *  PURPOSE:  TO CREATE A TWO-POINT POLYLINE FROM COORDINATES      *
' *                                                                 *
' *  GIVEN:    X1            = x coordinate of start point          *
' *            Y1            = y coordinate of start point          *
' *            X2            = x coordinate of end point            *
' *            Y2            = y coordinate of end point            *
' *                                                                 *
' *  RETURN:   avPolyline2Pt = the polyine feature                  *
' *                                                                 *
' *  Dim X1, Y1, X2, Y2 As Double                                   *
' *  Dim avPolyline2Pt As IPolyline                                 *
' *                                                                 *
' * * * * * * * * * * * * * * * * * * * * * * * * * * * * * * * * * *
'
Public Function avPolyline2Pt(X1, Y1, X2, Y2) As esriCore.IPolyline
'
' * * * * * * * * * * * * * * * * * * * * * * * * * * * * * * * * * *
' *                                                                 *
' *  Name: avPolylineMake                    File Name: avplinmk.bas *
' *                                                                 *
' *  PURPOSE:  TO CREATE A POLYLINE OBJECT FROM A POINT LIST        *
' *                                                                 *
' *  GIVEN:    shapeList      = the list of points comprising the   *
' *                             polyline                            *
' *                             structure of shapeList is:          *
' *                             Item 1: collection of points in part 1*
' *                             Repeat Item 1 for each part         *
' *                             So that, shapeList is a list of     *
' *                             collections with each collection    *
' *                             containing points                   *
' *                                                                 *
' *  RETURN:   avPolylineMake = the polyline feature                *
' *                                                                 *
' *  Dim shapeList As New Collection                                *
' *  Dim avPolylineMake As IPolyline                                *
' *                                                                 *
' * * * * * * * * * * * * * * * * * * * * * * * * * * * * * * * * * *
'
Public Function avPolylineMake(shapeList) As esriCore.IPolyline
'
```

```
' * * * * * * * * * * * * * * * * * * * * * * * * * * * * * * * * * * *
' *                                                                     *
' *  Name: avPolylineMake2                    File Name: avplinm2.bas  *
' *                                                                     *
' *  PURPOSE:  TO CREATE A POLYLINE OBJECT FROM A POINT LIST            *
' *                                                                     *
' *  GIVEN:    shapeList        = the list of points comprising the    *
' *                               polyline                              *
' *                               structure of shapeList is:            *
' *                               Item 1: number of parts               *
' *                               Item 2: number of points in part 1    *
' *                               Item 3: x value of point 1 in part 1 *
' *                               Item 4: y value of point 1 in part 1 *
' *                               Item 5: z value of point 1 in part 1 *
' *                               Item 6: m value of point 1 in part 1 *
' *                               Item 7: id value of point 1 in part 1*
' *                               Item 8: Repeat Items 3 - 7 for each  *
' *                                       point                         *
' *                               Repeat Items 2 - 8 for each part     *
' *                                                                     *
' *  RETURN:   avPolylineMake2 = the polyline feature                   *
' *                                                                     *
' *  Dim shapeList As New Collection                                    *
' *  Dim avPolylineMake2 As IPolyline                                   *
' *                                                                     *
' * * * * * * * * * * * * * * * * * * * * * * * * * * * * * * * * * * *
'
Public Function avPolylineMake2(shapeList) As esriCore.IPolyline
'
' * * * * * * * * * * * * * * * * * * * * * * * * * * * * * * * * * * *
' *                                                                     *
' *  Name: avQuantile                         File Name: avquntil.bas  *
' *                                                                     *
' *  PURPOSE:  TO SET THE LEGEND THAT IS ASSOCIATED WITH A THEME        *
' *           TO BE OF QUANTILE TYPE WITH THE CLASSES DETERMINED BY    *
' *            USING A QUANTILE METHOD                                  *
' *                                                                     *
' *  GIVEN:     pmxDoc     = the active view                            *
' *             theTheme   = theme to be processed                      *
' *            aField     = field name that theme is to be classified  *
' *                          upon                                       *
' *            numClass   = number of classes to be generated           *
' *                                                                     *
' *  RETURN:    nothing                                                 *
' *                                                                     *
' *  NOTE:     Divides the features in the theme into numClass          *
' *            classifications of equal size using the values in        *
' *            aField. This is only supported for numeric fields        *
' *                                                                     *
' *  Dim pmxDoc As IMxDocument                                          *
' *  Dim theTheme As Variant                                            *
' *  Dim aField As String                                               *
' *  Dim numClass As Long                                               *
' *                                                                     *
' * * * * * * * * * * * * * * * * * * * * * * * * * * * * * * * * * * *
'
Public Sub avQuantile(pmxDoc As esriCore.IMxDocument, _
                      theTheme, aField, numClass)
'
' * * * * * * * * * * * * * * * * * * * * * * * * * * * * * * * * * * *
' *                                                                     *
' *  Name: avQuantileX                        File Name: avquntlx.bas  *
' *                                                                     *
' *  PURPOSE:   TO SET THE LEGEND THAT IS ASSOCIATED WITH A THEME       *
' *             TO BE OF QUANTILE TYPE                                  *
' *                                                                     *
' *  GIVEN:     pmxDoc     = the active view                            *
' *             theTheme   = theme to be processed                      *
' *            aField     = field name that theme is to be classified  *
' *                          upon                                       *
' *            numClass   = number of classes to be generated           *
' *            method     = classification method                       *
' *                          1 : Defined Interval (not implemented)     *
```

```
' *                        2 : Equal Interval                        *
' *                        3 : Natural Breaks                        *
' *                        4 : Quantile                              *
' *                        5 : Standard Deviation (not implemented)  *
' *                                                                  *
' *  RETURN:   nothing                                               *
' *                                                                  *
' *  NOTE:     Divides the features in the theme into numClass       *
' *            classifications using the values in aField. This is   *
' *            only supported for numeric fields                     *
' *                                                                  *
' *  Dim pmxDoc As IMxDocument                                       *
' *  Dim theTheme As Variant                                         *
' *  Dim aField As String                                            *
' *  Dim numClass As Long                                            *
' *  Dim method As Long                                              *
' *                                                                  *
' * * * * * * * * * * * * * * * * * * * * * * * * * * * * * * * * * *
'
Public Sub avQuantileX(pmxDoc As esriCore.IMxDocument, _
                       theTheme, aField, numClass, method)
'
' * * * * * * * * * * * * * * * * * * * * * * * * * * * * * * * * * *
' *                                                                  *
' *  Name: avQuery                        File Name: avquery.bas     *
' *                                                                  *
' *  PURPOSE:  TO APPLY A QUERY TO A THEME OR A TABLE                *
' *                                                                  *
' *  GIVEN:    pmxDoc       = the active view                        *
' *            theTheme     = name of theme or table to be processed *
' *            aQueryString = query string to be applied             *
' *                           Sample String field query for a        *
' *                           Shapefile:                             *
' *                           aQueryStr = """PTCODE""" + " = 'BBBB'" *
' *                           for a Personal geodatabase:            *
' *                           aQueryStr = "PTCODE = 'bbbb'"          *
' *                           Sample Numeric field query for a       *
' *                           Shapefile and a Personal geodatabase   *
' *                           aQueryStr = "SLN >= 10"                *
' *            selSet       = theme selection set                    *
' *            setType      = type of selection desired              *
' *                           "NEW" : new selection set              *
' *                           "ADD" : add to selection set           *
' *                           "AND" : select from selection set      *
' *                                                                  *
' *  RETURN:   nothing                                               *
' *                                                                  *
' *  NOTE:     (a) Use avGetSelection to get the selection set that  *
' *                contains the result of this query                 *
' *            (b) The query is applied even if the theme or layer   *
' *                is set to be not selectable in ArcMap             *
' *            (c) String queries on shapefiles are case sensitive,  *
' *                while for personal geodatabases they are case     *
' *                insensitive                                       *
' *                                                                  *
' *  Dim pmxDoc As IMxDocument                                       *
' *  Dim theTheme As Variant                                         *
' *  Dim aQueryString As String                                      *
' *  Dim selSet As ISelectionSet                                     *
' *  Dim setType As String                                           *
' *                                                                  *
' * * * * * * * * * * * * * * * * * * * * * * * * * * * * * * * * * *
'
Public Sub avQuery(pmxDoc As esriCore.IMxDocument, _
                   theTheme, aQueryString, _
                   selSet As esriCore.ISelectionSet, setType)
'
```

```
'*****************************************************************
'*                                                               *
'*  Name: avRectMakeXY                   File Name: avr2ptmk.bas *
'*                                                               *
'*  PURPOSE:  TO CREATE A RECTANGULAR POLYGON FROM COORDINATES OF A *
'*            DIAGONAL OF THE POLYGON                            *
'*                                                               *
'*  GIVEN:    X1          = x coordinate of diagonal start point *
'*            Y1          = y coordinate of diagonal start point *
'*            X2          = x coordinate of diagonal end point   *
'*            Y2          = y coordinate of diagonal end point   *
'*                                                               *
'*  RETURN:   avRectMakeXY = the polygon feature                 *
'*                                                               *
'*  Dim X1, Y1, X2, Y2 As Double                                 *
'*  Dim avRectMakeXY As IPolygon                                 *
'*                                                               *
'*****************************************************************
'
Public Function avRectMakeXY(X1, Y1, X2, Y2) As esriCore.IPolygon
'
'*****************************************************************
'*                                                               *
'*  Name: avRemoveDoc                    File Name: avremdoc.bas *
'*                                                               *
'*  PURPOSE:  REMOVE THE SPECIFIED LAYER OR TABLE FROM THE TABLE OF *
'*            CONTENTS (DOES NOT DELETE ANY FILES FROM DISK)     *
'*                                                               *
'*  GIVEN:    aDocName = name of theme or table to be removed from *
'*                       the Table of Contents                   *
'*                                                               *
'*  RETURN:   nothing                                            *
'*                                                               *
'*  NOTE:     If the theme or table can not be found, the Table of *
'*            Contents is left unaltered and no error is generated *
'*                                                               *
'*  Dim aDocName As String                                       *
'*                                                               *
'*****************************************************************
'
Public Sub avRemoveDoc(aDocName)
'
'*****************************************************************
'*                                                               *
'*  Name: avRemoveDupStrings             File Name: avrmdups.bas *
'*                                                               *
'*  PURPOSE:  REMOVE DUPLICATE STRINGS OR NUMBERS FROM A COLLECTION *
'*                                                               *
'*  GIVEN:    theColl  = collection containing strings from which *
'*                       any duplicates will be removed          *
'*            caseFlag = flag denoting whether the collection is to *
'*                       be processed as case sensitive or case  *
'*                       insensitive (upper/lower case characters *
'*                       are treated the same)                   *
'*                       true = case sensitive, false = insensitive *
'*                                                               *
'*  RETURN:   nothing                                            *
'*                                                               *
'*  NOTE:     (a) theColl is changed by this subroutine          *
'*            (b) this subroutine will work for collections that *
'*                contain numbers, as well as, strings           *
'*            (c) if theColl contains numbers, not strings, set the *
'*                caseFlag to be true, if it is false an error will *
'*                be generated                                   *
'*            (d) if theColl contains numbers and strings, set the *
'*                caseFlag to be true, if it is false an error will *
'*                be generated                                   *
'*                                                               *
'*  Dim theColl As New Collection                                *
'*  Dim caseFlag As Boolean                                      *
'*                                                               *
'*****************************************************************
'
```

```
Public Sub avRemoveDupStrings(theColl, caseFlag)
'
' * * * * * * * * * * * * * * * * * * * * * * * * * * * * * * * * * * *
' *                                                                     *
' *  Name: avRemoveFields                      File Name: avremfld.bas  *
' *                                                                     *
' *  PURPOSE:  TO REMOVE FIELDS FROM A LAYER OR TABLE                   *
' *                                                                     *
' *  GIVEN:    pmxDoc        = the active view                          *
' *            theTheme      = the theme or table to be processed       *
' *            theFields     = list of fields to be removed, the        *
' *                            items in this list are index values      *
' *                            for the fields to be deleted, they       *
' *                            are numeric values not objects           *
' *                                                                     *
' *  RETURN:   avRemoveFields = error flag (0 = no error, 1 = error)    *
' *                                                                     *
' *  NOTE:     (a) In order to remove fields from a layer or table      *
' *                the editor can not be in an edit state, this         *
' *                routine will stop the editor, saving any changes     *
' *                that may have been made, prior to removing the       *
' *                fields                                               *
' *            (b) If an invalid index value appears in the list, -1,   *
' *                it will be ignored (an error is not generated)       *
' *            (c) Do not use this routine to delete the SHAPE field    *
' *                                                                     *
' *  Dim pmxDoc As IMxDocument                                          *
' *  Dim theTheme As Variant                                            *
' *  Dim theFields As New Collection                                    *
' *  Dim avRemoveFields As Integer                                      *
' *                                                                     *
' * * * * * * * * * * * * * * * * * * * * * * * * * * * * * * * * * * *
'
Public Function avRemoveFields(pmxDoc As esriCore.IMxDocument, _
                               theTheme, theFieldS)
'
' * * * * * * * * * * * * * * * * * * * * * * * * * * * * * * * * * * *
' *                                                                     *
' *  Name: avRemoveGraphic                     File Name: avremovg.bas  *
' *                                                                     *
' *  PURPOSE:  TO DELETE A GRAPHIC ELEMENT FROM THE DISPLAY             *
' *                                                                     *
' *  GIVEN:    pElement = graphic to be deleted                         *
' *                                                                     *
' *  RETURN:   nothing                                                  *
' *                                                                     *
' *  Dim pElement As IElement                                           *
' *                                                                     *
' * * * * * * * * * * * * * * * * * * * * * * * * * * * * * * * * * * *
'
Public Sub avRemoveGraphic(pElement As esriCore.IElement)
'
' * * * * * * * * * * * * * * * * * * * * * * * * * * * * * * * * * * *
' *                                                                     *
' *  Name: avRemoveRecord                      File Name: avrmrcrd.bas  *
' *                                                                     *
' *  PURPOSE:  DELETE A RECORD OR THE SELECTED FEATURES (ROWS) IN A     *
' *            LAYER OR TABLE                                           *
' *                                                                     *
' *  GIVEN:    pmxDoc   = the active view                               *
' *            theTheme = the theme or table to be processed            *
' *            theRcrd  = mode of deletion                              *
' *                       >= 0 : record of feature (row) for deletion   *
' *                        = -1 : delete selected features (rows) in a  *
' *                               theme or table, if there are no       *
' *                               selected features (rows), nothing     *
' *                               will be deleted                       *
' *                                                                     *
' *  RETURN:   nothing                                                  *
' *                                                                     *
' *  NOTE:     The theme or table must be editable prior to deleting    *
' *            any features (rows) from the theme or table              *
' *                                                                     *
```

```
' *  Dim pmxDoc As IMxDocument                                          *
' *  Dim theTheme As Variant                                            *
' *  Dim theRcrd As Long                                                *
' *                                                                     *
' * * * * * * * * * * * * * * * * * * * * * * * * * * * * * * * * * * * *
'
Public Sub avRemoveRecord(pmxDoc As esriCore.IMxDocument, _
                          theTheme, theRcrd)
'
' * * * * * * * * * * * * * * * * * * * * * * * * * * * * * * * * * * * *
' *                                                                     *
' *  Name: avResize                        File Name: avresize.bas     *
' *                                                                     *
' *  PURPOSE:  TO RESIZE A WINDOW OBJECT                                *
' *                                                                     *
' *  GIVEN:    aDoc    = the window object                              *
' *            aWidth  = the new width of the window                    *
' *            aHeight = the new height of the window                   *
' *                                                                     *
' *  RETURN:   nothing                                                  *
' *                                                                     *
' *  Dim aDoc As IUnknown                                               *
' *  Dim aWidth, aHeight As Long                                        *
' *                                                                     *
' * * * * * * * * * * * * * * * * * * * * * * * * * * * * * * * * * * * *
'
Public Sub avResize(aDoc As IUnknown, aWidth, aHeight)
'
' * * * * * * * * * * * * * * * * * * * * * * * * * * * * * * * * * * * *
' *                                                                     *
' *  Name: avReturnArea                    File Name: avrtarea.bas     *
' *                                                                     *
' *  PURPOSE:  GET THE AREA OF A GEOMETRY                               *
' *                                                                     *
' *  GIVEN:    theGeom      = the geometry to be processed              *
' *                                                                     *
' *  RETURN:   avReturnArea = the area of the geometry                  *
' *                                                                     *
' *  NOTE:     If invalid geometry is specified, avReturnArea will      *
' *            be set to zero                                           *
' *                                                                     *
' *  Dim theGeom As IGeometry                                           *
' *  Dim avReturnArea As Double                                         *
' *                                                                     *
' * * * * * * * * * * * * * * * * * * * * * * * * * * * * * * * * * * * *
'
Public Function avReturnArea(theGeom As esriCore.IGeometry) As Double
'
' * * * * * * * * * * * * * * * * * * * * * * * * * * * * * * * * * * * *
' *                                                                     *
' *  Name: avReturnCenter                  File Name: avrtcntr.bas     *
' *                                                                     *
' *  PURPOSE:  GET THE CENTROID OF A GEOMETRY                           *
' *                                                                     *
' *  GIVEN:    theGeom        = the geometry to be processed            *
' *                                                                     *
' *  RETURN:   avReturnCenter = the centroid of the geometry            *
' *                                                                     *
' *  NOTE:     If invalid geometry is specified, avReturnCenter will    *
' *            be set to NOTHING                                        *
' *                                                                     *
' *  Dim theGeom As IGeometry                                           *
' *  Dim avReturnCenter As IPoint                                       *
' *                                                                     *
' * * * * * * * * * * * * * * * * * * * * * * * * * * * * * * * * * * * *
'
Public Function avReturnCenter(theGeom As esriCore.IGeometry) _
                                                  As esriCore.IPoint
'
```

```
' * * * * * * * * * * * * * * * * * * * * * * * * * * * * * * * * * * *
' *                                                                     *
' *  Name: avReturnIntersection               File Name: avrtintr.bas   *
' *                                                                     *
' *  PURPOSE:  TO INTERSECT TWO SHAPES TO FORM A NEW SHAPE              *
' *                                                                     *
' *  GIVEN:    aShape1                = base shape                      *
' *            aShape2                = second shape to be intersected  *
' *                                     with the base shape             *
' *                                                                     *
' *  RETURN:   avReturnIntersection = new shape reflecting the          *
' *                                   intersection of the two shapes    *
' *                                                                     *
' *  NOTE:     If the shapes do not intersect an empty shape will be    *
' *            passed back                                              *
' *                                                                     *
' *  Dim aShape1 As IGeometry                                           *
' *  Dim aShape2 As IGeometry                                           *
' *  Dim avReturnIntersection As IGeometry                              *
' *                                                                     *
' * * * * * * * * * * * * * * * * * * * * * * * * * * * * * * * * * * *
'
Public Function avReturnIntersection(aShape1 As esriCore.IGeometry, _
                                     aShape2 As esriCore.IGeometry) _
                                     As esriCore.IGeometry
'
' * * * * * * * * * * * * * * * * * * * * * * * * * * * * * * * * * * *
' *                                                                     *
' *  Name: avReturnLength                     File Name: avrtlngt.bas   *
' *                                                                     *
' *  PURPOSE:  GET THE PERIMETER OR LENGTH OF A GEOMETRY                *
' *                                                                     *
' *  GIVEN:    theGeom         = the geometry to be processed           *
' *                                                                     *
' *  RETURN:   avReturnLength = the perimeter or length of the          *
' *                             geometry                                *
' *                                                                     *
' *  NOTE:     For multi-part features avReturnLength will be the       *
' *            total length, which includes all parts                   *
' *                                                                     *
' *  Dim theGeom As IGeometry                                           *
' *  Dim avReturnLength As Double                                       *
' *                                                                     *
' * * * * * * * * * * * * * * * * * * * * * * * * * * * * * * * * * * *
'
Public Function avReturnLength(theGeom As esriCore.IGeometry) As Double
'
' * * * * * * * * * * * * * * * * * * * * * * * * * * * * * * * * * * *
' *                                                                     *
' *  Name: avReturnMerged                     File Name: avrtmerg.bas   *
' *                                                                     *
' *  PURPOSE:  TO MERGE TWO SHAPES TOGETHER TO FORM A NEW SHAPE         *
' *                                                                     *
' *  GIVEN:    aShape1         = base shape                             *
' *            aShape2         = second shape to be merged with the     *
' *                              base shape                             *
' *                                                                     *
' *  RETURN:   avReturnMerged = new shape reflecting the merging        *
' *                                                                     *
' *  Dim aShape1 As IGeometry                                           *
' *  Dim aShape2 As IGeometry                                           *
' *  Dim avReturnMerged As IGeometry                                    *
' *                                                                     *
' * * * * * * * * * * * * * * * * * * * * * * * * * * * * * * * * * * *
'
Public Function avReturnMerged(aShape1 As esriCore.IGeometry, _
                              aShape2 As esriCore.IGeometry) _
                              As esriCore.IGeometry
'
```

```
' * * * * * * * * * * * * * * * * * * * * * * * * * * * * * * * * * * *
' *                                                                   *
' *  Name: avReturnUnion                     File Name: avrtunin.bas  *
' *                                                                   *
' *  PURPOSE:  TO UNION TWO SHAPES TOGETHER TO FORM A NEW SHAPE       *
' *                                                                   *
' *  GIVEN:    aShape1        = base shape                            *
' *            aShape2        = second shape to be unioned with the   *
' *                             base shape                            *
' *                                                                   *
' *  RETURN:   avReturnUnion = new shape reflecting the unioning      *
' *                                                                   *
' *  Dim aShape1 As IGeometry                                         *
' *  Dim aShape2 As IGeometry                                         *
' *  Dim avReturnUnion As IGeometry                                   *
' *                                                                   *
' * * * * * * * * * * * * * * * * * * * * * * * * * * * * * * * * * * *
'
Public Function avReturnUnion(aShape1 As esriCore.IGeometry, _
                         aShape2 As esriCore.IGeometry) _
                                  As esriCore.IGeometry
'
' * * * * * * * * * * * * * * * * * * * * * * * * * * * * * * * * * * *
' *                                                                   *
' *  Name: avReturnVisExtent                 File Name: avrtvext.bas  *
' *                                                                   *
' *  PURPOSE:  GET THE CURRENT EXTENT OF THE VIEW                     *
' *                                                                   *
' *  GIVEN:    pDT              = the screen display transformation   *
' *                                                                   *
' *  RETURN:   avReturnVisExtent = the current extent of the view     *
' *                                                                   *
' *  Dim pDT As IDisplayTransformation                                *
' *  Dim avReturnVisExtent As IEnvelope                               *
' *                                                                   *
' * * * * * * * * * * * * * * * * * * * * * * * * * * * * * * * * * * *
'
Public Function avReturnVisExtent( _
                         pDT As esriCore.IDisplayTransformation) _
                                  As esriCore.IEnvelope
'
' * * * * * * * * * * * * * * * * * * * * * * * * * * * * * * * * * * *
' *                                                                   *
' *  Name: avSelectByFTab                    File Name: avslftab.bas  *
' *                                                                   *
' *  PURPOSE:  TO SELECT FEATURES IN A THEME BASED UPON THE SELECTED  *
' *            FEATURES IN ANOTHER THEME                              *
' *                                                                   *
' *  GIVEN:    pmxDoc     = the active view                           *
' *            elmntTheme = name of theme to be processed             *
' *            seltrTheme = selector theme to be used                 *
' *            selMod     = selection mode of operation               *
' *                         "INTERSECTS"                              *
' *                         "ISWITHINDISTANCEOF"                      *
' *            selTol     = selection distance tolerance              *
' *            setType    = type of selection desired                 *
' *                         "NEW" : new selection set                 *
' *                         "ADD" : add to selection set              *
' *                         "AND" : select from selection set         *
' *                                                                   *
' *  RETURN:   nothing                                                *
' *                                                                   *
' *  NOTE:     The features that are selected by this command will be *
' *            highlighted (there is no need to use avUpdateSelection *
' *            to update the display)                                 *
' *                                                                   *
' *  Dim pmxDoc As IMxDocument                                        *
' *  Dim elmntTheme As Variant                                        *
' *  Dim seltrTheme As Variant                                        *
' *  Dim selMod As String                                             *
' *  Dim selTol As Double                                             *
' *  Dim setType As String                                            *
' *                                                                   *
```

```
' * * * * * * * * * * * * * * * * * * * * * * * * * * * * * * * * * * * * *
'
Public Sub avSelectByFTab(pmxDoc As esriCore.IMxDocument, _
                          elmntTheme, seltrTheme, _
                          selMod, selTol, setType)
'
' * * * * * * * * * * * * * * * * * * * * * * * * * * * * * * * * * * * * *
' *                                                                        *
' *  Name: avSelectByPoint                       File Name: avslpont.bas   *
' *                                                                        *
' *  PURPOSE:  TO SELECT FEATURES IN A THEME BASED UPON A POINT            *
' *                                                                        *
' *  GIVEN:    pmxDoc     = the active view                                *
' *            elmntTheme = name of theme to be processed                  *
' *            thePoint   = selector theme to be used                      *
' *            selTol     = selection distance tolerance                   *
' *            setType    = type of selection desired                      *
' *                         "NEW" : new selection set                      *
' *                         "ADD" : add to selection set                   *
' *                         "AND" : select from selection set              *
' *                                                                        *
' *  RETURN:   nothing                                                     *
' *                                                                        *
' *  NOTE:     The features that are selected by this command will be     *
' *            highlighted (there is no need to use avUpdateSelection      *
' *            to update the display)                                      *
' *                                                                        *
' *  Dim pmxDoc As IMxDocument                                             *
' *  Dim elmntTheme As Variant                                             *
' *  Dim thePoint As IPoint                                                *
' *  Dim selTol As Double                                                  *
' *  Dim setType As String                                                 *
' *                                                                        *
' * * * * * * * * * * * * * * * * * * * * * * * * * * * * * * * * * * * * *
'
Public Sub avSelectByPoint(pmxDoc As esriCore.IMxDocument, elmntTheme, _
                           thePoint As esriCore.IPoint, selTol, setType)
'
' * * * * * * * * * * * * * * * * * * * * * * * * * * * * * * * * * * * * *
' *                                                                        *
' *  Name: avSelectByPolygon                     File Name: avslpoly.bas   *
' *                                                                        *
' *  PURPOSE:  TO SELECT FEATURES IN A THEME BASED UPON A POLYGON          *
' *                                                                        *
' *  GIVEN:    pmxDoc     = the active view                                *
' *            elmntTheme = name of theme to be processed                  *
' *            theGeom    = geometry to be used                            *
' *            setType    = type of selection desired                      *
' *                         "NEW" : new selection set                      *
' *                         "ADD" : add to selection set                   *
' *                         "AND" : select from selection set              *
' *                                                                        *
' *  RETURN:   nothing                                                     *
' *                                                                        *
' *  NOTE:     The features that are selected by this command will be     *
' *            highlighted (there is no need to use avUpdateSelection      *
' *            to update the display)                                      *
' *                                                                        *
' *  Dim pmxDoc As IMxDocument                                             *
' *  Dim elmntTheme As Variant                                             *
' *  Dim theGeom As IGeometry                                              *
' *  Dim setType As String                                                 *
' *                                                                        *
' * * * * * * * * * * * * * * * * * * * * * * * * * * * * * * * * * * * * *
'
Public Sub avSelectByPolygon(pmxDoc As esriCore.IMxDocument, _
                             elmntTheme, _
                             theGeom As esriCore.IGeometry, setType)
'
```

```
' * * * * * * * * * * * * * * * * * * * * * * * * * * * * * * * * * * *
' *                                                                     *
' *  Name: avSetActive                      File Name: avsetact.bas     *
' *                                                                     *
' *  PURPOSE:  TO MAKE A THEME SELECTABLE OR NOT                        *
' *                                                                     *
' *  GIVEN:    pmxDoc   = the active view                               *
' *            theTheme = theme to be processed                         *
' *            sStatus  = selectable status (True  = selectable)        *
' *                                         (False = not selectable)    *
' *                                                                     *
' *  RETURN:   nothing                                                  *
' *                                                                     *
' *  Dim pmxDoc As IMxDocument                                          *
' *  Dim theTheme As Variant                                            *
' *  Dim sStatus As Boolean                                             *
' *                                                                     *
' * * * * * * * * * * * * * * * * * * * * * * * * * * * * * * * * * * *
'
Public Sub avSetActive(pmxDoc As esriCore.IMxDocument, theTheme, _
                       sStatus)
'
' * * * * * * * * * * * * * * * * * * * * * * * * * * * * * * * * * * *
' *                                                                     *
' *  Name: avSetAll                         File Name: avseltal.bas     *
' *                                                                     *
' *  PURPOSE:  TO SELECT ALL OF THE FEATURES OR ROWS FOR A LAYER OR     *
' *            TABLE                                                    *
' *                                                                     *
' *  GIVEN:    pmxDoc     = the active view                             *
' *            theTheme   = the theme or table to be processed          *
' *                                                                     *
' *  RETURN:   psTableSel = the selection set for the theme or table    *
' *                         with all features or rows selected          *
' *                                                                     *
' *  NOTE:     Following the call to this subroutine make a call to     *
' *            avUpdateSelection to update the display so that the      *
' *            selected features can be seen                            *
' *                                                                     *
' *  Dim pmxDoc As IMxDocument                                          *
' *  Dim theTheme As Variant                                            *
' *  Dim psTableSel As ISelectionSet                                    *
' *                                                                     *
' * * * * * * * * * * * * * * * * * * * * * * * * * * * * * * * * * * *
'
Public Sub avSetAll(pmxDoc As esriCore.IMxDocument, theTheme, _
                    psTableSel As esriCore.ISelectionSet)
'
' * * * * * * * * * * * * * * * * * * * * * * * * * * * * * * * * * * *
' *                                                                     *
' *  Name: avSetEditable                    File Name: avsetedt.bas     *
' *                                                                     *
' *  PURPOSE:  START OR TERMINATE THE EDITING ON A LAYER OR TABLE       *
' *                                                                     *
' *  GIVEN:    pmxDoc   = the active view                               *
' *            theTheme = theme or table to be processed                *
' *            eStatus  = editing status (True  = start editing)        *
' *                                      (False = stop editing)         *
' *                                                                     *
' *  RETURN:   nothing                                                  *
' *                                                                     *
' *  NOTE:     (a) For layers the editing is not terminated but         *
' *                rather any buffered writes are simply flushed.       *
' *                For tables the editing is terminated.  To            *
' *                terminate the editing on layers use the subroutine   *
' *                avStopEditing.                                       *
' *            (b) If the layer or table is to be made editable and     *
' *                the layer or table is already editable, no action    *
' *                will be taken and the layer or table will remain     *
' *                editable                                             *
' *                                                                     *
' *  Dim pmxDoc As IMxDocument                                          *
' *  Dim theTheme As Variant                                            *
```

```
' *  Dim eStatus As Boolean                                          *
' *                                                                  *
' * * * * * * * * * * * * * * * * * * * * * * * * * * * * * * * * * * *
'
Public Sub avSetEditable(pmxDoc As esriCore.IMxDocument, _
                         theTheme, eStatus)
'
' * * * * * * * * * * * * * * * * * * * * * * * * * * * * * * * * * * *
' *                                                                  *
' *  Name: avSetEditableTheme                 File Name: avsedthm.bas *
' *                                                                  *
' *  PURPOSE:  SCRIPT TO SET THE TYPE OF TASK FOR EDITING A THEME     *
' *                                                                  *
' *  GIVEN:    pmxDoc   = the active view                            *
' *            theTheme = the theme to be processed                  *
' *                       if NULL, editor will be stopped saving any *
' *                       edits that may have been made              *
' *            theType  = the type of task to be performed, if not   *
' *                       equal to zero will set the current task of *
' *                       the editor                                 *
' *                       0    = stop sketch session                 *
' *                       1    = modify feature (this will start a   *
' *                              sketch session), for polylines and  *
' *                              polygons, handles will be drawn at  *
' *                              the vertices composing the feature  *
' *                       2    = create new feature                  *
' *                       NULL = do nothing                          *
' *                                                                  *
' *  RETURN:   nothing                                               *
' *                                                                  *
' *  NOTE:     The global variable ugSketch is used to keep track of *
' *            whether a sketch session is active or not. If the     *
' *            value of ugSketch = 0, a sketch session is not active,*
' *            if ugSketch = 1, a sketch session is active           *
' *                                                                  *
' *  Dim pmxDoc As IMxDocument                                       *
' *  Dim theTheme As Variant                                         *
' *  Dim theType As Variant                                          *
' *                                                                  *
' * * * * * * * * * * * * * * * * * * * * * * * * * * * * * * * * * * *
'
Public Sub avSetEditableTheme(pmxDoc As esriCore.IMxDocument, _
                              theTheme, theType)
'
' * * * * * * * * * * * * * * * * * * * * * * * * * * * * * * * * * * *
' *                                                                  *
' *  Name: avSetExtension                     File Name: avsnmext.bas *
' *                                                                  *
' *  PURPOSE:  SET THE FILE EXTENSION IN A BASE NAME OR A PATH NAME   *
' *                                                                  *
' *  GIVEN:    aPath          = a base name or a full path name to be *
' *                             processed, the base name may or may  *
' *                             not contain an extension             *
' *            aExt           = extension to be set on the base name, *
' *                             should not contain a period, just the *
' *                             desired three character extension, if *
' *                             the base name has an extension, it   *
' *                             will be changed to aExt, if not, it  *
' *                             will be added                        *
' *                                                                  *
' *  RETURN:   avSetExtension = the new base name or full path name   *
' *                             with the specified extension applied *
' *                                                                  *
' *  Dim aPath, aExt As String                                       *
' *  Dim avSetExtension As String                                    *
' *                                                                  *
' * * * * * * * * * * * * * * * * * * * * * * * * * * * * * * * * * * *
'
Public Function avSetExtension(aPath, aExt) As String
'
```

```
' * * * * * * * * * * * * * * * * * * * * * * * * * * * * * * * * * * *
' *                                                                   *
' *  Name: avSetExtent                      File Name: avstvext.bas  *
' *                                                                   *
' *  PURPOSE:  SET THE CURRENT EXTENT OF THE VIEW                    *
' *                                                                   *
' *  GIVEN:    pActiveView = the active view                         *
' *            pDT         = the screen display transformation       *
' *            newRect     = view extent rectangle                   *
' *                                                                   *
' *  RETURN:   nothing                                               *
' *                                                                   *
' *  Dim pActiveView As IActiveView                                  *
' *  Dim pDT As IDisplayTransformation                               *
' *  Dim newRect As IEnvelope                                        *
' *                                                                   *
' * * * * * * * * * * * * * * * * * * * * * * * * * * * * * * * * * * *
'
Public Sub avSetExtent(pActiveView As esriCore.IActiveView, _
                       pDT As esriCore.IDisplayTransformation, _
                       newRect As esriCore.IEnvelope)
'
' * * * * * * * * * * * * * * * * * * * * * * * * * * * * * * * * * * *
' *                                                                   *
' *  Name: avSetGraphicsLayer               File Name: avstglyr.bas  *
' *                                                                   *
' *  PURPOSE:  TO SET THE CURRENT ANNOTATION TARGET LAYER AS THE     *
' *            BASIC GRAPHICS LAYER OR CREATE A NEW USER DEFINED     *
' *            GRAPHICS LAYER                                        *
' *                                                                   *
' *  GIVEN:    theGLayer  = graphics layer to contain the graphics   *
' *                         that are subsequently created, if NULL   *
' *                         is specified for this argument this will *
' *                         indicate that the basic graphics layer is *
' *                         to get the graphics that are subsequently *
' *                         created                                  *
' *                                                                   *
' *  RETURN:   pCurGraLyr = graphics layer that will contain the user *
' *                         programmed graphics                      *
' *                                                                   *
' *  NOTE:     If theGLayer name specified exists, it will not be    *
' *            deleted, but rather, will become the current graphics *
' *            layer, a new graphics layer will not be created thus  *
' *            any graphics generated will be added to the layer     *
' *                                                                   *
' *  Dim theGLayer As Variant                                        *
' *  Dim pCurGraLyr As IGraphicsLayer                                *
' *                                                                   *
' * * * * * * * * * * * * * * * * * * * * * * * * * * * * * * * * * * *
'
Public Sub avSetGraphicsLayer(theGLayer, _
                              pCurGraLyr As esriCore.IGraphicsLayer)
'
' * * * * * * * * * * * * * * * * * * * * * * * * * * * * * * * * * * *
' *                                                                   *
' *  Name: avSetName                        File Name: avsetnam.bas  *
' *                                                                   *
' *  PURPOSE:  TO SET THE CAPTION OF THE APPLICATION                 *
' *                                                                   *
' *  GIVEN:    aTitle  = name of the application to appear in the    *
' *                      upper left corner of the application window *
' *                                                                   *
' *  RETURN:   nothing                                               *
' *                                                                   *
' *  NOTE:     To set the name for a layer or table the user should  *
' *            use the subroutine avObjSetName                       *
' *                                                                   *
' *  Dim aTitle As String                                            *
' *                                                                   *
' * * * * * * * * * * * * * * * * * * * * * * * * * * * * * * * * * * *
'
Public Sub avSetName(aTitle)
'
```

```
' * * * * * * * * * * * * * * * * * * * * * * * * * * * * * * * * * * *
' *                                                                     *
' *  Name: avSetSelFeatures                   File Name: avstsfet.bas  *
' *                                                                     *
' *  PURPOSE:  TO SET THE SELECTED FEATURES FOR A SET OF THEMES         *
' *                                                                     *
' *  GIVEN:    pmxDoc     = the active view                             *
' *            selThmList = list of themes with selected features       *
' *            selRecList = list of selected features record numbers    *
' *                                                                     *
' *  RETURN:   nothing                                                  *
' *                                                                     *
' *  NOTE:     (a) The records selected here will be added to the       *
' *                current selected set for the theme                   *
' *            (b) Following the call to this subroutine make a call    *
' *                to avUpdateSelection to update the display so that   *
' *                the selected features can be seen                    *
' *            (c) structure of selThmList is:                          *
' *                Item 1: name of theme 1                              *
' *                Item 2: number of selected features in theme 1       *
' *                Item 3: name of theme 2                              *
' *                Item 4: number of selected features in theme 2       *
' *                Repeat Items 1 and 2 for each theme                  *
' *            (d) structure of selRecList is:                          *
' *                Item 1: selected feature 1 OID in theme 1            *
' *                Item 2: selected feature 2 OID in theme 1            *
' *                Repeat Item 1 for each selected feature in theme 1   *
' *                Item 3: selected feature 1 OID in theme 2            *
' *                Item 4: selected feature 2 OID in theme 2            *
' *                Repeat Item 3 for each selected feature in theme 2   *
' *                                                                     *
' *  Dim pmxDoc As IMxDocument                                          *
' *  Dim selThmList As New Collection                                   *
' *  Dim selRecList As New Collection                                   *
' *                                                                     *
' * * * * * * * * * * * * * * * * * * * * * * * * * * * * * * * * * * *
'
Public Sub avSetSelFeatures(pmxDoc As esriCore.IMxDocument, _
                            selThmList, selRecList)
'
' * * * * * * * * * * * * * * * * * * * * * * * * * * * * * * * * * * *
' *                                                                     *
' *  Name: avSetSelection                     File Name: avstselt.bas  *
' *                                                                     *
' *  PURPOSE:  SET THE SELECTED SET FOR A LAYER OR TABLE                *
' *                                                                     *
' *  GIVEN:    pmxDoc     = the active view                             *
' *            theTheme   = the theme or table to be processed          *
' *            psTableSel = the selection set for the theme or table    *
' *                                                                     *
' *  RETURN:   nothing                                                  *
' *                                                                     *
' *  Dim pmxDoc As IMxDocument                                          *
' *  Dim theTheme As Variant                                            *
' *  Dim psTableSel As ISelectionSet                                    *
' *                                                                     *
' * * * * * * * * * * * * * * * * * * * * * * * * * * * * * * * * * * *
'
Public Sub avSetSelection(pmxDoc As esriCore.IMxDocument, theTheme, _
                          psTableSel As esriCore.ISelectionSet)
'
' * * * * * * * * * * * * * * * * * * * * * * * * * * * * * * * * * * *
' *                                                                     *
' *  Name: avSetValueG                        File Name: avsetvlg.bas  *
' *                                                                     *
' *  PURPOSE:  TO STORE A SHAPE IN THE SHAPE FIELD OF A SPECIFIC ROW    *
' *            FOR A LAYER                                              *
' *                                                                     *
' *  GIVEN:    pmxDoc   = the active view                               *
' *            theTheme = the theme to be processed                     *
' *            aField   = the shape field                               *
' *            aRecord  = record of theme or table to be processed      *
```

```
' *           aShape   = shape to be stored (not attribute data but  *
' *                      only geometry, see note below)               *
' *                                                                   *
' *  RETURN:   nothing                                                *
' *                                                                   *
' *  NOTE:      Do not use this routine to store attribute data, use  *
' *             this routine to store geometry only.  Use the routine *
' *             avSetValue to store attribute data.                   *
' *                                                                   *
' *  Dim pmxDoc As IMxDocument                                        *
' *  Dim theTheme As Variant                                          *
' *  Dim aField, aRecord As Long                                      *
' *  Dim aShape As IGeometry                                          *
' *                                                                   *
' * * * * * * * * * * * * * * * * * * * * * * * * * * * * * * * * * * *
'
Public Sub avSetValueG(pmxDoc As esriCore.IMxDocument, theTheme, _
                       aField, aRecord, aShape As esriCore.IGeometry)
'
' * * * * * * * * * * * * * * * * * * * * * * * * * * * * * * * * * * *
' *                                                                   *
' *  Name: avSetValue                          File Name: avsetval.bas *
' *                                                                   *
' *  PURPOSE:  TO STORE A VALUE IN A SPECIFIC FIELD OF A SPECIFIC     *
' *            ROW FOR A LAYER OR TABLE                               *
' *                                                                   *
' *  GIVEN:    pmxDoc   = the active view                             *
' *            theTheme = the theme or table to be processed          *
' *            aField   = field to be written to                      *
' *            aRecord  = record of theme or table to be processed    *
' *            anObj    = object to be stored (not geometry but only  *
' *                       attribute information, see note below)      *
' *                                                                   *
' *  RETURN:   nothing                                                *
' *                                                                   *
' *  NOTE:      Do not use this routine to store geometry in the SHAPE *
' *             field, use this routine to store attribute information *
' *             only.  Use avSetValueG to store geometry in the SHAPE *
' *             field.                                                *
' *                                                                   *
' *  Dim pmxDoc As IMxDocument                                        *
' *  Dim theTheme As Variant                                          *
' *  Dim aField, aRecord As Long                                      *
' *  Dim anObj As Variant                                             *
' *                                                                   *
' * * * * * * * * * * * * * * * * * * * * * * * * * * * * * * * * * * *
'
Public Sub avSetValue(pmxDoc As esriCore.IMxDocument, theTheme, _
                      aField, aRecord, anObj)
'
' * * * * * * * * * * * * * * * * * * * * * * * * * * * * * * * * * * *
' *                                                                   *
' *  Name: avSetVisible                       File Name: avstvibl.bas *
' *                                                                   *
' *  PURPOSE:  TO SET THE VISIBILITY STATUS OF AN OBJECT              *
' *                                                                   *
' *  GIVEN:    name     = name of input object for which its          *
' *                       visibility status is to be defined          *
' *            aStatus = the visible state of the input object        *
' *                       true = visible, false = not visible         *
' *                                                                   *
' *  RETURN:   nothing                                                *
' *                                                                   *
' *  Dim name As Variant                                              *
' *  Dim aStatus As Boolean                                           *
' *                                                                   *
' * * * * * * * * * * * * * * * * * * * * * * * * * * * * * * * * * * *
'
Public Sub avSetVisible(name, aStatus)
'
```

```
' * * * * * * * * * * * * * * * * * * * * * * * * * * * * * * * * * * * *
' *                                                                       *
' *  Name: avSetWorkDir                        File Name: avstwdir.bas    *
' *                                                                       *
' *  PURPOSE:  SET THE CURRENT WORKING DIRECTORY                          *
' *                                                                       *
' *  GIVEN:    theWorkDir = the new working directory                     *
' *                                                                       *
' *  RETURN:   nothing                                                    *
' *                                                                       *
' *  Dim theWorkDir As String                                             *
' *                                                                       *
' * * * * * * * * * * * * * * * * * * * * * * * * * * * * * * * * * * * *
'
Public Sub avSetWorkDir(theWorkDir)
'
' * * * * * * * * * * * * * * * * * * * * * * * * * * * * * * * * * * * *
' *                                                                       *
' *  Name: avShowMsg                           File Name: avshwmsg.bas    *
' *                                                                       *
' *  PURPOSE:  DISPLAY A MESSAGE IN THE STATUS BAR AREA                   *
' *                                                                       *
' *  GIVEN:    aMessage = the message to be displayed                     *
' *                                                                       *
' *  RETURN:   nothing                                                    *
' *                                                                       *
' *  Dim aMessage As String                                               *
' *                                                                       *
' * * * * * * * * * * * * * * * * * * * * * * * * * * * * * * * * * * * *
'
Public Sub avShowMsg(aMessage)
'
' * * * * * * * * * * * * * * * * * * * * * * * * * * * * * * * * * * * *
' *                                                                       *
' *  Name: avShowStopButton                    File Name: avshwstb.bas    *
' *                                                                       *
' *  PURPOSE:  DISPLAY THE STOP BUTTON ON THE PROGRESS BAR                *
' *                                                                       *
' *  GIVEN:    nothing                                                    *
' *                                                                       *
' *  RETURN:   nothing                                                    *
' *                                                                       *
' *  NOTE:     Use of this command will result in the progress bar        *
' *            appearing in the middle of the display and not in the     *
' *            status bar area                                            *
' *                                                                       *
' * * * * * * * * * * * * * * * * * * * * * * * * * * * * * * * * * * * *
'
Public Sub avShowStopButton()
'
' * * * * * * * * * * * * * * * * * * * * * * * * * * * * * * * * * * * *
' *                                                                       *
' *  Name: avSingleSymbol                      File Name: avsngsym.bas    *
' *                                                                       *
' *  PURPOSE:  TO SET THE LEGEND THAT IS ASSOCIATED WITH A THEME TO      *
' *            BE OF SINGLE SYMBOL TYPE                                   *
' *                                                                       *
' *  GIVEN:    pmxDoc   = the active view                                 *
' *            theTheme = theme to be processed                           *
' *            pDesc    = renderer description                            *
' *            pLabel   = label (appears in the Table of Contents)        *
' *            pSym     = symbol used to draw every feature in theme      *
' *                                                                       *
' *  RETURN:   nothing                                                    *
' *                                                                       *
' *  NOTE:     (a) All features in the theme will be classified such      *
' *                that every feature is drawn using the same             *
' *                symbology:  color, etc.                                *
' *            (b) pDesc and pLabel can be specified as NULL and pSym     *
' *                as NOTHING if default values are to be used for        *
' *                these parameters                                       *
' *                                                                       *
' *  Dim pmxDoc As IMxDocument                                            *
```

```
' *  Dim theTheme As Variant                                       *
' *  Dim pDesc As String                                           *
' *  Dim pLabel As String                                          *
' *  Dim pSym As ISymbol                                           *
' *                                                                *
' * * * * * * * * * * * * * * * * * * * * * * * * * * * * * * * * * *
'
Public Sub avSingleSymbol(pmxDoc As esriCore.IMxDocument, _
                          theTheme, pDesc, pLabel, _
                          pSym As esriCore.ISymbol)
'
' * * * * * * * * * * * * * * * * * * * * * * * * * * * * * * * * * *
' *                                                                *
' *  Name: avSplit                          File Name: avsplit.bas *
' *                                                                *
' *  PURPOSE:  TO SPLIT A SHAPE USING A SECOND SHAPE AS THE SPLITTER *
' *                                                                *
' *  GIVEN:     aShape1    = shape to be split                     *
' *             aShape2    = shape to be used as the split line    *
' *                                                                *
' *  RETURN:    shapeList = list of new shapes created as a result of *
' *                         the splitting process                  *
' *                                                                *
' *  Dim aShape1 As IGeometry                                      *
' *  Dim aShape2 As IGeometry                                      *
' *  Dim shapeList As New Collection                               *
' *                                                                *
' * * * * * * * * * * * * * * * * * * * * * * * * * * * * * * * * * *
'
Public Sub avSplit(aShape1 As esriCore.IGeometry, _
                   aShape2 As esriCore.IGeometry, shapeList)
'
' * * * * * * * * * * * * * * * * * * * * * * * * * * * * * * * * * *
' *                                                                *
' *  Name: avStartOperation               File Name: avstropr.bas *
' *                                                                *
' *  PURPOSE:  TO START AN OPERATION WITHIN AN EDIT SESSION        *
' *                                                                *
' *  GIVEN:     nothing                                            *
' *                                                                *
' *  RETURN:    nothing                                            *
' *                                                                *
' *  NOTE:      (a) The global variable ugEditMode is used to keep *
' *                 track of if an operation is or is not in progress *
' *                 If the value of ugEditMode is 0, an operation has *
' *                 not been started, if ugEditMode is 1, an operation *
' *                 has been started. A new operation can not be   *
' *                 started if one is currently in progress        *
' *             (b) The theme or table must be editable prior to using *
' *                 this subroutine                                *
' *                                                                *
' * * * * * * * * * * * * * * * * * * * * * * * * * * * * * * * * * *
'
Public Sub avStartOperation()
'
' * * * * * * * * * * * * * * * * * * * * * * * * * * * * * * * * * *
' *                                                                *
' *  Name: avStopEditing                  File Name: avstpedt.bas *
' *                                                                *
' *  PURPOSE:  TERMINATE THE EDITING ON ALL LAYERS AND TABLES      *
' *                                                                *
' *  GIVEN:     nothing                                            *
' *                                                                *
' *  RETURN:    nothing                                            *
' *                                                                *
' *  NOTE:      (a) This command when used will empty the Undo list so *
' *                 that the user will not be able to use the Edit sub *
' *                 menu item Undo (all edits are committed to disk) *
' *             (b) If the editor is not in an edit state, an error *
' *                 message will not be generated, but rather, no  *
' *                 action will take place                         *
' *                                                                *
' *  Dim pmxDoc As IMxDocument                                     *
```

```
' *  Dim theTheme As Variant                                                *
' *                                                                         *
' * * * * * * * * * * * * * * * * * * * * * * * * * * * * * * * * * * * * * *
'
Public Sub avStopEditing()
'
' * * * * * * * * * * * * * * * * * * * * * * * * * * * * * * * * * * * * * *
' *                                                                         *
' *  Name: avStopOperation                      File Name: avstpopr.bas    *
' *                                                                         *
' *  PURPOSE:  TO STOP AN OPERATION WITHIN AN EDIT SESSION                  *
' *                                                                         *
' *  GIVEN:    oprMssg = edit operation message that will appear to        *
' *                      the right of the Undo menu item under the         *
' *                      Edit menu item                                     *
' *                                                                         *
' *  RETURN:   nothing                                                      *
' *                                                                         *
' *  NOTE:     This subroutine does not stop the editor, it will only      *
' *            terminate an operation. When the editor is stopped, it      *
' *            is not possible to use the Undo command under the Edit      *
' *            menu item, so that, if the Undo command is to be used,      *
' *            the Editor must be active (in use)                          *
' *                                                                         *
' *  Dim oprMssg As Variant                                                 *
' *                                                                         *
' * * * * * * * * * * * * * * * * * * * * * * * * * * * * * * * * * * * * * *
'
Public Sub avStopOperation(oprMssg)
'
' * * * * * * * * * * * * * * * * * * * * * * * * * * * * * * * * * * * * * *
' *                                                                         *
' *  Name: avSummarize                          File Name: avsumriz.bas    *
' *                                                                         *
' *  PURPOSE:  TO SUMMARIZE A THEME OR A TABLE ON A SPECIFIC FIELD         *
' *            THE RECORDS PROCESSED ARE THOSE THAT ARE SELECTED           *
' *                                                                         *
' *  GIVEN:    pmxDoc      = the active view                               *
' *            theTheme    = theme or table to be processed                *
' *            aFileName   = name of the output table to be created,       *
' *                          table will be stored in the workspace of      *
' *                          the theme or table that is summarized so      *
' *                          do not specify a full pathname and do         *
' *                          not include an extension such as .dbf         *
' *                          If an extension appears in the name it        *
' *                          will be removed with no error generated       *
' *            aType       = type of output table                          *
' *                          "dBase"                                        *
' *            aField      = field that theme or table summarized on       *
' *            fieldList   = list of fields to be summarized               *
' *            sumryList   = type of summary to be performed on items      *
' *                          in the fieldList (operation codes)            *
' *                          Dissolve (for use on the Shape field)         *
' *                          Count                                          *
' *                          Minimum                                        *
' *                          Maximum                                        *
' *                          Sum                                            *
' *                          Average                                        *
' *                          Variance                                       *
' *                          StdDev                                         *
' *                                                                         *
' *  RETURN:   avSummarize = list of attributes in summarized table,       *
' *                          will be set to NOTHING if an error was        *
' *                          encountered during the processing             *
' *                                                                         *
' *  NOTE:     (a) Since this routine passes avSummarize as NOTHING        *
' *                if an error is detected, make sure to check for         *
' *                this in the code that calls this function               *
' *            (b) If fieldList and sumryList are empty lists or           *
' *                passed in as NOTHING default values will be used,       *
' *                that is, Count.aField and Maximum.aField                *
```

```
' *           (c) If the table to be created exists on disk, the      *
' *               routine will overwrite the existing table without   *
' *               asking or informing the user                        *
' *           (d) If the table contains selected records, then only   *
' *               the selected records will be processed, if there    *
' *               are no selected records, then the entire table      *
' *               will be processed                                   *
' *                                                                   *
' *  Dim pmxDoc As IMxDocument                                        *
' *  Dim theTheme As Variant                                          *
' *  Dim aFileName, aType, aField As String                           *
' *  Dim fieldList, sumryList As New Collection                       *
' *  Dim avSummarize As ITable                                        *
' *                                                                   *
' * * * * * * * * * * * * * * * * * * * * * * * * * * * * * * * * * * *
'
Public Function avSummarize(pmxDoc As esriCore.IMxDocument, theTheme, _
                            aFileName, aType, aField, _
                            fieldList, sumryList) As esriCore.ITable
'
' * * * * * * * * * * * * * * * * * * * * * * * * * * * * * * * * * * *
' *                                                                   *
' *  Name: avSymbolGetAngle                 File Name: avsygang.bas   *
' *                                                                   *
' *  PURPOSE:  TO GET THE ANGLE ASSIGNED TO A GRAPHIC SYMBOL          *
' *                                                                   *
' *  GIVEN:    aSymTyp          = type of symbol to be processed      *
' *                               PEN    : line symbol                *
' *                               MARKER : point symbol               *
' *                               FILL   : polygon symbol             *
' *            pSymbol          = symbol to be processed              *
' *                                                                   *
' *  RETURN:   avSymbolGetAngle = angle assigned to symbol (degrees)  *
' *                                                                   *
' *  NOTE:     (a) This routine processes only MARKER symbols, the    *
' *                PEN and FILL symbols will result in a value of     *
' *                zero for avSymbolGetAngle                          *
' *            (b) For text symbols use avGraphicTextGetAngle to get  *
' *                the text angle                                     *
' *                                                                   *
' *  Dim aSymTyp As String                                            *
' *  Dim pSymbol As ISymbol                                           *
' *  Dim avSymbolGetAngle As Double                                   *
' *                                                                   *
' * * * * * * * * * * * * * * * * * * * * * * * * * * * * * * * * * * *
'
Public Function avSymbolGetAngle(aSymTyp, _
                                 pSymbol As esriCore.ISymbol) As Double
'
' * * * * * * * * * * * * * * * * * * * * * * * * * * * * * * * * * * *
' *                                                                   *
' *  Name: avSymbolGetColor                 File Name: avsygclr.bas   *
' *                                                                   *
' *  PURPOSE:  TO GET THE COLOR ASSIGNED TO A GRAPHIC SYMBOL          *
' *                                                                   *
' *  GIVEN:    aSymTyp          = type of symbol to be processed      *
' *                               PEN    : line symbol                *
' *                               MARKER : point symbol               *
' *                               FILL   : polygon symbol             *
' *                               TEXT   : text symbol                *
' *            pSymbol          = symbol to be processed              *
' *                                                                   *
' *  RETURN:   avSymbolGetColor = color assigned to symbol            *
' *                                                                   *
' *  NOTE:     It is possible for avSymbolGetColor to be NOTHING so   *
' *            make sure to check for this condition before using the *
' *            result (i.e. some polygon fills have no color, so that *
' *            avSymbolGetColor will be NOTHING in those instances)   *
' *                                                                   *
' *  Dim aSymTyp As String                                            *
' *  Dim pSymbol As ISymbol                                           *
' *  Dim avSymbolGetColor As iColor                                   *
' *                                                                   *
```

```
' * * * * * * * * * * * * * * * * * * * * * * * * * * * * * * * * * * * *
'
Public Function avSymbolGetColor(aSymTyp, _
                  pSymbol As esriCore.ISymbol) As esriCore.iColor
'
' * * * * * * * * * * * * * * * * * * * * * * * * * * * * * * * * * * * *
' *                                                                       *
' *  Name: avSymbolGetOLColor                     File Name: avsygolc.bas *
' *                                                                       *
' *  PURPOSE:  TO GET THE OUTLINE COLOR ASSIGNED TO A GRAPHIC SYMBOL      *
' *                                                                       *
' *  GIVEN:    aSymTyp            = type of symbol to be processed        *
' *                                 PEN    : line symbol                  *
' *                                 MARKER : point symbol                 *
' *                                 FILL   : polygon symbol               *
' *            pSymbol            = symbol to be processed                *
' *                                                                       *
' *  RETURN:   avSymbolGetOLColor = color assigned to symbol              *
' *                                                                       *
' *  NOTE:     It is possible for avSymbolGetOLColor to be NOTHING so     *
' *            make sure to check for this condition before using the     *
' *            result (i.e. some polygon fills have no color, so that     *
' *            avSymbolGetOLColor will be NOTHING in those instances)     *
' *                                                                       *
' *  Dim aSymTyp As String                                                *
' *  Dim pSymbol As ISymbol                                               *
' *  Dim avSymbolGetOLColor As iColor                                     *
' *                                                                       *
' * * * * * * * * * * * * * * * * * * * * * * * * * * * * * * * * * * * *
'
Public Function avSymbolGetOLColor(aSymTyp, _
                  pSymbol As esriCore.ISymbol) As esriCore.iColor
'
' * * * * * * * * * * * * * * * * * * * * * * * * * * * * * * * * * * * *
' *                                                                       *
' *  Name: avSymbolGetOLWidth                     File Name: avsygolw.bas *
' *                                                                       *
' *  PURPOSE:  TO GET THE OUTLINE WIDTH ASSIGNED TO A GRAPHIC SYMBOL      *
' *                                                                       *
' *  GIVEN:    aSymTyp            = type of symbol to be processed        *
' *                                 PEN    : line symbol                  *
' *                                 MARKER : point symbol                 *
' *                                 FILL   : polygon symbol               *
' *            pSymbol            = symbol to be processed                *
' *                                                                       *
' *  RETURN:   avSymbolGetOLWidth = outline width assigned to symbol      *
' *                                                                       *
' *  NOTE:     For PEN symbols the width of the symbol is assigned to     *
' *            avSymbolGetOLWidth, for MARKER symbols if the outline      *
' *            is to be drawn the outline size is returned, otherwise     *
' *            the size of the marker will be returned                    *
' *                                                                       *
' *  Dim aSymTyp As String                                                *
' *  Dim pSymbol As ISymbol                                               *
' *  Dim avSymbolGetOLWidth As Double                                     *
' *                                                                       *
' * * * * * * * * * * * * * * * * * * * * * * * * * * * * * * * * * * * *
'
Public Function avSymbolGetOLWidth(aSymTyp, _
                          pSymbol As esriCore.ISymbol) As Double
'
' * * * * * * * * * * * * * * * * * * * * * * * * * * * * * * * * * * * *
' *                                                                       *
' *  Name: avSymbolGetSize                        File Name: avsygsiz.bas *
' *                                                                       *
' *  PURPOSE:  TO GET THE SIZE ASSIGNED TO A GRAPHIC SYMBOL               *
' *                                                                       *
' *  GIVEN:    aSymTyp         = type of symbol to be processed           *
' *                              PEN    : line symbol                     *
' *                              MARKER : point symbol                    *
' *                              FILL   : polygon symbol                  *
' *                              TEXT   : text symbol                     *
' *            pSymbol         = symbol to be processed                   *
```

```
' *                                                                   *
' *  RETURN:    avSymbolGetSize = size assigned to symbol             *
' *                                                                   *
' *  NOTE:      For PEN and FILL symbols the width of the symbol is   *
' *             assigned to avSymbolGetSize                           *
' *                                                                   *
' *  Dim aSymTyp As String                                            *
' *  Dim pSymbol As ISymbol                                           *
' *  Dim avSymbolGetSize As Double                                    *
' *                                                                   *
' * * * * * * * * * * * * * * * * * * * * * * * * * * * * * * * * * * *
'
Public Function avSymbolGetSize(aSymTyp, _
                               pSymbol As esriCore.ISymbol) As Double
'
' * * * * * * * * * * * * * * * * * * * * * * * * * * * * * * * * * * *
' *                                                                   *
' *  Name: avSymbolGetStipple                 File Name: avsygstp.bas *
' *                                                                   *
' *  PURPOSE:  TO GET THE STIPPLE ASSIGNED TO A GRAPHIC SYMBOL        *
' *                                                                   *
' *  GIVEN:    aSymTyp            = type of symbol to be processed    *
' *                                 PEN    : line symbol              *
' *                                 MARKER : point symbol             *
' *                                 FILL   : polygon symbol           *
' *            pSymbol            = symbol to be processed            *
' *                                                                   *
' *  RETURN:   avSymbolGetStipple = IMultiLayerFillSymbol interface   *
' *                                 for the symbol if it is of this   *
' *                                 type, otherwise, NOTHING          *
' *                                                                   *
' *  NOTE:     Since there is no direct correlation between ArcView   *
' *            .Stipple request and an ArcObject method or property, *
' *            we will use this macro to return an object of type    *
' *            IMultiLayerFillSymbol provided the symbol is of that  *
' *            type, otherwise, NOTHING will be passed back          *
' *                                                                   *
' *  Dim aSymTyp As String                                            *
' *  Dim pSymbol As ISymbol                                           *
' *  Dim avSymbolGetStipple As IMultiLayerFillSymbol                  *
' *                                                                   *
' * * * * * * * * * * * * * * * * * * * * * * * * * * * * * * * * * * *
'
Public Function avSymbolGetStipple(aSymTyp, _
       pSymbol As esriCore.ISymbol) As esriCore.IMultiLayerFillSymbol
'
' * * * * * * * * * * * * * * * * * * * * * * * * * * * * * * * * * * *
' *                                                                   *
' *  Name: avSymbolGetStyle                   File Name: avsygsty.bas *
' *                                                                   *
' *  PURPOSE:  TO GET THE STYLE ASSIGNED TO A GRAPHIC SYMBOL          *
' *                                                                   *
' *  GIVEN:    aSymTyp          = type of symbol to be processed      *
' *                               PEN    : line symbol                *
' *                               MARKER : point symbol               *
' *                               FILL   : polygon symbol             *
' *            pSymbol          = symbol to be processed              *
' *                                                                   *
' *  RETURN:   avSymbolGetStyle = style assigned to symbol,           *
' *                               varies depending upon symbol type   *
' *                               for PEN symbols                     *
' *                               0 : Solid                           *
' *                               1 : Dashed                          *
' *                               2 : Dotted                          *
' *                               3 : dashes & dots                   *
' *                               4 : dashes & double dots            *
' *                               5 : Is invisible                    *
' *                               6 : Fit into bounding rectangle     *
' *                               for MARKER symbols                  *
' *                               0 : Circle                          *
' *                               1 : Square                          *
' *                               2 : Cross                           *
' *                               3 : X                               *
```

```
' *                              4 : Diamond                          *
' *                              for FILL symbols                     *
' *                              0 : Solid                            *
' *                              1 : Empty                            *
' *                              2 : Horizontal hatch                 *
' *                              3 : Vertical hatch                   *
' *                              4 : 45° left-to-right hatch          *
' *                              5 : 45° left-to-right hatch          *
' *                              6 : Horz. and vert. crosshatch       *
' *                              7 : 45° crosshatch                   *
' *                                                                   *
' *  NOTE:     This routine processes only ISimpleLineSymbol,         *
' *            ISimpleMarkerSymbol and ISimpleFillSymbol type symbols *
' *                                                                   *
' *  Dim aSymTyp As String                                            *
' *  Dim pSymbol As ISymbol                                           *
' *  Dim avSymbolGetStyle As Variant                                  *
' *                                                                   *
' * * * * * * * * * * * * * * * * * * * * * * * * * * * * * * * * * * *
'
Public Function avSymbolGetStyle(aSymTyp, _
                                 pSymbol As esriCore.ISymbol) As Variant
'
' * * * * * * * * * * * * * * * * * * * * * * * * * * * * * * * * * * *
' *                                                                   *
' *  Name: avSymbolMake                    File Name: avsymmak.bas    *
' *                                                                   *
' *  PURPOSE:  TO CREATE A NEW GRAPHIC SYMBOL                         *
' *                                                                   *
' *  GIVEN:    aSymTyp      = type of graphic to be created           *
' *                           PEN    : line symbol                    *
' *                           MARKER : point symbol                   *
' *                           FILL   : polygon symbol                 *
' *                                                                   *
' *  RETURN:   avSymbolMake = symbol describing a graphic that can be *
' *                           added to the graphics layer             *
' *                                                                   *
' *  NOTE:     (a) This routine will create only ISimpleLineSymbol,   *
' *                ISimpleMarkerSymbol and ISimpleFillSymbol type     *
' *                symbols                                            *
' *            (b) For text symbols use MakeTextSymbol to create a    *
' *                text symbol or avGraphicTextMake to create a text  *
' *                element                                            *
' *                                                                   *
' *  Dim aSymTyp As String                                            *
' *  Dim avSymbolMake As ISymbol                                      *
' *                                                                   *
' * * * * * * * * * * * * * * * * * * * * * * * * * * * * * * * * * * *
'
Public Function avSymbolMake(aSymTyp) As esriCore.ISymbol
'
' * * * * * * * * * * * * * * * * * * * * * * * * * * * * * * * * * * *
' *                                                                   *
' *  Name: avSymbolSetAngle                File Name: avsymang.bas    *
' *                                                                   *
' *  PURPOSE:  TO SET THE ANGLE OF A SYMBOL                           *
' *                                                                   *
' *  GIVEN:    aSymTyp = type of symbol to be processed               *
' *                      PEN    : line symbol                         *
' *                      MARKER : point symbol                        *
' *                      FILL   : polygon symbol                      *
' *            pSymbol = symbol to be processed                       *
' *            aAngle  = angle to be assigned (degrees)               *
' *                                                                   *
' *  RETURN:   nothing                                                *
' *                                                                   *
' *  NOTE:     For text symbols use avGraphicTextSetAngle to define   *
' *            the text angle                                         *
' *                                                                   *
' *  Dim aSymTyp As String                                            *
' *  Dim pSymbol As ISymbol                                           *
' *  Dim aAngle As Variant                                            *
' *                                                                   *
```

```
' * * * * * * * * * * * * * * * * * * * * * * * * * * * * * * * * * * *
'
Public Sub avSymbolSetAngle(aSymTyp, _
                            pSymbol As esriCore.ISymbol, aAngle)
'
' * * * * * * * * * * * * * * * * * * * * * * * * * * * * * * * * * * *
' *                                                                   *
' *  Name: avSymbolSetColor                   File Name: avsymclr.bas *
' *                                                                   *
' *  PURPOSE:  TO SET THE COLOR FOR A SYMBOL                          *
' *                                                                   *
' *  GIVEN:    aSymTyp = type of symbol to be processed               *
' *                      PEN    : line symbol                         *
' *                      MARKER : point symbol                        *
' *                      FILL   : polygon symbol                      *
' *                      TEXT   : text symbol                         *
' *            pSymbol = symbol to be processed                       *
' *            aColor  = color to be assigned, if numeric will refer  *
' *                      to a RGB color index value, otherwise one of *
' *                      the predefined values listed below           *
' *                      BLUE                                         *
' *                      RED                                          *
' *                      GREEN                                        *
' *                      YELLOW                                       *
' *                      MAGENTA                                      *
' *                      BLACK                                        *
' *                      BROWN                                        *
' *                      CYAN                                         *
' *                      GRAY                                         *
' *                      ORANGE                                       *
' *                      WHITE                                        *
' *                                                                   *
' *  RETURN:   nothing                                                *
' *                                                                   *
' *  Dim aSymTyp As String                                            *
' *  Dim pSymbol As ISymbol                                           *
' *  Dim aColor As Variant                                            *
' *                                                                   *
' * * * * * * * * * * * * * * * * * * * * * * * * * * * * * * * * * * *
'
Public Sub avSymbolSetColor(aSymTyp, _
                            pSymbol As esriCore.ISymbol, aColor)
'
' * * * * * * * * * * * * * * * * * * * * * * * * * * * * * * * * * * *
' *                                                                   *
' *  Name: avSymbolSetOLColor                 File Name: avsymolc.bas *
' *                                                                   *
' *  PURPOSE:  TO SET THE OUTLINE COLOR FOR A SYMBOL                  *
' *                                                                   *
' *  GIVEN:    aSymTyp = type of symbol to be processed               *
' *                      PEN    : line symbol                         *
' *                      MARKER : point symbol                        *
' *                      FILL   : polygon symbol                      *
' *            pSymbol = symbol to be processed                       *
' *            aColor  = color to be assigned, if numeric will refer  *
' *                      to a RGB color index value, otherwise one of *
' *                      the predefined values listed below           *
' *                      BLUE                                         *
' *                      RED                                          *
' *                      GREEN                                        *
' *                      YELLOW                                       *
' *                      MAGENTA                                      *
' *                      BLACK                                        *
' *                      BROWN                                        *
' *                      CYAN                                         *
' *                      GRAY                                         *
' *                      ORANGE                                       *
' *                      WHITE                                        *
' *                                                                   *
' *  RETURN:   nothing                                                *
' *                                                                   *
' *  NOTE:     For PEN symbols the color of the symbol is set to the  *
' *            value of aColor, for MARKER symbols the outline        *
```

```
' *            property is set to be true, denoting that the outline  *
' *            for the marker is to be drawn, and the outline color   *
' *            is set to the value of aColor                          *
' *                                                                   *
' *  Dim aSymTyp As String                                            *
' *  Dim pSymbol As ISymbol                                           *
' *  Dim aColor As Variant                                            *
' *                                                                   *
' * * * * * * * * * * * * * * * * * * * * * * * * * * * * * * * * * * *
'
Public Sub avSymbolSetOLColor(aSymTyp, _
                             pSymbol As esriCore.ISymbol, aColor)
'
' * * * * * * * * * * * * * * * * * * * * * * * * * * * * * * * * * * *
' *                                                                   *
' *  Name: avSymbolSetOLWidth                File Name: avsymolw.bas  *
' *                                                                   *
' *  PURPOSE:  TO SET THE OUTLINE WIDTH FOR A SYMBOL                  *
' *                                                                   *
' *  GIVEN:    aSymTyp = type of symbol to be processed               *
' *                      PEN    : line symbol                         *
' *                      MARKER : point symbol                        *
' *                      FILL   : polygon symbol                      *
' *            pSymbol = symbol to be processed                       *
' *            aWidth  = outline width to be assigned, a value of     *
' *                      zero denotes no outline is to be drawn       *
' *                                                                   *
' *  RETURN:   nothing                                                *
' *                                                                   *
' *  NOTE:     For PEN symbols the width of the symbol is set to the  *
' *            value of aWidth, for MARKER symbols the outline        *
' *            property is set to be true, denoting that the outline  *
' *            for the marker is to be drawn, and the outline width   *
' *            is set to the value of aWidth                          *
' *                                                                   *
' *  Dim aSymTyp As String                                            *
' *  Dim pSymbol As ISymbol                                           *
' *  Dim aWidth As Variant                                            *
' *                                                                   *
' * * * * * * * * * * * * * * * * * * * * * * * * * * * * * * * * * * *
'
Public Sub avSymbolSetOLWidth(aSymTyp, _
                             pSymbol As esriCore.ISymbol, aWidth)
'
' * * * * * * * * * * * * * * * * * * * * * * * * * * * * * * * * * * *
' *                                                                   *
' *  Name: avSymbolSetSize                   File Name: avsymsiz.bas  *
' *                                                                   *
' *  PURPOSE:  TO SET THE SIZE OF A SYMBOL                            *
' *                                                                   *
' *  GIVEN:    aSymTyp = type of symbol to be processed               *
' *                      PEN    : line symbol                         *
' *                      MARKER : point symbol                        *
' *                      FILL   : polygon symbol                      *
' *                      TEXT   : text symbol                         *
' *            pSymbol = symbol to be processed                       *
' *            aSize   = size to be assigned (greater than zero)      *
' *                                                                   *
' *  RETURN:   nothing                                                *
' *                                                                   *
' *  NOTE:     For PEN and FILL symbols this routine works the same   *
' *            as avSymbolSetOLWidth                                  *
' *                                                                   *
' *  Dim aSymTyp As String                                            *
' *  Dim pSymbol As ISymbol                                           *
' *  Dim aSize As Variant                                             *
' *                                                                   *
' * * * * * * * * * * * * * * * * * * * * * * * * * * * * * * * * * * *
'
Public Sub avSymbolSetSize(aSymTyp, pSymbol As esriCore.ISymbol, aSize)
'
```

```
' * * * * * * * * * * * * * * * * * * * * * * * * * * * * * * * * * * *
' *                                                                     *
' *  Name: avSymbolSetStipple                 File Name: avsymstp.bas  *
' *                                                                     *
' *  PURPOSE:  TO SET THE STIPPLE OF A SYMBOL                           *
' *                                                                     *
' *  GIVEN:     aSymTyp  = type of symbol to be processed               *
' *                        PEN    : line symbol                         *
' *                        MARKER : point symbol                        *
' *                        FILL   : polygon symbol                      *
' *             pSymbol  = symbol to be processed                       *
' *             aStipple = stipple to be assigned                       *
' *                                                                     *
' *  RETURN:    nothing                                                 *
' *                                                                     *
' *  NOTE:      Since there is no direct correlation between ArcView    *
' *             .Stipple request and an ArcObject method or property,   *
' *             we will use this macro to allow the user to change a    *
' *             ISymbol object into a IMultiLayerFillSymbol provided    *
' *             a valid IMultiLayerFillSymbol object is given           *
' *                                                                     *
' *  Dim aSymTyp As String                                              *
' *  Dim pSymbol As ISymbol                                             *
' *  Dim aStipple As IMultiLayerFillSymbol                              *
' *                                                                     *
' * * * * * * * * * * * * * * * * * * * * * * * * * * * * * * * * * * *
'
Public Sub avSymbolSetStipple(aSymTyp, pSymbol As esriCore.ISymbol, _
                        aStipple As esriCore.IMultiLayerFillSymbol)
'
' * * * * * * * * * * * * * * * * * * * * * * * * * * * * * * * * * * *
' *                                                                     *
' *  Name: avSymbolSetStyle                   File Name: avsymsty.bas  *
' *                                                                     *
' *  PURPOSE:  TO SET THE STYLE FOR A SYMBOL                            *
' *                                                                     *
' *  GIVEN:     aSymTyp = type of symbol to be processed                *
' *                       PEN    : line symbol                          *
' *                       MARKER : point symbol                         *
' *                       FILL   : polygon symbol                       *
' *             pSymbol = symbol to be processed                        *
' *             aStyle  = style to be assigned, varies depending upon   *
' *                       the type of symbol                            *
' *                       for PEN symbols                               *
' *                       0 : Solid                                     *
' *                       1 : Dashed                                    *
' *                       2 : Dotted                                    *
' *                       3 : Has alternating dashes and dots           *
' *                       4 : Has alternating dashes and double dots    *
' *                       5 : Is invisible                              *
' *                       6 : Will fit into it's bounding rectangle     *
' *                       for MARKER symbols                            *
' *                       0 : Circle                                    *
' *                       1 : Square                                    *
' *                       2 : Cross                                     *
' *                       3 : X                                         *
' *                       4 : Diamond                                   *
' *                       for FILL symbols                              *
' *                       0 : Solid                                     *
' *                       1 : Empty                                     *
' *                       2 : Horizontal hatch                          *
' *                       3 : Vertical hatch                            *
' *                       4 : 45-degree downward, left-to-right hatch   *
' *                       5 : 45-degree upward, left-to-right hatch     *
' *                       6 : Horizontal and vertical crosshatch        *
' *                       7 : 45-degree crosshatch                      *
' *                                                                     *
' *  RETURN:    nothing                                                 *
' *                                                                     *
' *  Dim aSymTyp As String                                              *
' *  Dim pSymbol As ISymbol                                             *
' *  Dim aStyle As Variant                                              *
' *                                                                     *
```

```
' * * * * * * * * * * * * * * * * * * * * * * * * * * * * * * * * * * * *
'
Public Sub avSymbolSetStyle(aSymTyp, _
                            pSymbol As esriCore.ISymbol, aStyle)
'
' * * * * * * * * * * * * * * * * * * * * * * * * * * * * * * * * * * * *
' *                                                                     *
' *  Name: avThemeInvalidate                    File Name: avthminv.bas  *
' *                                                                     *
' *  PURPOSE:  REDRAW A THEME                                           *
' *                                                                     *
' *  GIVEN:    pmxDoc   = the active view                               *
' *            theTheme = theme to be processed                         *
' *            rdStatus = redraw status (True = redraw entire view)     *
' *                                     (False = redraw theme only)     *
' *                                                                     *
' *  RETURN:   nothing                                                  *
' *                                                                     *
' *  NOTE:     If False is specified for rdStatus it may be necessary *
' *            to follow the call to avThemeInvalidate with a call to *
' *            pActiveView.Refresh or avGetDisplayFlush in order to    *
' *            refresh the display so that the changes made to the     *
' *            theme are properly displayed, these calls are not made *
' *            here thereby eliminating multiple screen redraws        *
' *                                                                     *
' *  Dim pmxDoc As IMxDocument                                          *
' *  Dim theTheme As Variant                                            *
' *  Dim rdStatus As Boolean                                            *
' *                                                                     *
' * * * * * * * * * * * * * * * * * * * * * * * * * * * * * * * * * * * *
'
Public Sub avThemeInvalidate(pmxDoc As esriCore.IMxDocument, _
                             theTheme, rdStatus)
'
' * * * * * * * * * * * * * * * * * * * * * * * * * * * * * * * * * * * *
' *                                                                     *
' *  Name: avUnique                             File Name: avunique.bas  *
' *                                                                     *
' *  PURPOSE:  TO SET THE LEGEND THAT IS ASSOCIATED WITH A THEME        *
' *            TO BE OF UNIQUE TYPE                                     *
' *                                                                     *
' *  GIVEN:    pmxDoc     = the active view                             *
' *            theTheme   = theme to be processed                       *
' *            aField     = field name that theme is to be classified   *
' *                         upon                                        *
' *            showNulls = flag denoting whether features that have     *
' *                         not been assigned a value for aField        *
' *                         should be drawn or not (true, false)        *
' *                                                                     *
' *  RETURN:   nothing                                                  *
' *                                                                     *
' *  NOTE:     All features in the theme will be classified such that *
' *            features having a unique value, within a field, will    *
' *            be drawn in a unique or different symbol from the       *
' *            other unique values within the field                    *
' *                                                                     *
' *  Dim pmxDoc As IMxDocument                                          *
' *  Dim theTheme As Variant                                            *
' *  Dim aField As String                                               *
' *  Dim showNulls As Boolean                                           *
' *                                                                     *
' * * * * * * * * * * * * * * * * * * * * * * * * * * * * * * * * * * * *
'
Public Sub avUnique(pmxDoc As esriCore.IMxDocument, _
                    theTheme, aField, showNulls)
'
```

```
' * * * * * * * * * * * * * * * * * * * * * * * * * * * * * * * * * *
' *                                                                  *
' *  Name: avUpdateAnno                     File Name: avupdann.bas  *
' *                                                                  *
' *  PURPOSE:  TO TRANSFORM AN EXISTING ANNOTATION FEATURE           *
' *                                                                  *
' *  GIVEN:    pFeature   = annotation feature to be processed       *
' *            oldX       = x coordinate of feature control point    *
' *            oldY       = y coordinate of feature control point    *
' *            newX       = x coordinate control point to be moved to *
' *            newY       = y coordinate control point to be moved to *
' *            rotang     = rotation angle to be applied (degrees)   *
' *                          0.0 : do not rotate feature             *
' *                          <>0 : add rotation angle to feature     *
' *            scaleX     = X scale factor (can not be <= 0.0)       *
' *            scaleY     = Y scale factor (can not be <= 0.0)       *
' *                                                                  *
' *  RETURN:   newFeature = feature after transformation applied     *
' *                                                                  *
' *  NOTE:     (a) The rotation angle is added to the existing angle *
' *                of the annotation (positive value denotes counter- *
' *                clockwise rotation, negative value clockwise)     *
' *            (b) Scale factor greater than 1.0 increases the size, *
' *                while a value less than 1.0 decreases the size    *
' *            (c) The X scale factor is always used in the scaling  *
' *                process, the Scale method does not seem to work as *
' *                it should on Annotation features when the X and Y *
' *                scale factors are different                       *
' *            (d) The layer that the feature resides must be in an  *
' *                editable state                                    *
' *                                                                  *
' *  Dim pFeature As IFeature                                        *
' *  Dim oldX, oldY, newX, newY, rotang, scaleX, scaleY As Double    *
' *  Dim newFeature As IFeature                                      *
' *                                                                  *
' * * * * * * * * * * * * * * * * * * * * * * * * * * * * * * * * * *
'
Public Sub avUpdateAnno(pFeature As esriCore.IFeature, _
                        oldX, oldY, newX, newY, rotang, _
                        scaleX, scaleY, _
                        newFeature As esriCore.IFeature)
'
' * * * * * * * * * * * * * * * * * * * * * * * * * * * * * * * * * *
' *                                                                  *
' *  Name: avUpdateLegend                   File Name: avuplgnd.bas  *
' *                                                                  *
' *  PURPOSE:  TO UPDATE A THEME TO REFLECT ANY CHANGES MADE TO ITS  *
' *            LEGEND, BOTH THE THEME AND THE TABLE OF CONTENTS WILL *
' *            BE UPDATED (REDRAWN)                                  *
' *                                                                  *
' *  GIVEN:    pmxDoc   = the active view                            *
' *            theTheme = theme to be processed                      *
' *                                                                  *
' *  RETURN:   nothing                                               *
' *                                                                  *
' *  NOTE:     It may be necessary to follow the call to the routine *
' *            avUpdateLegend with a call to avDisplayInvalidate or  *
' *            avGetDisplayFlush in order to refresh the display so  *
' *            that the changes made to the theme are properly       *
' *            displayed, these calls are not made here thereby      *
' *            eliminating multiple screen redraws                   *
' *                                                                  *
' *  Dim pmxDoc As IMxDocument                                       *
' *  Dim theTheme As Variant                                         *
' *                                                                  *
' * * * * * * * * * * * * * * * * * * * * * * * * * * * * * * * * * *
'
Public Sub avUpdateLegend(pmxDoc As esriCore.IMxDocument, theTheme)
'
```

```
' * * * * * * * * * * * * * * * * * * * * * * * * * * * * * * * * * * *
' *                                                                     *
' *  Name: avUpdateSelection                  File Name: avupselt.bas  *
' *                                                                     *
' *  PURPOSE:  TO UPDATE THE ATTRIBUTE TABLE FOR A THEME TO REFLECT     *
' *            THE CURRENT SELECTION SET FOR THE THEME                  *
' *                                                                     *
' *  GIVEN:    pmxDoc   = the active view                               *
' *            theTheme = theme to be processed                         *
' *                                                                     *
' *  RETURN:   nothing                                                  *
' *                                                                     *
' *  Dim pmxDoc As IMxDocument                                          *
' *  Dim theTheme As Variant                                            *
' *                                                                     *
' * * * * * * * * * * * * * * * * * * * * * * * * * * * * * * * * * * *
'
Public Sub avUpdateSelection(pmxDoc As esriCore.IMxDocument, theTheme)
'
' * * * * * * * * * * * * * * * * * * * * * * * * * * * * * * * * * * *
' *                                                                     *
' *  Name: avVTabMakeNew                      File Name: avvtabmk.bas  *
' *                                                                     *
' *  PURPOSE:  CREATE A NEW TABLE THAT IS OF dBASE OR TEXT FILE TYPE   *
' *                                                                     *
' *  GIVEN:    aFileName     = name of the table to be created,         *
' *                            if the name does not contain a           *
' *                            complete pathname the current working    *
' *                            directory will be used, some examples    *
' *                            of name include:                         *
' *                               c:\project\test\atable                *
' *                               c:\project\test\atable.dbf            *
' *                               atable                                *
' *                               atable.dbf                            *
' *                            the name can or can not contain the      *
' *                            extension .dbf or .txt                   *
' *            aClass        = type of table to be created              *
' *                            dBase                                    *
' *                            TEXT                                     *
' *                                                                     *
' *  RETURN:   avVTabMakeNew = table object that is created             *
' *                                                                     *
' *  NOTE:     (a) Two fields called OID and ID will be created by      *
' *                this routine, the function avAddDoc can be used to  *
' *                add the table to the map, if need be                 *
' *            (b) If the table to be created exists on disk, the       *
' *                routine will abort the existing table will not be    *
' *                overwritten                                          *
' *                                                                     *
' *  Dim aFileName, aClass As String                                    *
' *  Dim avVTabMakeNew As ITable                                        *
' *                                                                     *
' * * * * * * * * * * * * * * * * * * * * * * * * * * * * * * * * * * *
'
Public Function avVTabMakeNew(aFileName, aclass) As esriCore.ITable
'
' * * * * * * * * * * * * * * * * * * * * * * * * * * * * * * * * * * *
' *                                                                     *
' *  Name: avViewAddGraphic                   File Name: avvwagra.bas  *
' *                                                                     *
' *  PURPOSE:  TO ADD A GRAPHIC INTO THE ACTIVE GRAPHICS LAYER          *
' *                                                                     *
' *  GIVEN:    pElement = graphic to be added                           *
' *                                                                     *
' *  RETURN:   nothing                                                  *
' *                                                                     *
' *  NOTE:     Use the subroutine avSetGraphicsLayer to set the         *
' *            active graphics layer (annotation target layer)          *
' *                                                                     *
' *  Dim pElement As IElement                                           *
' *                                                                     *
' * * * * * * * * * * * * * * * * * * * * * * * * * * * * * * * * * * *
'
```

```
Public Sub avViewAddGraphic(pElement As esriCore.IElement)
'
' * * * * * * * * * * * * * * * * * * * * * * * * * * * * * * * * * *
' *                                                                  *
' *  Name: avViewGetGraphics                File Name: avvgtgra.bas  *
' *                                                                  *
' *  PURPOSE:  TO GET A LIST OF ALL OF THE GRAPHICS IN THE MAP       *
' *                                                                  *
' *  GIVEN:    nothing                                               *
' *                                                                  *
' *  RETURN:   graList = list of all graphic elements in the map     *
' *                                                                  *
' *  Dim graList As New Collection                                   *
' *                                                                  *
' * * * * * * * * * * * * * * * * * * * * * * * * * * * * * * * * * *
'
Public Sub avViewGetGraphics(graList)
'
' * * * * * * * * * * * * * * * * * * * * * * * * * * * * * * * * * *
' *                                                                  *
' *  Name: avXORSelection                   File Name: avxorsel.bas  *
' *                                                                  *
' *  PURPOSE:  PERFORM AN EXCLUSIVE XOR ON A LAYER OR TABLE SELECTION *
' *            SET                                                   *
' *                                                                  *
' *  GIVEN:    pmxDoc    = the active view                           *
' *            theTheme  = the theme or table to be processed        *
' *            orgSelSet = the selection set to be XOR               *
' *                                                                  *
' *  RETURN:   xorSelSet = the selection set after XOR was performed *
' *                                                                  *
' *  NOTE:     The current selection set for the theme or table is   *
' *            used in the XORing with the set that is passed in     *
' *            (orgSelSet). If theTheme has no selection set, the    *
' *            command will select all features (rows) in theTheme   *
' *            and use this set in the XOR process                   *
' *                                                                  *
' *  Dim pmxDoc As IMxDocument                                       *
' *  Dim theTheme As Variant                                         *
' *  Dim orgSelSet As ISelectionSet                                  *
' *  Dim xorSelSet As ISelectionSet                                  *
' *                                                                  *
' * * * * * * * * * * * * * * * * * * * * * * * * * * * * * * * * * *
'
Public Sub avXORSelection(pmxDoc As esriCore.IMxDocument, theTheme, _
                          orgSelSet As esriCore.ISelectionSet, _
                          xorSelSet As esriCore.ISelectionSet)
'
' * * * * * * * * * * * * * * * * * * * * * * * * * * * * * * * * * *
' *                                                                  *
' *  Name: avZoomToSelected                 File Name: avzm2sel.bas  *
' *                                                                  *
' *  PURPOSE:  TO ZOOM TO THE EXTENT OF THE SELECTED SET FOR A THEME *
' *            OR THE EXTENT OF ALL SELECTED FEATURES IN THE MAP     *
' *                                                                  *
' *  GIVEN:    pmxDoc   = the active view                            *
' *            theTheme = theme for which its selected features will *
' *                       be zoomed to, if NULL is specified all     *
' *                       selected features will be zoomed to        *
' *                                                                  *
' *  RETURN:   nothing                                               *
' *                                                                  *
' *  NOTE:     If a theme is specified and the theme does not contain *
' *            any selected features, the command will zoom to the   *
' *            full extent of the theme (all features processed in   *
' *            this condition)                                       *
' *                                                                  *
' *  Dim pmxDoc As IMxDocument                                       *
' *  Dim theTheme As Variant                                         *
' *                                                                  *
' * * * * * * * * * * * * * * * * * * * * * * * * * * * * * * * * * *
'
Public Sub avZoomToSelected(pmxDoc As esriCore.IMxDocument, theTheme)
```

```
'
' * * * * * * * * * * * * * * * * * * * * * * * * * * * * * * * * * * *
' *                                                                     *
' *  Name: avZoomToTheme                     File Name: avzm2thm.bas    *
' *                                                                     *
' *  PURPOSE:  TO ZOOM TO THE EXTENT OF A THEME                         *
' *                                                                     *
' *  GIVEN:    pmxDoc   = the active view                               *
' *            theTheme = theme to be processed                         *
' *                                                                     *
' *  RETURN:   nothing                                                  *
' *                                                                     *
' *  Dim pmxDoc As IMxDocument                                          *
' *  Dim theTheme As Variant                                            *
' *                                                                     *
' * * * * * * * * * * * * * * * * * * * * * * * * * * * * * * * * * * *
'
Public Sub avZoomToTheme(pmxDoc As esriCore.IMxDocument, theTheme)
'
' * * * * * * * * * * * * * * * * * * * * * * * * * * * * * * * * * * *
' *                                                                     *
' *  Name: avZoomToThemes                    File Name: avzm2ths.bas    *
' *                                                                     *
' *  PURPOSE:  TO ZOOM TO THE EXTENT OF A GROUP OF THEMES               *
' *                                                                     *
' *  GIVEN:    pmxDoc  = the active view                                *
' *            thmList = list of themes to be processed                 *
' *                                                                     *
' *  RETURN:   nothing                                                  *
' *                                                                     *
' *  Dim pmxDoc As IMxDocument                                          *
' *  Dim thmList As New Collection                                      *
' *                                                                     *
' * * * * * * * * * * * * * * * * * * * * * * * * * * * * * * * * * * *
'
Public Sub avZoomToThemes(pmxDoc As esriCore.IMxDocument, thmList)
'
' * * * * * * * * * * * * * * * * * * * * * * * * * * * * * * * * * * *
' *                                                                     *
' *  Name: Calc_Callback               File Name: Calc_Callback.cls     *
' *                                                                     *
' *  PURPOSE:  Initialize, report on, and terminate the reporting of    *
' *            progress bar for the Calculate method                    *
' *                                                                     *
' *  GIVEN:    nothing                                                  *
' *                                                                     *
' *  RETURN:   nothing                                                  *
' *                                                                     *
' * * * * * * * * * * * * * * * * * * * * * * * * * * * * * * * * * * *
'
Private Sub Class_Initialize()
'
' * * * * * * * * * * * * * * * * * * * * * * * * * * * * * * * * * * *
' *                                                                     *
' *  Name: ChangeView                        File Name: chngview.bas    *
' *                                                                     *
' *  PURPOSE:  SCRIPT TO ALTER THE DISPLAY OF THE VIEW                  *
' *                                                                     *
' *  GIVEN:    pmxDoc  = the active view                                *
' *            opmode  = mode of operation                              *
' *                        1 : zoom scale factor to be applied          *
' *                        2 : panning values to be applied             *
' *                        3 : a new extent to be defined               *
' *                        4 : center display about a point             *
' *            sclFctr = scale factor to be applied to view             *
' *            panXval = distance in world units to pan along x axis    *
' *                      or display center x coordinate if opmode = 4   *
' *            panYval = distance in world units to pan along y axis    *
' *                      or display center y coordinate if opmode = 4   *
' *            usrView = user-defined view extent rectangle, can be     *
' *                      either an IPolygon or IEnvelope object         *
' *                                                                     *
```

```
' *  RETURN:   iok     = error flag (0 = no error, 1 = error)          *
' *            newRect = view extent rectangle                          *
' *                                                                     *
' *  Dim pmxDoc As IMxDocument                                          *
' *  Dim opmode As Integer                                              *
' *  Dim sclFctr, panXval, panYval As Double                            *
' *  Dim usrView As IUnknown                                            *
' *  Dim iok As Integer                                                 *
' *  Dim newRect As IEnvelope                                           *
' *                                                                     *
' * * * * * * * * * * * * * * * * * * * * * * * * * * * * * * * * * * * *
'
Public Sub ChangeView(pmxDoc As esriCore.IMxDocument, opmode, _
                     sclFctr, panXval, panYval, usrView As IUnknown, _
                     iok, newRect As esriCore.IEnvelope)
'
' * * * * * * * * * * * * * * * * * * * * * * * * * * * * * * * * * * * *
' *                                                                     *
' *  Name: CopyList                          File Name: copylist.bas   *
' *                                                                     *
' *  PURPOSE:  COPY A COLLECTION INTO ANOTHER COLLECTION AND THEN      *
' *            INITIALIZE OR CLEAR THE ORIGINAL COLLECTION              *
' *                                                                     *
' *  GIVEN:    origList = list to be copied and then cleared           *
' *                                                                     *
' *  RETURN:   newList  = copy of the original list                    *
' *                                                                     *
' *  Dim origList As New Collection                                     *
' *  Dim newList As New Collection                                      *
' *                                                                     *
' * * * * * * * * * * * * * * * * * * * * * * * * * * * * * * * * * * * *
'
Public Sub CopyList(origList, newList)
'
' * * * * * * * * * * * * * * * * * * * * * * * * * * * * * * * * * * * *
' *                                                                     *
' *  Name: CopyList2                         File Name: copylst2.bas   *
' *                                                                     *
' *  PURPOSE:  COPY A COLLECTION OF OBJECTS INTO ANOTHER COLLECTION    *
' *            AND THEN INITIALIZE OR CLEAR THE ORIGINAL COLLECTION    *
' *                                                                     *
' *  GIVEN:    origList = list to be copied and then cleared           *
' *                                                                     *
' *  RETURN:   newList  = copy of the original list                    *
' *                                                                     *
' *  NOTE:     Objects are not variables (i.e. doubles, longs, etc.)   *
' *                                                                     *
' *  Dim origList As New Collection                                     *
' *  Dim newList As New Collection                                      *
' *                                                                     *
' * * * * * * * * * * * * * * * * * * * * * * * * * * * * * * * * * * * *
'
Public Sub CopyList2(origList, newList)
'
' * * * * * * * * * * * * * * * * * * * * * * * * * * * * * * * * * * * *
' *                                                                     *
' *  Name: CreateAccessDB                    File Name: creaccdb.bas   *
' *                                                                     *
' *  PURPOSE:  CREATE A PERSONAL GEODATABASE                            *
' *                                                                     *
' *  GIVEN:    sDir            = directory location                     *
' *            sDBName         = geodatabase name                       *
' *            bOverWrite      = flag denoting whether to overwrite     *
' *                              the geodatabase if it exists           *
' *                              true = overwrite, false = do not       *
' *                                                                     *
' *  RETURN:   CreateAccessDB = geodatabase that is created, will      *
' *                             be set to NOTHING if an error was      *
' *                             encountered during the processing      *
' *                                                                     *
' *  NOTE:     The geodatabase created will contain no dataset or      *
' *            feature class. These will have to be added later on     *
' *            if need be (only the .mdb file is created)              *
```

```
' *                                                                    *
' *   Dim sDir As String                                               *
' *   Dim sDBName As String                                            *
' *   Dim bOverWrite As Boolean                                        *
' *   Dim CreateAccessDB As IWorkspace                                 *
' *                                                                    *
' * * * * * * * * * * * * * * * * * * * * * * * * * * * * * * * * * * *
'
Public Function CreateAccessDB(sDir, sDBName, _
                               bOverWrite) As esriCore.IWorkspace
'
' * * * * * * * * * * * * * * * * * * * * * * * * * * * * * * * * * * *
' *                                                                    *
' *   Name: CreateAnnoClass                   File Name: creancls.bas  *
' *                                                                    *
' *   PURPOSE:  CREATE AN ANNOTATION CLASS IN A PERSONAL GEODATABASE   *
' *                                                                    *
' *   GIVEN:    pWorkspace      = connection to the geodatabase        *
' *             theName         = annotation feature class name        *
' *             pFields         = shapefile attributes                 *
' *             dRefScale       = reference scale of the current view  *
' *             dUnits          = units setting of the current view    *
' *                                                                    *
' *   RETURN:   CreateAnnoClass = feature class that is created, will  *
' *                               be set to NOTHING if an error was    *
' *                               encountered during the processing    *
' *                                                                    *
' *   Dim pWorkspace As IWorkspace                                     *
' *   Dim theName As String                                            *
' *   Dim pFields As esriCore.IFields                                  *
' *   Dim dRefScale As Double                                          *
' *   Dim dUnits As esriUnits                                          *
' *   Dim CreateAnnoClass As IFeatureClass                             *
' *                                                                    *
' * * * * * * * * * * * * * * * * * * * * * * * * * * * * * * * * * * *
'
Public Function CreateAnnoClass(pWorkspace As esriCore.IWorkspace, _
                        theName, pFields As esriCore.IFields, _
                        dRefScale, dUnits) As esriCore.IFeatureClass
'
' * * * * * * * * * * * * * * * * * * * * * * * * * * * * * * * * * * *
' *                                                                    *
' *   Name: CreateFeatClass                   File Name: creftcls.bas  *
' *                                                                    *
' *   PURPOSE:  CREATE A FEATURE CLASS WITHIN A FEATURE DATASET IN A   *
' *             GEODATABASE                                            *
' *                                                                    *
' *   GIVEN:    pFeatureDataset = feature dataset to be processed      *
' *             theName         = featureclass name                    *
' *             geomType        = featureclass geometry type, such as: *
' *                               esriGeometryPoint                    *
' *                               esriGeometryPolyline                 *
' *                               esriGeometryPolygon                  *
' *             pFields         = feature attributes (optional)        *
' *                                                                    *
' *   RETURN:   CreateFeatClass = feature class that is created, will  *
' *                               be set to NOTHING if an error was    *
' *                               encountered during the processing    *
' *                                                                    *
' *   NOTE:     (a) The function CreateNewShapefile can be used to     *
' *                 create the IFeatureDataset object, if appropriate  *
' *             (b) If pFields contains any geometry fields they will  *
' *                 be ignored, only valid attribute fields will be    *
' *                 processed                                          *
' *             (c) If pFields is not specified only the OID and SHAPE *
' *                 fields will be added to the featureclass           *
' *                                                                    *
' *   Dim pFeatureDataset As IFeatureDataset                           *
' *   Dim theName As String                                            *
' *   Dim geomType As esriGeometryType                                 *
' *   Dim pFields As IFields                                           *
' *   Dim CreateFeatClass As IFeatureClass                             *
' *                                                                    *
```

```
' * * * * * * * * * * * * * * * * * * * * * * * * * * * * * * * * * * * *
'
Public Function CreateFeatClass(pFeatureDataset _
                    As esriCore.IFeatureDataset, _
                    theName, geomType As esriCore.esriGeometryType, _
                    Optional pFields As esriCore.IFields) _
                                            As esriCore.IFeatureClass
'
' * * * * * * * * * * * * * * * * * * * * * * * * * * * * * * * * * * * *
' *                                                                      *
' *  Name: CreateList                          File Name: crealist.bas  *
' *                                                                      *
' *  PURPOSE:  TO CREATE A NEW COLLECTION MAKING SURE IT IS EMPTY       *
' *                                                                      *
' *  GIVEN:    nothing                                                   *
' *                                                                      *
' *  RETURN:   newList = new list                                        *
' *                                                                      *
' *  Dim newList As New Collection                                       *
' *                                                                      *
' * * * * * * * * * * * * * * * * * * * * * * * * * * * * * * * * * * * *
'
Public Sub CreateList(newList)
'
' * * * * * * * * * * * * * * * * * * * * * * * * * * * * * * * * * * * *
' *                                                                      *
' *  Name: CreateNewGeoDB                      File Name: cregeodb.bas  *
' *                                                                      *
' *  PURPOSE:  CREATE A PERSONAL GEODATABASE ALLOWING THE USER TO       *
' *            SPECIFY THE DIRECTORY AND NAME USING A FILE DIALOG BOX *
' *                                                                      *
' *  GIVEN:    pFieldsI       = attributes to be stored in the          *
' *                             new personal geodatabase                *
' *            geomType       = shapefile geometry type                 *
' *            defName        = default filename                        *
' *            aTitle         = file dialog message box title           *
' *                                                                      *
' *  RETURN:   CreateNewGeoDB = feature class that is created, will     *
' *                             be set to NOTHING if an error was       *
' *                             encountered during the processing       *
' *                                                                      *
' *  NOTE:     (a) A stand-alone annotation feature class is created    *
' *                in the geodatabase by this function                  *
' *            (b) Use CreateNewShapefile specifying the .mdb file      *
' *                name extension in the default filename to create     *
' *                a geodatabase that contains a feature class and      *
' *                not an annotation feature class                      *
' *            (c) The new annotation feature class is automatically    *
' *                added to the map once it has been created            *
' *            (d) If an existing .mdb file is selected, the user can *
' *                either abort the command (CANCEL), add to the .mdb *
' *                file (NO) or overwrite the existing file (YES)       *
' *            (e) When an existing .mdb file is appended the root      *
' *                name of the default filename is used as the name     *
' *                of the new annotation class that is created          *
' *            (f) When an existing .mdb file is to be overwritten,     *
' *                if the file exists in the map the function will      *
' *                not delete the file but will inform the user and     *
' *                abort the function                                   *
' *                                                                      *
' *  Dim pFieldsI As esriCore.IFields                                    *
' *  Dim geomType As esriCore.esriGeometryType                           *
' *  Dim defName As String                                               *
' *  Dim aTitle As String                                                *
' *  Dim CreateNewGeoDB As IFeatureClass                                 *
' *                                                                      *
' * * * * * * * * * * * * * * * * * * * * * * * * * * * * * * * * * * * *
'
Public Function CreateNewGeoDB(pFieldsI As esriCore.IFields, _
                          geomType As esriCore.esriGeometryType, _
                          defname As String, _
                          aTitle As String) As esriCore.IFeatureClass
'
```

```
' * * * * * * * * * * * * * * * * * * * * * * * * * * * * * * * * * * * * *
' *                                                                         *
' *  Name: CreateNewShapefile                  File Name: creatshp.bas      *
' *                                                                         *
' *  PURPOSE:  USING A FILE DIALOG BOX PROMPT THE USER TO SPECIFY A         *
' *            DATASET TO BE CREATED (THIS CAN BE EITHER A SHAPEFILE        *
' *            OR A PERSONAL GEODATABASE)                                   *
' *                                                                         *
' *  GIVEN:    pFieldsI           = attributes to be stored in the          *
' *                                 new shapefile                           *
' *            geomType           = shapefile geometry type, such as:       *
' *                                 esriGeometryPoint                       *
' *                                 esriGeometryPolyline                    *
' *                                 esriGeometryPolygon                     *
' *            defName            = default filename, this may or may       *
' *                                 not contain a filename extension        *
' *                                 (see the note below)                    *
' *            aTitle             = file dialog message box title           *
' *                                                                         *
' *  RETURN:   CreateNewShapefile = feature class that is created,          *
' *                                 will be set to NOTHING if an            *
' *                                 error was encountered during the        *
' *                                 processing                              *
' *                                                                         *
' *  NOTE:     (a) If the defName argument contains the .shp filename       *
' *                extension, the dataset type that will be created         *
' *                will be a shapefile. If the .mdb filename extension*
' *                is found, the type of dataset created will be a          *
' *                personal geodatabase. If no filename extension is        *
' *                given both types will appear in the list of              *
' *                available types and the user can pick the desired        *
' *                type.                                                    *
' *            (b) The new shapefile or geodatabase is automatically        *
' *                added to the map once it has been created                *
' *            (c) If an existing .mdb file is selected, the user can       *
' *                either abort the command (CANCEL), add to the .mdb       *
' *                file (NO) or overwrite the existing file (YES)           *
' *            (d) When an existing .mdb file is appended the root          *
' *                name of the default filename is used as the name         *
' *                of the new feature class that is created                 *
' *            (e) When an existing .mdb file is to be overwritten,         *
' *                if the file exists in the map the function will          *
' *                not delete the file but will inform the user and         *
' *                abort the function                                       *
' *                                                                         *
' *  Dim pFieldsI As esriCore.IFields                                       *
' *  Dim geomType As esriCore.esriGeometryType                              *
' *  Dim defName As String                                                  *
' *  Dim aTitle As String                                                   *
' *  Dim CreateNewShapefile As esriCore.IFeatureClass                       *
' *                                                                         *
' * * * * * * * * * * * * * * * * * * * * * * * * * * * * * * * * * * * * *
'
Public Function CreateNewShapefile(pFieldsI As esriCore.IFields, _
                          geomType As esriCore.esriGeometryType, _
                          defname As String, _
                          aTitle As String) As esriCore.IFeatureClass
'
' * * * * * * * * * * * * * * * * * * * * * * * * * * * * * * * * * * * * *
' *                                                                         *
' *  Name: CreateShapeFile                     File Name: creashap.bas      *
' *                                                                         *
' *  PURPOSE:  CREATE A NEW SHAPEFILE USING INFORMATION EXPLICITLY          *
' *            DEFINED IN THE CALLING ARGUMENTS (NO USER INTERACTION)       *
' *                                                                         *
' *  GIVEN:    featWorkspace    = directory location                        *
' *            strName          = shapefile name                            *
' *            geomType         = shapefile geometry type                   *
' *                               esriGeometryPoint                         *
' *                               esriGeometryPolyline                      *
' *                               esriGeometryPolygon                       *
' *            pFields          = shapefile attributes (optional)           *
' *            pCLSID           = geometry type subclass (optional)         *
```

```
' *                                                                  *
' *  RETURN:   CreateShapeFile = feature class that is created       *
' *                                                                  *
' *  NOTE:     (a) The name of the shapefile should not contain the  *
' *                .shp extension, if it does it will be stripped off *
' *            (b) If the pFields argument is not specified a default *
' *                shape field with a default spatial reference will *
' *                be assigned and one attribute called ID will be   *
' *                added to the shapefile                            *
' *                                                                  *
' *  Dim featWorkspace As IFeatureWorkspace                          *
' *  Dim strName As String                                           *
' *  Dim geomType As esriGeometryType                                *
' *  Dim pFields As IFields                                          *
' *  Dim pCLSID As UID                                               *
' *  Dim CreateShapeFile As IFeatureClass                            *
' *                                                                  *
' * * * * * * * * * * * * * * * * * * * * * * * * * * * * * * * * * * *
'
Public Function CreateShapeFile(featWorkspace _
                               As esriCore.IFeatureWorkspace, _
                               strName As String, _
                               geomType As esriCore.esriGeometryType, _
                               Optional pFields As esriCore.IFields, _
                               Optional pCLSID As esriCore.UID) _
                                             As esriCore.IFeatureClass
'
' * * * * * * * * * * * * * * * * * * * * * * * * * * * * * * * * * * *
' *                                                                  *
' *  Name: Dformat                          File Name: dformat.bas   *
' *                                                                  *
' *  PURPOSE:  This is a subroutine type script used to format a data *
' *            field of an output file according to a Fortran  Fa.b  *
' *            format.                                               *
' *            The input argument list contains three elements which *
' *            are expected to be read as numbers                    *
' *                                                                  *
' *  GIVEN:    theNumber   = the real number to be formatted         *
' *            TotalDigits = the number of the data field characters *
' *                          including leading spaces, decimal point *
' *                          and decimal digits                      *
' *            DigitsRight = digits to right of decimal point        *
' *                                                                  *
' *  RETURN:   Dformat     = string representing the number in the   *
' *                          specified format                        *
' *                                                                  *
' *  NOTE:     If the number will not fit within the specified data  *
' *            field length as specified by TotalDigits, then the    *
' *            data field will be expanded to accomodate the number. *
' *                                                                  *
' *  Dim theNumber As Double                                         *
' *  Dim TotalDigits, DigitsRight As Integer                         *
' *  Dim Dformat As String                                           *
' *                                                                  *
' * * * * * * * * * * * * * * * * * * * * * * * * * * * * * * * * * * *
'
Public Function Dformat(theNumber, TotalDigits, DigitsRight)
'
' * * * * * * * * * * * * * * * * * * * * * * * * * * * * * * * * * * *
' *                                                                  *
' *  Name: ExportVBAcode                   File Name: exptvbac.bas   *
' *                                                                  *
' *  PURPOSE:  EXPORT VBA COMPONENTS FROM THE CURRENT PROJECT INTO A *
' *            SPECIFIED DIRECTORY                                   *
' *                                                                  *
' *  GIVEN:    nothing                                               *
' *                                                                  *
' *  RETURN:   nothing                                               *
' *                                                                  *
' *  NOTE:     (a) The directory is a hardcoded path stored in aDIR, *
' *                if the directory does not exist, it is created    *
' *            (b) The Avenue Wraps: avFileExists and                *
' *                                  avFileDelete                    *
```

```
' *              must appear in the project file in order for this  *
' *               macro to execute                                  *
' *           (c) If the component being exported exists on disk, it *
' *              will be deleted and the component in the project   *
' *              file will replace whatever was on disk previously  *
' *           (d) A file called index.txt will be created which will *
' *              contain a list of the components that were written *
' *               to disk                                           *
' *                                                                 *
' * * * * * * * * * * * * * * * * * * * * * * * * * * * * * * * * * *
'
Public Sub ExportVBAcode()
'
' * * * * * * * * * * * * * * * * * * * * * * * * * * * * * * * * * *
' *                                                                 *
' *  Name: FindLayer                       File Name: findlayr.bas  *
' *                                                                 *
' *  PURPOSE:  FIND A LAYER IN A MAP                                *
' *                                                                 *
' *  GIVEN:    map       = map to be searched                       *
' *            name      = name of layer to be found                *
' *                                                                 *
' *  RETURN:   FindLayer = the layer in the map                     *
' *                                                                 *
' *  Dim map As IMap                                                *
' *  Dim name As Variant                                            *
' *  Dim FindLayer As ILayer                                        *
' *                                                                 *
' * * * * * * * * * * * * * * * * * * * * * * * * * * * * * * * * * *
'
Public Function FindLayer(map As esriCore.IMap, _
                          name As Variant) As esriCore.iLayer
'
' * * * * * * * * * * * * * * * * * * * * * * * * * * * * * * * * * *
' *                                                                 *
' *  Name: FindTheme                       File Name: findthem.bas  *
' *                                                                 *
' *  PURPOSE:  FIND A THEME IN A MAP                                *
' *                                                                 *
' *  GIVEN:    map       = map to be searched                       *
' *            nameIN    = name of theme to be found                *
' *                                                                 *
' *  RETURN:   FindTheme = the theme in the map                     *
' *                                                                 *
' *  Dim map As IMap                                                *
' *  Dim nameIN As Variant                                          *
' *  Dim FindTheme As Variant                                       *
' *                                                                 *
' * * * * * * * * * * * * * * * * * * * * * * * * * * * * * * * * * *
'
Public Function FindTheme(map As esriCore.IMap, _
                          nameIN As Variant) As Variant
'
' * * * * * * * * * * * * * * * * * * * * * * * * * * * * * * * * * *
' *                                                                 *
' *  Name: GetTextFont                     File Name: gettxfnt.bas  *
' *                                                                 *
' *  PURPOSE:  TO DETERMINE THE CURRENT ACTIVE TEXT FONT IN ARCMAP  *
' *                                                                 *
' *  GIVEN:    pMxDoc    = current active document                  *
' *                                                                 *
' *  RETURN:   fontStrg = name of current active font               *
' *            currSize = font size                                 *
' *            defTFINC = font style (1 = normal, 2 = italic)       *
' *            defPMODE = font style (1 = normal, 3 = bold)         *
' *            defCOLOR = font color, RGB color index value         *
' *                                                                 *
' *  Dim pMxDoc As IMxDocument                                      *
' *  Dim fontStrg As String                                         *
' *  Dim currSize As Double                                         *
' *  Dim defTFINC, defPMODE As Integer                              *
' *  Dim defCOLOR As Long                                           *
' *                                                                 *
```

```
' * * * * * * * * * * * * * * * * * * * * * * * * * * * * * * * * * * * *
'
Public Sub GetTextFont(pmxDoc As esriCore.IMxDocument,
                       fontStrg, currSize, defTFINC, defPMODE, defCOLOR)
'
' * * * * * * * * * * * * * * * * * * * * * * * * * * * * * * * * * * * *
' *                                                                     *
' *  Name: GetTextRect                         File Name: gttxtrec.bas  *
' *                                                                     *
' *  PURPOSE:  TO GET THE ANGLE, HEIGHT AND WIDTH OF A TEXT ELEMENT     *
' *                                                                     *
' *  GIVEN:    pTextElement   = text element representing the text      *
' *            pScreenDisplay = display which text element appears in  *
' *                                                                     *
' *  RETURN:   x1,y1          = low left corner coordinates of text     *
' *            aAngle         = text angle (degrees)                    *
' *            aWidth         = text width (along the text angle)       *
' *            aHeight        = text height (perpendicular to angle)    *
' *                                                                     *
' *  NOTE:     The attributes passed passed back reflect those of an    *
' *            inclined, not an orthogonal, enclosing rectangle that    *
' *            circumscribes the text element                           *
' *                                                                     *
' *  Dim pTextElement As ITextElement                                   *
' *  Dim pScreenDisplay As IScreenDisplay                               *
' *  Dim x1, y1, aAngle, aWidth, aHeight As Double                      *
' *                                                                     *
' * * * * * * * * * * * * * * * * * * * * * * * * * * * * * * * * * * * *
'
Public Sub GetTextRect(pTextElement As esriCore.ITextElement, _
                       pScreenDisplay As esriCore.IScreenDisplay, _
                       X1, Y1, aAngle, aWidth, aHeight)
'
' * * * * * * * * * * * * * * * * * * * * * * * * * * * * * * * * * * * *
' *                                                                     *
' *  Name: HDBbuild                            File Name: hdbbuild.bas  *
' *                                                                     *
' *  PURPOSE:  BUILD A CUSTOMIZABLE HORIZONTAL DIALOG BOX               *
' *                                                                     *
' *  GIVEN:    instruct    = message box instruction                    *
' *            Heading     = message box heading (title)                *
' *            labelList   = list of column labels to be displayed      *
' *            defaultInfo = list of default values for each column     *
' *            typeList    = list of item data types for each column    *
' *                            1 = data line, 2 = combo box             *
' *            colmnList   = list of column widths                      *
' *            nRows       = number of rows in the dialog box           *
' *                                                                     *
' *  RETURN:   userInfo    = list of user responses for data items      *
' *                                                                     *
' *  Dim instruct, Heading As String                                    *
' *  Dim labelList As New Collection                                    *
' *  Dim defaultInfo As New Collection                                  *
' *  Dim typeList As New Collection                                     *
' *  Dim colmnList As New Collection                                    *
' *  Dim nRows As Integer                                               *
' *  Dim userInfo As New Collection                                     *
' *                                                                     *
' * * * * * * * * * * * * * * * * * * * * * * * * * * * * * * * * * * * *
'
Public Sub HDBbuild(instruct, Heading, LabelList, defaultInfo, _
                    typeList, colmnList, nRows, userInfo)
'
' * * * * * * * * * * * * * * * * * * * * * * * * * * * * * * * * * * * *
' *                                                                     *
' *  Name: icasinan                            File Name: icasinan.bas  *
' *                                                                     *
' *  PURPOSE:  COMPUTE THE ARCSIN OF A VALUE                            *
' *                                                                     *
' *  GIVEN:    angleX   = sin of an angle in radians                    *
' *                                                                     *
' *  RETURN:   icasinan = angle in radians whose sin is angleX          *
' *                                                                     *
```

```
' *  Dim angleX, icasinan As Double                                       *
' *                                                                       *
' * * * * * * * * * * * * * * * * * * * * * * * * * * * * * * * * * * * * *
'
Public Function icasinan(ANGLEX) As Double
'
' * * * * * * * * * * * * * * * * * * * * * * * * * * * * * * * * * * * *
' *                                                                       *
' *  Name: icatan                              File Name: icatan.bas      *
' *                                                                       *
' *  PURPOSE:  COMPUTE THE ARC TANGENT OF A NUMBER                        *
' *                                                                       *
' *  GIVEN:    D      = a number                                          *
' *                                                                       *
' *  RETURN:   icatan = arc tangent of a number in radians                *
' *                                                                       *
' *  Dim D, icatan As Double                                              *
' *                                                                       *
' * * * * * * * * * * * * * * * * * * * * * * * * * * * * * * * * * * * * *
'
Public Function icatan(D) As Double
'
' * * * * * * * * * * * * * * * * * * * * * * * * * * * * * * * * * * * *
' *                                                                       *
' *  Name: iccomdis                          File Name: iccomdis.bas      *
' *                                                                       *
' *  PURPOSE:  TO COMPUTE THE DISTANCE BETWEEN TWO POINTS                 *
' *                                                                       *
' *  GIVEN:    X1,Y1    = coordinates of the first point                  *
' *            X2,Y2    = coordinates of the second point                 *
' *                                                                       *
' *  RETURN:   iccomdis = distance between the two points                 *
' *                                                                       *
' *  Dim X1, Y1, X2, Y2, iccomdis As Double                               *
' *                                                                       *
' * * * * * * * * * * * * * * * * * * * * * * * * * * * * * * * * * * * * *
'
Public Function iccomdis(X1, Y1, X2, Y2) As Double
'
' * * * * * * * * * * * * * * * * * * * * * * * * * * * * * * * * * * * *
' *                                                                       *
' *  Name: iccomppt                          File Name: iccomppt.bas      *
' *                                                                       *
' *  PURPOSE:  TO CHECK IF A POINT IS WITHIN A TOLERANCE, THAT VARIES     *
' *            BASED UPON THE DISPLAY OF THE VIEW, OF ANOTHER POINT       *
' *                                                                       *
' *  GIVEN:    XCORD,YCORD = coordinates of point to be checked           *
' *            X2,Y2       = coordinates of the base point                *
' *                                                                       *
' *  RETURN:   XCRD2,YCRD2 = input values if NOFND = 0                    *
' *                        = X2,Y2 values if NOFND = 1                    *
' *            NOFND       = 0 : no match was found                       *
' *                        = 1 : match was found within the point         *
' *                              snapping tolerance.                      *
' *                                                                       *
' *  Dim XCORD, YCORD, X2, Y2, XCRD2, YCRD2 As Double                     *
' *  Dim NOFND As Integer                                                 *
' *                                                                       *
' * * * * * * * * * * * * * * * * * * * * * * * * * * * * * * * * * * * * *
'
Public Sub iccomppt(XCORD, YCORD, X2, Y2, XCRD2, YCRD2, noFnd)
'
' * * * * * * * * * * * * * * * * * * * * * * * * * * * * * * * * * * * *
' *                                                                       *
' *  Name: icdegrad                          File Name: icdegrad.bas      *
' *                                                                       *
' *  PURPOSE:  TO CONVERT FROM DEGREES TO RADIANS                         *
' *                                                                       *
' *  GIVEN:    ANGLE    = angle in degrees (decimal)                      *
' *                                                                       *
' *  RETURN:   icdegrad = angle in radians                                *
' *                                                                       *
```

```
' *  Dim ANGLE As Double                                                   *
' *  Dim icdegrad As Double                                                *
' *                                                                        *
' * * * * * * * * * * * * * * * * * * * * * * * * * * * * * * * * * * * *
'
Public Function icdegrad(angle) As Double
'
' * * * * * * * * * * * * * * * * * * * * * * * * * * * * * * * * * * * *
' *                                                                        *
' *  Name: icforce                              File Name: icforce.bas     *
' *                                                                        *
' *  PURPOSE:  INVERSE FROM POINT 1 TO POINT 2                             *
' *                                                                        *
' *  GIVEN:    PTN1,PTE1 = north-east coordinates of point 1               *
' *            PTN2,PTE2 = north-east coordinates of point 2               *
' *                                                                        *
' *  RETURN:   D         = distance from point 1 to point 2                *
' *            az        = azimuth (radians) from point 1 to 2             *
' *                                                                        *
' *  Dim PTN1, PTE1, PTN2, PTE2, D, AZ As Double                           *
' *                                                                        *
' * * * * * * * * * * * * * * * * * * * * * * * * * * * * * * * * * * * *
'
Public Sub icforce(PTN1, PTE1, PTN2, PTE2, D, AZ)
'
' * * * * * * * * * * * * * * * * * * * * * * * * * * * * * * * * * * * *
' *                                                                        *
' *  Name: icmakdir                             File Name: icmakdir.bas    *
' *                                                                        *
' *  PURPOSE:  COMPUTE THE CARTESIAN DIRECTION OF TWO POINTS               *
' *                                                                        *
' *  GIVEN:    X1,Y1    = coordinates of the first point                   *
' *            X2,Y2    = coordinates of the second point                  *
' *                                                                        *
' *  RETURN:   icmakdir = direction of the two points in radians           *
' *                                                                        *
' *  Dim X1, Y1, X2, Y2, icmakdir As Double                                *
' *                                                                        *
' * * * * * * * * * * * * * * * * * * * * * * * * * * * * * * * * * * * *
'
Public Function icmakdir(X1, Y1, X2, Y2) As Double
'
' * * * * * * * * * * * * * * * * * * * * * * * * * * * * * * * * * * * *
' *                                                                        *
' *  Name: icraddeg                             File Name: icraddeg.bas    *
' *                                                                        *
' *  PURPOSE:  TO CONVERT FROM RADIANS TO DEGREES                          *
' *                                                                        *
' *  GIVEN:    ANGLE    = angle in radians                                 *
' *                                                                        *
' *  RETURN:   icraddeg = angle in degrees (decimal)                       *
' *                                                                        *
' *  Dim ANGLE As Double                                                   *
' *  Dim icraddeg As Double                                                *
' *                                                                        *
' * * * * * * * * * * * * * * * * * * * * * * * * * * * * * * * * * * * *
'
Public Function icraddeg(angle) As Double
'
' * * * * * * * * * * * * * * * * * * * * * * * * * * * * * * * * * * * *
' *                                                                        *
' *  Name: LoadVBAcode                          File Name: loadvbac.bas    *
' *                                                                        *
' *  PURPOSE:  LOAD VBA COMPONENTS FROM A DIRECTORY INTO THE CURRENT       *
' *            ACTIVE PROJECT                                              *
' *                                                                        *
' *  GIVEN:    nothing                                                     *
' *                                                                        *
' *  RETURN:   nothing                                                     *
' *                                                                        *
' *  NOTE:     (a) The directory is a hardcoded path stored in aDIR,       *
' *            (b) Avenue Wraps: avGetWorkDir, avListFiles,                *
' *                              avSetWorkDir, and CreateList              *
```

```
' *              must appear in the project file in order for this *
' *               macro to execute                                  *
' *                                                                 *
' * * * * * * * * * * * * * * * * * * * * * * * * * * * * * * * * * *
'
Public Sub LoadVBAcode()
'
' * * * * * * * * * * * * * * * * * * * * * * * * * * * * * * * * * *
' *                                                                 *
' *  Name: MakeTextElement                 File Name: mktxtele.bas  *
' *                                                                 *
' *  PURPOSE:  TO CREATE A TEXT ELEMENT                             *
' *                                                                 *
' *  GIVEN:    sText          = text string to appear               *
' *            dX,dY          = low left corner coordinates         *
' *            dAngle         = text angle of inclination (degrees) *
' *            pTextSymbol    = text symbol reflecting font, size   *
' *                             and color                           *
' *                                                                 *
' *  RETURN:   MakeTextElement = text element representing the text *
' *                                                                 *
' *  Dim sText As String                                            *
' *  Dim dX, dY, dAngle As Double                                   *
' *  Dim pTextSymbol As ITextSymbol                                 *
' *  Dim MakeTextElement As ITextElement                            *
' *                                                                 *
' * * * * * * * * * * * * * * * * * * * * * * * * * * * * * * * * * *
'
Public Function MakeTextElement(sText, DX, DY, dAngle, _
                          pTextSymbol As esriCore.ITextSymbol) _
                                         As esriCore.ITextElement
'
' * * * * * * * * * * * * * * * * * * * * * * * * * * * * * * * * * *
' *                                                                 *
' *  Name: MakeTextSymbol                  File Name: mktxtsym.bas  *
' *                                                                 *
' *  PURPOSE:  TO CREATE A TEXT SYMBOL                              *
' *                                                                 *
' *  GIVEN:    strFont        = text font                           *
' *            dFontSize      = font size                           *
' *            iItalic        = font style (1 = normal, 2 = italic) *
' *            iBold          = font style (1 = normal, 3 = bold)   *
' *            iColor         = RGB color index value               *
' *                                                                 *
' *  RETURN:   MakeTextSymbol = text symbol representing the text   *
' *                                                                 *
' *  Dim strFont As String                                          *
' *  Dim dFontSize As Double                                        *
' *  Dim iItalic, iBold As Integer                                  *
' *  Dim iColor As Long                                             *
' *  Dim MakeTextSymbol As ITextSymbol                              *
' *                                                                 *
' * * * * * * * * * * * * * * * * * * * * * * * * * * * * * * * * * *
'
Public Function MakeTextSymbol(strFont, dFontSize, iItalic, iBold, _
                          iColor) As esriCore.ITextSymbol
'
' * * * * * * * * * * * * * * * * * * * * * * * * * * * * * * * * * *
' *                                                                 *
' *  Name: RunProgress                     File Name: runprgrs.bas  *
' *                                                                 *
' *  PURPOSE:  Initialize, report on, and terminate the reporting of *
' *            the progress for a processing operation.             *
' *                                                                 *
' *  GIVEN:    xyzRec    = the current unit of measure of progress  *
' *                      = 0: initiate the progress report phase    *
' *                      > 0: report on the progress                *
' *                      < 0: terminate the progress report phase   *
' *            totRecs   = the total unit of measure of progress    *
' *            aMessage  = identification of the progress reporting *
' *                                                                 *
' *  RETURN:   nothing                                              *
' *                                                                 *
```

```
' *  NOTE:     The progress bar can appear in one of two forms, the    *
' *            first is when no stop button is displayed. In this      *
' *            form the progress bar and message appear in the status *
' *            bar area and remain visible until the progress bar is   *
' *            terminated. In the second form, the progress bar will   *
' *            appear in the middle of the display in a dialog box     *
' *            containing the cancel button. Selecting the cancel      *
' *            button will set the global variable ugpProCancel to be *
' *            TRUE.  By testing the value of ugpProCancel, the user   *
' *            can detect if the operation should be canceled or not. *
' *            In addition, the ugpProDesc variable can be used to     *
' *            display additional information about the operation.     *
' *                                                                    *
' *  Dim xyzRec, totRecs As Long                                       *
' *  Dim aMessage As Variant                                           *
' *                                                                    *
' * * * * * * * * * * * * * * * * * * * * * * * * * * * * * * * * * * *
'
Public Sub RunProgress(xyzRec, totRecs, aMessage)
'
' * * * * * * * * * * * * * * * * * * * * * * * * * * * * * * * * * * *
' *                                                                    *
' *  Name: SetViewSnapTol                     File Name: setsnapt.bas  *
' *                                                                    *
' *  PURPOSE:  SCRIPT TO SET THE SNAP TOLERANCE FOR THE CURRENT VIEW   *
' *                                                                    *
' *  GIVEN:     theView  = the current active view                     *
' *             xP       = x coordinate of given point                 *
' *             yP       = y coordinate of given point                 *
' *                                                                    *
' *  RETURN:    viewRect = the width of the current visible display    *
' *            thePoint = point in the projected coordinate system,    *
' *                       will be xP,yP if no projection applied, if   *
' *                       a projection is applied will be different    *
' *                        from xP,yP                                  *
' *            difxxx   = tolerance as a percentage of the view width *
' *                       based upon user-defined value for ugsnapTol *
' *             difzzz   = smaller tolerance (difxxx * 0.1)            *
' *             difwww   = tolerance which will be:                    *
' *                        (a) the same as difxxx if the tolerance is  *
' *                             defined as a percentage                *
' *                             (ugsnapTolMode = "P"), or              *
' *                        (b) equal to the absolute tolerance value   *
' *                             (ugsnapTol), which is converted into   *
' *                             the projected environment, if the      *
' *                             tolerance is defined to be absolute    *
' *                             (ugsnapTolMode = "A")                  *
' *                                                                    *
' *  Dim theView As IMxDocument                                        *
' *  Dim xP, yP As Double                                              *
' *  Dim viewRect, difxxx, difzzz, difwww As Double                    *
' *  Dim thePoint As IPoint                                            *
' *                                                                    *
' * * * * * * * * * * * * * * * * * * * * * * * * * * * * * * * * * * *
'
Public Sub SetViewSnapTol(theView, xP, yP, _
                          viewRect, thePoint, difxxx, difzzz, difwww)
'
' * * * * * * * * * * * * * * * * * * * * * * * * * * * * * * * * * * *
' *                                                                    *
' *  Name: SortTwoLists                       File Name: srt2list.bas  *
' *                                                                    *
' *  PURPOSE:  SCRIPT TO SORT UP TO TWO DIFFERENT LISTS, SORTING THE   *
' *            SECOND LIST BASED UPON THE SORT OF THE FIRST LIST       *
' *                                                                    *
' *  GIVEN:     list1    = first list of items to be sorted            *
' *             list2    = second list of items to be sorted           *
' *             aMssg    = progress bar message                        *
' *             anOrder = the sort order as a Boolean                  *
' *                       True = ascending, False = Descending         *
' *                                                                    *
' *  RETURN:    nothing                                                *
' *                                                                    *
```

```
' *  NOTE:     (a) The order of the lists passed in are changed by    *
' *                this script to reflect the effects of the sort     *
' *            (b) If only one list is to be sorted the collection    *
' *                list2 can be an empty list or passed in as NOTHING *
' *            (c) If NULL is specified for aMssg, no progress bar    *
' *                will be displayed                                  *
' *                                                                   *
' *  Dim list1 As New Collection                                      *
' *  Dim list2 As New Collection                                      *
' *  Dim aMssg As Variant                                             *
' *  Dim anOrder As Boolean                                           *
' *                                                                   *
' * * * * * * * * * * * * * * * * * * * * * * * * * * * * * * * * * * *
'
Public Sub SortTwoLists(list1, list2, aMssg, anOrder)
'
' * * * * * * * * * * * * * * * * * * * * * * * * * * * * * * * * * * *
' *                                                                   *
' *  Name: VDBbuild                       File Name: vdbbuild.bas     *
' *                                                                   *
' *  PURPOSE:  BUILD A CUSTOMIZABLE VERTICAL DIALOG BOX               *
' *                                                                   *
' *  GIVEN:     instruct    = message box instruction                 *
' *             Heading     = message box heading (title)             *
' *             labelList   = list of labels for data items           *
' *             defaultInfo = list of default values for data items   *
' *             typeList    = type of data item to be displayed       *
' *                           1 = data line, 2 = combo box,           *
' *                           3 = text box with multiselect           *
' *                                                                   *
' *  RETURN:    userInfo    = list of user responses for data items   *
' *                                                                   *
' *  NOTE:      Only use a text box control when a single data item   *
' *             is to be displayed such as done with avMsgBoxMultiList *
' *                                                                   *
' *  Dim instruct, Heading As String                                  *
' *  Dim labelList As New Collection                                  *
' *  Dim defaultInfo As New Collection                                *
' *  Dim typeList As New Collection                                   *
' *  Dim userInfo As New Collection                                   *
' *                                                                   *
' * * * * * * * * * * * * * * * * * * * * * * * * * * * * * * * * * * *
'
Public Sub VDBbuild(instruct, Heading, LabelList, defaultInfo, _
                   typeList, userInfo)
'
' * * * * * * * * * * * * * * * * * * * * * * * * * * * * * * * * * * *
' *                                                                   *
' *  Name: VDBbuild2                      File Name: vdbbuld2.bas     *
' *                                                                   *
' *  PURPOSE:  BUILD A CUSTOMIZABLE VERTICAL DIALOG BOX WITH A BACK   *
' *            BUTTON                                                 *
' *                                                                   *
' *  GIVEN:     instruct    = message box instruction                 *
' *             Heading     = message box heading (title)             *
' *             labelList   = list of labels for data items           *
' *             defaultInfo = list of default values for data items   *
' *             typeList    = type of data item to be displayed       *
' *                           1 = data line, 2 = combo box            *
' *                                                                   *
' *  RETURN:    userInfo    = list of user responses for data items   *
' *                                                                   *
' *  Dim instruct, Heading As String                                  *
' *  Dim labelList As New Collection                                  *
' *  Dim defaultInfo As New Collection                                *
' *  Dim typeList As New Collection                                   *
' *  Dim userInfo As New Collection                                   *
' *                                                                   *
' * * * * * * * * * * * * * * * * * * * * * * * * * * * * * * * * * * *
'
Public Sub VDBbuild2(instruct, Heading, LabelList, defaultInfo, _
                    typeList, userInfo)
```

INDEX

Symbols

A

B

C

D